Springer Series on
Atoms+Plasmas

14

Editor: I. I. Sobel'man

Springer Series on

Atoms+Plasmas

Editors: G. Ecker   P. Lambropoulos   I. I. Sobel'man   H. Walther

Managing Editor: H. K. V. Lotsch

V. S. Lisitsa

Atoms in Plasmas

With 65 Figures

Springer-Verlag

Berlin Heidelberg New York
London Paris Tokyo
Hong Kong Barcelona
Budapest

Dr. Valery S. Lisitsa

Kurchatov Institute of Atomic Energy, 123182 Moscow, Russia

Series Editors:

Professor Dr. Günter Ecker

Ruhr-Universität Bochum, Fakultät für Physik und Astronomie,
Lehrstuhl Theoretische Physik I, Universitätsstrasse 150,
D-44801 Bochum, Germany

Professor Peter Lambropoulos, Ph. D.

Max-Planck-Institut für Quantenoptik
D-85748 Garching, Germany, and
Foundation for Research and Technology – Hellas (FO.R.T.H.),
Institute of Electronic Structure & Laser (IESL),
University of Crete, PO Box 1527, Heraklion, Crete 71110, Greece

Professor Igor I. Sobel'man

Lebedev Physical Institute, Russian Academy of Sciences,
Leninsky Prospekt 53, 117333 Moscow, Russia

Professor Dr. Herbert Walther

Sektion Physik der Universität München, Am Coulombwall 1,
D-85748 Garching/München, Germany

Managing Editor: Dr.-Ing. Helmut K.V. Lotsch

Springer-Verlag, Tiergartenstrasse 17, D-69121 Heidelberg, Germany

ISBN-13: 978-3-642-78728-7   e-ISBN-13: 978-3-642-78726-3
DOI: 10.1007/978-3-642-78726-3

Library of Congress Cataloging-in-Publication Data. Lisitsa, V. S. (Valery S.), 1945- Atoms in plasmas/V. S. Lisitsa. p. cm. – (Springer series on atoms + plasmas; 14) Includes bibliographical references and index. Atoms. 2. Atomic spectra. 3. Plasma (Ionized gases) 4. Radiation. 5. Collisions (Nuclear physics) I. Title. II. Series. QC173.3.L57 1994 539.7–dc20 93-46645

Typesetting: K +V Fotosatz, D-64743 Beerfelden
SPIN 10066058   54/3140 - 5 4 3 2 1 0 - Printed on acid-free paper

Preface

The monograph is the result of more than 20 years activity of the author in the field of radiation-collisional phenomena during his work in the I. V. Kurchatov Institute of Atomic Energy (Moscow, Russia). When investigating the specific applications in the fields of high- and low-temperature plasmas, atomic collisions, nonlinear laser spectroscopy and other fields, the author has faced a wide spectrum of problems of a fundamental nature which are of general physical interest. These problems have led to the conception of a "perturbed atom," that is, an atom that is under the simultaneous action of a number of perturbations in a medium. Such perturbations are the electric and magnetic fields, the fields resulting from laser and Planck's radiation, plasma oscillations, collisions of an atom with other particles, stochastic acceleration of the atom in a dense medium, etc. Solving fundamental problems connected with the aforementioned phenomena is the goal of this monograph.

The main approach to the above problems involves quasiclassical or purely classical methods which have unexpectedly revealed great potential in recent years. The results obtained on this basis are of great clarity and simplicity. I hope that the methods and results considered in the monograph will be useful for many investigators in their research work.

I am very grateful to my colleagues Profs. V. I. Kogan, I. I. Sobelman, G. V. Sholin, A. N. Starostin, S. I. Yakovlenko, Drs. A. V. Demura, A. B. Kukushkin, and others for long-term fruitful cooperation and helpful discussions.

Moscow, February 1994   *Valery S. Lisitsa*

Contents

1. Introduction. General Problems of Description of Atomic Spectra in Plasmas

1.1 Atomic Physics and Plasma Physics. Quasiclassical Methods for Atomic Processes

As a rule, numerous reviews and monographs on atomic spectra theory [1.1 – 10] are devoted to a specific type of external influence on the atom with no regard to any other influence that could exist. It is true of both radiation and autoionization decay, influence of constant electrical and magnetic fields, dynamic electrical fields, collision of atoms with particles and other processes.

However, in actual research work, one has to deal with the simultaneous influence of many factors on the atom. This is the case with plasma applications when the atom experiences the simultaneous combined influence of both electrical and magnetic fields, constant and dynamic fields, collisions in the presence of external fields, and so on.

Recording the combined action of numerous factors on an atom is far from being a trivial problem. In many instances it is not even clear how to approach the problem and to choose the correct approximation. This is the range of problems which will be considered in the present monograph. It should be noted that only the dynamics of atomic systems subjected to different disturbances will be examined here and no detailed calculation of their spectral radiation in plasma should be expected. In this sense, plasmas are only the source of disturbances of different types. The results obtained are of rather universal importance and therefore can be used for other applications.

The following is a description of the atom's dynamics in rarefied plasma which does not strongly distort its basic energetic structure. In this sense the influence of the plasma is weak. However, it does not allow the application of perturbation theory *a priori* since a weak influence is present only in terms of Z^2Ry atomic energy units and has nothing to do with close-lying energy levels.

The importance of a large variety of atomic processes is very well known in the physics of both low and high temperature plasmas. The atomic characteristics are important from the point of view of hot plasma diagnostics and their influence on energy balance, thermodynamic properties, radiation ability etc. [1.3 – 11]. Atomic physics and plasma physics substantially differ in the methods of achieving their objectives. To solve atomic physics problems, the quantum mechanical approach is used; whereas for plasma physics, the motion of particles may be considered as purely classical. However, unexpected possibilities of quasiclassical and purely classical methods for describing collisional and radiation atomic processes in plasmas [1.12] have been brought to light in recent years. These possibilities are conditioned by what is called the "moderate" domain of electron energies in a plasma which are on one hand fast enough to efficiently ionize the medium, and on the other hand slow enough as compared with atomic velocities of electrons on inner shells. In this way the plasma with multicharged ions becomes a "natural" object for the application of quasiclassical methods [1.12]. One more reason for using classical methods is the Coulomb potential typical of multicharged ions where the quantum and classical approaches to scattering problems are known to lead to identical results.

Classical methods prove very efficient even for such very well known problems as Stark and Zeeman effects, especially in the case of highly excited atomic states. As a matter of fact, we are now arriving at a re-evaluation of the role of classical methods in atomic physics. The success of these methods brought during the last decades apparently account for the scare-tactic exaggeration of "old" classical atomic mechanics ousted by new quantum theory [1.13].

The advantages of classical methods are due to the universal results they yield. In fact, quantum mechanics is "a physics of matrix element" results, and more often than not, is a collection of numerical data for describing this or that process. For actual research work with plasmas, these data are not an end in itself, as it is the case of theoretical spectroscopy, but part of more complete plasma models where the data must be coordinated with numerous parameters changing within a wide range.

Thus, it is desirable that proper plasma applications should have available universal analytical process probabilities and constituent parameter dependences which would allow their realization and enable the connection between atomic and plasma parameters. Classical methods cater exactly to this. In this respect, methods of description of particle motion both in plasmas and atoms are drawn together.

This monograph connects the main features between the atomic and plasma parameters in two ways. First, the combined influence of many perturbations on the atom are considered; and, second, a wide application of quasiclassical or purely classical methods are used for the description of the atom.

1.2 General Problems of Atomic-State Mixing in a Plasma Medium. Density Matrix Method

The atom immersed in a plasma experiences various perturbations from the plasma, which leads to a definite distribution of the atom with regard to excited states as well as to changing the excited states wave functions themselves. As a result of such perturbations the atomic states turn out to be "mixed," strongly different from "pure," unperturbed atomic states. Consequently, the spectral characteristics of radiation emission or absorption by the atom in plasmas are also substantially different from the spectra of an unperturbed atom. The atomic state "mixing" mechanisms can easily be imagined by considering a plasma as some external system of classically moving particles interacting with the atom [1.14−16]. If we look at a plasma in this way, the plasma action is composed of fast changing perturbations due to electrons on the one hand, and slow ones due to ions on the other. It is convenient to define these last perturbations by the electric micro-fields $F(t)$ caused by the ions. These border on other types of electrical and magnetic fields in plasmas, such as plasma oscillation fields, fields of magnetic systems, etc.

As far as the electron-atomic interaction is concerned, it may be taken into account by the introduction of additional atomic relaxation, described by the S-matrix. Thus, the role of plasma electrons comes at a nonelastic drop of an atomic electron from one sublevel to another. The action of an ion electric field F as well as other fields lead to periodic oscillations of the electron between different atomic states. Properly speaking, it is exactly these oscillations that express the fact of atomic wave function "mixing" in an external field.

The usual approach to the atomic spectral description in plasmas is divided into two main problems (for example, [1.3−5]): 1) the investigation of excited atomic state dynamics in external fields and 2) the investigation of atomic level population kinetics. For example, such is the approach to the description of the Stark effect [1.1−3], where at first the true atomic states are defined (atomic dynamics), and then their population and relaxation processes are analyzed. In reality, the values of the relaxation (kinetic) parameters may be compared to atomic interaction with external fields, and the whole character of the atomic state evolution is found to be more complicated. The situation is typical, for example, of multicharged ions in dense plasmas [1.9, 10]. Thus, it is more consistent to take into account both dynamic and kinetic parameters simultaneously.

This very approach is used in laser physics [1.7, 16] where the atomic dynamics in external electromagnetic fields is described by kinetic equations, including a relaxation parameter. Although we shall investigate dynamic and kinetic processes separately, here we shall try to formulate a more general approach to the description of atomic excited state evolution based just as in laser physics [1.7] on general kinetic equations [1.16, 17]. The corresponding kinetic equation is written for the atomic density matrix $[\varrho]$, the diagonal elements of which determine the atomic state populations, and the non-diagonal ones determine the polarization at a corresponding atomic transition. Processes leading to atomic excitation as well as to atomic state mixing induce the media polarization at the transition considered. The calculation of the polarization spectral distribution determines the characteristics of the radiation emitted or absorbed by atoms in plasmas. The main sources of atomic excitation in a plasma are electrons and light quanta. The electron action on the atom is characterized in the simplest cases by incoherent pumping of excited atomic states Q, being the rate of an atomic electron appearing at a given atomic level in a unit of time and a unit of volume. The excitation of atomic electrons by light quanta is of importance in optically dense plasmas [1.18] and requires more detailed information for investigation of the spectral parameter. In fact, for excitation by incoherent pumping (electrons), the only spectral characteristics which define the media polarization is the line shape $I(\omega)$ of spontaneous radiation.

However, for excitation by a light quantum with frequency ω', the emitted radiation with frequency ω "remembers" the initial quantum ω' inducing the polarization at the transition considered. As a result, the polarization distribution induced by the quanta in the media is described by the more complicated frequency redistribution functions $R(\omega, \omega')$ determining the probability of the emission of a quantum with frequency ω when the absorption of the initial quantum with frequency ω' has taken place [1.18–19].

The calculation of spectral functions $I(\omega)$ and $R(\omega, \omega')$ in plasmas is a very complicated problem connected with both atomic state dynamics in external fields and the population kinetics of these very states. Let us consider a general approach [1.19] to the atomic spectral calculations by a model example of a three-level system which consists of ground level 1, radiated level 2 and a near-lying metastable polarized level 3 (Fig. 1.1).

The model correctly describes the basic properties of excited atomic state evolution such as the mixing of excited states 2 and 3 due to the interaction $d \cdot F(t)$ of an atom dipole momentum d with ion field F, spontaneous radiation level relaxation described by matrix $[R]$ and the aforementioned collision relaxation S due to the nonelastic collision transitions between sublevels 2 and 3. The sources of polarization near the transition $2 \rightarrow 1$ are the incoherent pumping Q and interaction V_1 of the atom with a light quantum with frequency Ω_1 impinging on it. The general kinetic equation for the atomic density matrix is as follows:

$$[\dot{\varrho}] = -i[V^{\text{tot}}(t), [\varrho]] + S + R + Q \tag{1.1}$$

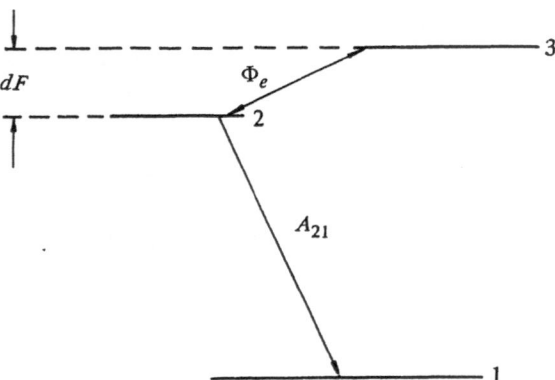

Fig. 1.1. The scheme of the three level system: the radiative (A_{21}) and collision (Φ_e) transition probabilities as well as the interaction $d \cdot F$ with the electric field are shown [1.19]

Here H_0 is the unperturbed Hamiltonian. $V^{\text{tot}}(t)$ is the total interaction operator and is the sum of the interactions with the plasma (ionic) field:

$$V(t) = -\exp(-iH_0 t)\, d \cdot F(t) \exp(iH_0 t)$$

and with the fields of falling V_1 and scattered $V_2(t)$ radiations:

$$V^{\text{tot}}(t) = V(t) + V_1(t) + V_2(t) \tag{1.2}$$

$$V_1(t) = -G_1 \exp(-i\Omega_1 t)\beta_1 \beta_2^+ \ ,$$
$$V_2(t) = -G_2^+ \exp(i\Omega_2 t)\beta_1^+ \beta_2 \ , \tag{1.3}$$

where G_1, G_2 are the interactions with the field's amplitudes, $\Omega_1 = \omega' - \omega_{21}$, $\Omega_2 = \omega - \omega_{21}$ are the frequencies counted from the unperturbed frequencies of the allowed transition $\omega_{21} = (E_2 - E_1)$; β, β^+ are the operators of the annihilation and creation of an atom in the corresponding states. In accordance with the redistribution radiation problem, the field G_1 is only absorbed and field G_2 is only emitted. Therefore, the commutators of $V_{1,2}$ are defined with the Hermite conjugation: $[V_{1,2}, [\varrho]] = V_{1,2}[\varrho] - [\varrho] V_{1,2}^+$.

In fact, (1.1) with the field interaction Hamiltonian (1.2, 3) defines the density matrix of not merely an isolated atom but for a compound-system "atom plus spontaneous electromagnetic field" in accordance with choosing the selection rules for perturbations (1.3). The nondiagonal element of the matrix $[\varrho]_{21}$ after multiplying it by G_2^* yields the medium's spontaneous radiation power [1.7]

$$P(\omega) = -2\hbar\omega\, \text{Re}\{\langle iG_2^* \exp(i\Omega_2 t)[\varrho]_{21}\rangle\} \ . \tag{1.4}$$

In this connection let us call the nondiagonal matrix element of the matrix a medium "spontaneous polarization."

The symbol $\Diamond$ in (1.4) means the ensemble average of radiative ions, that is, of the whole collection of stochastic variables entering into the equation for $[\varrho]$, first of which is the plasma microfields distribution F and the Maxwell velocity distribution of radiative ions, which determines the Doppler line broadening.

Operation S is expressed in terms of the well-known impact broadening operator in the following way [1.5]:

$$
\begin{aligned}
S_{aa'}(\varrho) &= \sum_{a_1,a'} \exp(i\varepsilon t)\langle aa_1'^+ | \Phi | a'a_1^+\rangle [\varrho]_{a_1,a_1'} \\
&= -\frac{4}{3}\pi N_e \frac{e^4}{\hbar^2}\left\langle \frac{1}{v}\right\rangle \Lambda \sum_{a_1,a_1'} \exp(i\varepsilon t)[\varrho]_{a,a_1'} \cdot \{(r\cdot r)_{aa_1}\delta_{a'a_1'} \\
&\quad + (r\cdot r)_{a'a_1'}\delta_{aa_1} - 2(r)_{aa_1}\cdot(r)_{a'a_1'}\} \ ,
\end{aligned}
\tag{1.5}
$$

where N_e, v_{T_e}, e, m are the electron density, thermal velocity, charge and mass, respectively, and r is an operator of the atomic electron coordinate.

The solution of the system (1.1–3) for the spontaneous emission (or redistribution) of the field with frequency Ω_2 is obtained by perturbation theory according to the parameter G_2. In the zeroth order approximation it reads as follows:

$$\mathcal{L}[\varrho]^0 + i[V_1, [\varrho]^0]_+ = Q \; ; \quad \mathcal{L}[\varrho] \equiv \frac{\partial[\varrho]}{\partial t} - S - R + i[V(t), [\varrho]] \qquad (1.6)$$

The interaction with the incident field V_1 may be neglected for the calculation of the line shape $I(\omega)$, determined by pumping Q,

$$\mathcal{L}[\varrho]_0^0 = Q \; . \qquad (1.7)$$

On the contrary, when determining the function $R(\omega, \omega')$ it is essential to take into account the interaction V_1, whereas the contribution of pumping Q to the excited atomic levels may be neglected. Accounting for term V_1 in (1.6) may be also done on the basis of perturbation theory, resulting in the following:

$$\mathcal{L}[\varrho]_1^0 = -i[V_1, [\varrho]_0^0]_+ \; ; \quad \mathcal{L}[\varrho]_2^0 = -i[V_1, [\varrho]_1^0]_+ \; . \qquad (1.8)$$

Equation (1.8) supposes that pumping Q (that is matrix $[\varrho]_0^0$) determines the population N_0 for the lower ground state 1 only, the 0 indices denote that the perturbation V_1 is taken into account as usual [1.7], in the first order for polarization and in the second order for populations. Thus, in the zeroth order approximation, according to the G_2 matrix $[\varrho]^0$ is the sum $[\varrho]^0 = [\varrho]_0^0 + [\varrho]_1^0 + [\varrho]_2^0$ of the solutions of (1.7–8), where the separate items in the calculations of $I(\omega)$ or $R(\omega, \omega')$ are determined as stated above. For the zeroth order approximation below, we simply use the designation $[\varrho]^0$ without working out the separate contributions in detail.

What we seek is an equation for the density matrix ϱ which determines the radiation or redistribution and it arises in the first order for the interaction G_2

$$\mathcal{L}[\varrho] = -i[V_2, [\varrho]^0(t)]_+ \; . \qquad (1.9)$$

Solving (1.9) and substituting $[\varrho]$ into (1.4) we find the radiated power $P(\omega)$. This calls for a preliminary solution to the equations for $[\varrho]^0$, namely (1.7) in the case of the line shape $I(\omega)$ calculations or (1.8) in the case of the calculation of the redistribution function $R(\omega, \omega')$. Matrix $[\varrho]^0$ may be called a matrix of the initial populations of excited atomic states caused by pumping Q or G_1. It should be noted that the formation of these populations (as well as of the polarization $[\varrho]_{12}$) occurs under the influence of a mixing ion field, as in (1.6). Moreover, matrix $[\varrho]^0$ is not diagonal in general: it not only has the elements $[\varrho]_{22}^0$ and $[\varrho]_{33}^0$ determining the proper population of sublevels 2 and 3 but also the nondiagonal element $[\varrho]^0_{23}$ due to the mixing effects leading to the state interference. Thus, the determination of the excited state populations in plasmas is a nonlinear (with regard to field F) effect of the state interference considering their evolution from a spherical basis, when $F \to 0$, to a parabolic one when $F \to \infty$. In accordance with the terminology adopted in nonlinear laser spectroscopy [1.7] it is appropriate to call the mentioned mixing effects nonlinear interference effects (NIEF). For the model system under consideration, the effects connected with the contribution of the nondiagonal matrix elements $[\varrho]^{23}_0$ may be selected in clear analytical form [1.19]. The practical NIEF contribution can be determined from [1.19] where this contribution is picked for transitions $2^1S - 4^1P$, 4^1D of the helium-like ion Al XIII in a plasma with the parameters $N_e = 20^{20} \text{ cm}^{-3}$, $T_e = T_i = 350 \text{ eV}$.

To summarize the above, we may say that the atomic spectral formation in plasmas is a complex interference process in which on a similar basis there are both relaxation (elements of the $[S]$- and $[R]$-matrix) and dynamic parameters of atomic interaction with a plasma. The state interference is essential for both the state population formation and the polarization of the medium. That is why the selection of purely dynamic effects in atomic spectral systems with subsequent relaxation parameters calculations is a conditional procedure to a considerable extent.

2. Classical Motion in an Atomic Potential. Atomic Structure

2.1 Classical Radiation Spectra in a Coulomb Field. Peculiarities of the High-Frequency Domain. Kramers' Electrodynamics

The foundations of radiation theory for a classically moving particle (electron) in a given potential $V(r)$ are stated in numerous books on classical electrodynamics [2.1,2]. In accordance with [2.3−6], we shall dwell on a number of classical spectral peculiarities with the attractive potential $V(r) = -|V(r)|$ playing an important role in the applicability of the classical method to atomic physics. The essence of the problem involves the situation when an emitting electron in an attractive field experiences an acceleration and may obtain the kinetic energy $W = E + |V(r)|$, considerably exceeding its initial energy E at infinity. In this case the classical nature of electron motion is preserved even when the quantum energy $\hbar\omega$ emitted by the electron exceeds its initial energy E. The circumstance essentially expands the applicability domain of classical description methods for atomic processes, including the inelastic domain $\hbar\omega \gtrsim E$. Below, we will focus on the Coulomb field case playing an important part in atomic processes in plasmas. Atomic potentials of a more general type are investigated in [2.4]. The results of following consideration will be used later in the quasiclassical approximation constructions for radiation transition probabilities.

Classical electrodynamics (CED) operates as is known [2.1,2] with an effective spectral radiation yield $d\kappa(\omega)$, cross section $d\sigma(\omega)$ and energy $dE(\varrho,\omega)$ emitted during the time of the collision with an impact parameter ϱ in a frequency interval $d\omega$. These quantities are connected by the relation

$$\frac{d\kappa(\omega)}{d\omega} = \hbar\omega \frac{d\sigma}{d\omega} = \int_0^\infty 2\pi\varrho \, d\varrho \, [dE(\varrho,\omega)/d\omega] \ . \tag{2.1}$$

The spectral distribution of the emitted energy is defined by the Fourier-coefficients $\dot{x}(\varrho,\omega)$ and $\dot{y}(\varrho,\omega)$ of the electron velocity components $\dot{x}(\varrho,t)$ and $\dot{y}(\varrho,t)$:

$$\frac{dE(\varrho,\omega)}{d\omega} = \frac{e^2}{3\pi c^3} \omega^2 [\,|\dot{x}(\varrho,\omega) + i\dot{y}(\varrho,\omega)|^2 + |\dot{x}(\varrho,\omega) - i\dot{y}(\varrho,\omega)|^2\,] \ . \tag{2.2}$$

These Fourier components for motion in a Coulomb field are expressed in terms of the Hankel functions $H_{i\nu}^{(1)}$ ($i\nu\varepsilon$) and their derivatives $H_{i\nu}^{(1)'}$ ($i\nu\varepsilon$):

$$\frac{dE(\varrho,\omega)}{d\omega} = \frac{2\pi}{3}\frac{Z^2 e^6 \omega^2}{m^2 v^4 c^3}f(\nu,\varepsilon) \ , \tag{2.3}$$

$$f(\nu,\varepsilon) = \left\{ [H_{i\nu}^{(1)'}(i\nu\varepsilon)]^2 - \left(1-\frac{1}{\varepsilon^2}\right)\ [H_{i\nu}^{(1)}(i\nu\varepsilon)]^2 \right\} \ . \tag{2.4}$$

Here $\nu = \omega/\tilde{\omega}$ is a dimensionless frequency in units of the "classical" Coulomb frequency, $\tilde{\omega} = v/a = m v^3/Ze^2$ ($a = Ze^2/mv^2$ is the Coulomb length), $\varepsilon^2 = 1+\varrho^2/a^2$ is the eccentricity of the hyperbolic trajectory (ϱ is the impact parameter): parameters z, e, m, v and c are the standard designations for nuclear charge, mass, electron velocity and light velocity, respectively.

The function $f(\nu,\varepsilon)$ is as well known [2.2] as the complete derivative of the function

$$g(\nu) = \frac{\pi\sqrt{3}}{4}i\nu H_{i\nu}(i\nu)H_{i\nu}'(i\nu) \ , \tag{2.5}$$

that makes it possible to perform the integration over $d\varrho$ in (2.1) and to obtain the cross section

$$d\sigma(\omega) = \frac{16\pi}{3\sqrt{3}}\frac{e^2}{\hbar c}\frac{v^2}{c^2}a^2\frac{d\nu}{\nu}g(\nu) \ . \tag{2.6}$$

The function $g(\nu)$ is named the Gaunt-factor. For a large radiation frequency $\omega/\tilde{\omega} = \nu \gg 1$, the factor $g(\nu)$ approaches unity, and the rest factor before $g(\nu)$ in (2.6) is termed the "Kramers" bremsstrahlung cross section.

The total (integral over ω) effective radiation κ is expressed in terms of the total radiation energy loss during the collision $\Delta E(\varrho)$:

$$\kappa = \int \hbar\omega\frac{d\sigma}{d\omega}d\omega = \int_0^\infty 2\pi\varrho\, d\varrho\,\Delta E(\varrho) \ . \tag{2.7}$$

The quantity $\Delta E(\varrho)$ may be expressed, in turn, with the help of a time integral from the square of electron acceleration $w(\varrho,t)$:

$$\Delta E(\varrho) = \frac{2e^2}{3c^3}\int_{-\infty}^\infty [w(\varrho,t)]^2 dt \ . \tag{2.8}$$

For the central field, (2.8) is rewritten in the form

$$\Delta E(\varrho) = \frac{4e^2}{3m^2c^3}\int_{r_0(\varrho)}^\infty \left(\frac{dV}{dr}\right)^2\frac{dr}{v_r} \ , \tag{2.9}$$

where $v_r(\varrho)$ is the radial velocity, and $r_0(\varrho)$ is the classical turning point defined from the relation

$$1 = \frac{\varrho^2}{r_0^2} - \frac{|V(r_0)|}{E} \ . \tag{2.10}$$

Let us write down the spectral distributions of the emitted energy in the domain of large and small frequencies. Following [2.4, 5], let us use the normalized spectral functions as the ratio of the spectral distribution $dE(\varrho, \omega)$ to the total radiation $\Delta E(\varrho)$

$$v \ll 1, \ \varrho \gg a: \ \frac{dE(\varrho, \omega)}{\Delta E(\varrho)} = \frac{8}{\pi^2} \{[sK_0(s)]^2 + [sK_1(s)]^2\} ds \tag{2.11}$$

$$s = M\omega/2E;$$

$$v \gg 1, \ \varrho \ll a: \ \frac{dE(\varrho, \omega)}{\Delta E(\varrho)} = \frac{12}{\pi^2} G(u) u \, du \ ; \tag{2.12}$$

$$u = M^3 \omega/3Z^2 me^4; \ G(u) \equiv u \, [K_{1/3}^2(u) + K_{2/3}^2(u)] \ ; \tag{2.13}$$

where $M = mv\varrho$ is the electron orbital momentum, and $K_v(x)$ are the McDonald functions.

Let us analyze in more detail the high frequency case (2.12, 13). First, it is obvious that the spectral distribution described by the variable u does not depend on the initial electron energy E. This "switch-off of the energy integral" (SEI) is due to the aforementioned electron acceleration in an attractive potential $V(r) = -Ze^2/r$. As a matter of fact, the radiation of large frequencies $\omega \gg \tilde{\omega} = v/a$ takes place with a sharp curvature of the impact electron trajectory at the radiating section where its acceleration is maximum. It is obvious that the largest acceleration is observed near the trajectory turning point r_0 (2.9–10). In this domain, the potential energy $V(r_0)$ is much larger compared with the initial energy $(|V(r_0)| \gg E)$, and that is the reason why the latter does not influence the spectral distribution of the emitted energy. This domain for the Coulomb field was first indicated by *Kramers* [2.7], so the non-Coulomb generalization of the SEI-approximation forms the basis of the "*Kramers'* electrodynamics" concept introduced by V. *Kogan* [2.5].

According to (2.2), the spectral distribution consists of the two polarizations corresponding to the rotation directions along and against the electron trajectory. In accordance with the total intensity (2.13), the sum of the two contributions from the two polarizations mentioned is

$$\frac{dE(\varrho, \omega)}{\Delta E(\varrho)} = \frac{6}{\pi^2} u^2 \{[K_{1/3}(u) + K_{2/3}(u)]^2 + [K_{1/3}(u) - K_{2/3}(u)]^2\}$$

$$\propto [F_-(u)^2 + F_+(u)^2] \ . \tag{2.14}$$

It is easy to see that in almost all of the variable domains, the change of the function $F_-(u)$ (corresponding to the sum of functions $K_{1/3}$ and $K_{2/3}$) substantially exceeds the function $F_+(u)$ (corresponding to their difference). This particular circumstance is caused mathematically by the compensation of these same functions $K_{1/3}$ and $K_{2/3}$, and reflects an important feature of radiation formation in the high-frequency domain, namely: the radiation is basically caused by the electron rotation near the turning point r_0 of the trajectory. The angular velocity of such a rotation $\omega_R(r_0)$ is defined by the relation:

$$\omega_R(r_0) = M/mr_0^2 = \sqrt{2(E+|V(r_0)|)/mr_0^2} = \frac{v_{max}}{r_0} , \qquad (2.15)$$

where v_{max} is the maximum electron velocity.

The aforementioned nature of the spectral formation becomes apparent if one writes down the functions $F_\pm(\omega, E, M)$ for an arbitrary potential $V(r)$ in the form of the Fourier-components of the electron trajectory [2.4−6].

$$F_\pm(\omega, E, M) = \int_{r_0(E,M)}^{\infty} \left(E+|V(r)| - \frac{M\omega_R(r)}{2} \right)^{-1/2} r\,dr$$

$$\times \cos \left(\int_{r_0(E,M)}^{r} [\omega \pm \omega_R(r')] \left\{ \frac{2}{m} \left[E+|V(r)| \right. \right.\right.$$

$$\left.\left.\left. - \frac{M\omega_R(r')}{2} \right] \right\}^{-1/2} dr' \right) . \qquad (2.16)$$

For large frequencies $\omega \gg \tilde{\omega}$ the integrated expression in (2.16) promptly oscillates everywhere, excluding the points of oscillation compensation $\omega \simeq \omega_R(r_\omega)$. The compensation takes place only for the function F_- (which explains the definition of its index) but not for the function F_+. Therefore, the more considerable the inequality of the F_- and F_+ contributions to the intensity $I(\omega)$, the higher the frequency ω becomes. This circumstance follows from pure classical mechanics and also manifests itself in quantum calculations of transition probabilities, being known as the Bethe rule (Chap. 3).

The above analysis of the radiation mechanism in the high frequency domain reveals a means of approximating a universal spectral description [2.6]. The description is reached by the replacement of the real electron motion by its rotation along a circle with angular velocity $\omega_R(r)$. This approximation is obtained by the introduction of the delta-function $\delta[\omega - \omega_R(r)]$ into (2.9), leading to the spectral distribution [2.4, 5]:

$$\left(\frac{d\kappa}{d\omega} \right)_R \simeq \frac{8\pi e^2}{3c^3 m^2 v} \int_0^{\infty} \left(\frac{\partial V}{\partial r} \right)^2 \sqrt{1 - \frac{V(r)}{E}} r^2 \delta[\omega - \omega_R(r)]\, dr . \qquad (2.17)$$

Calculation of the integral in (2.17) leads to the following expression for the Gaunt-factor (2.6) of *Bremsstrahlung* [2.4−6]

$$g_{rot}(\omega) = \frac{6}{Z^2 e^4} \frac{D_\omega}{2 + D_\omega^2} \frac{[E + |V(r_\omega)|]^3}{m\omega^2} \tag{2.18}$$

$$D_\omega = -d \ln [E + |V(r_\omega)|]/d \ln r_\omega ,$$

where in correspondence with the ideas presented, the radiation radius r_ω is defined by the relationship

$$\omega_R(r_\omega) = \omega \quad \text{or} \quad \frac{E + |V(r_\omega)|}{r_\omega^2} = \frac{m\omega^2}{2} . \tag{2.19}$$

The "rotational" approximation (2.18,19) is of high precision. For example, for a Coulomb potential the error of the approximation does not exceed 5% even with a frequency as low as $\omega = \tilde{\omega}/2$. The detailed analysis of the rotational approximation results in a more general class of atomic potentials, and is reduced to [2.4, 6].

The SEI-effect and the peculiarities of the radiation spectra connected with it work as pointed out in the high frequency domain $\omega \gg \tilde{\omega}$. The domain makes the main contribution to the total Bremsstrahlung intensity. As far as the low frequency domain $\omega \ll \tilde{\omega}$ is concerned, (2.11) shows that there is no compensation of K_0 and K_1 and consequently no domination of spectral function F_- upon F_+.

The independence of the radiation characteristics of the energy (SEI-effect) indicates the universal nature of the radiation spectral dependence on frequency, not only for the infinite motion ($E > 0$) considered above but for the finite motion ($E < 0$) of the electron along an elliptical trajectory, as well. It is easy to verify this by analyzing the finite motion intensity distribution $I_n(\omega)$, being the sum of harmonics $n = \omega/\omega_0$, where

$$\omega_0 = (2|E|)^{3/2}/Ze^2 \sqrt{m} \tag{2.20}$$

is the typical frequency of the finite motion (the analog of the frequency $\tilde{\omega}$ in continuum spectrum). The intensity I_n of the given harmonic is equal to [2.5]:

$$I_n \propto n^2 E^4 (1 - \tilde{\varepsilon}^2)^2 \left\{ K_{1/3}^2 \left[\frac{n}{3}(1 - \tilde{\varepsilon}^2)^{3/2} \right] + K_{2/3}^2 \left[\frac{n}{3}(1 - \tilde{\varepsilon}^2)^{3/2} \right] \right\} , \tag{2.21}$$

where $\tilde{\varepsilon} = (1 - 2|E| M^2/Z^2 me^4)^{1/2}$ is the eccentricity of the elliptical trajectory.

It is simple to ensure that the argument of the K-functions in (2.21) is reduced, as in the continuum spectrum case, to the universal variable $u \sim M^3 \omega/Z^2$. Independence of the energy spectrum is realized for the radiation intensity of the classical motion averaged over the period $T = 2\pi/\omega_0$, namely for the quantity [2.5]

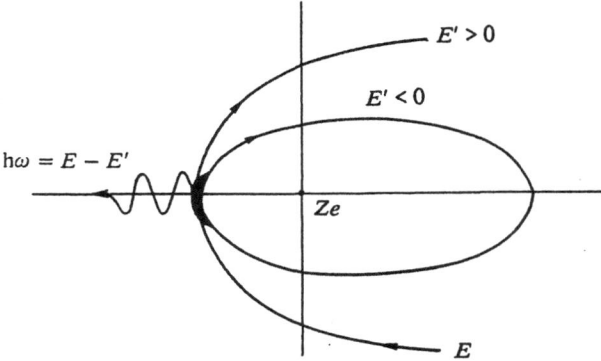

Fig. 2.1. Electron trajectories for free-free ($E'>0$) and free-bound ($E'<0$) radiative transitions in a Coulomb field. The mutual radiative part of the trajectories is marked

$$T\,dI = TI_n\,dn = \frac{2\pi}{\omega_0}\,I_n\,\frac{d\omega}{\omega_0} \propto \frac{I_n}{\omega_0^2}\,d\omega \ . \tag{2.22}$$

One can see that the quantity (2.22) becomes independent of the electron energy after substitution of (2.21).

To summarize, it should be noted that the high frequency "Kramers" spectral domain possesses a universal intensity distribution for transitions in both continuum and discrete spectra. The universality is connected with the SEI-effect realized for attractive atomic potentials. The essence of the "universality" effect for spectral distributions is demonstrated in Fig. 2.1 where there are the hyperbolic (infinite) and elliptical (finite) electron trajectories with identical radiating sections of trajectories independent of the initial energy ($E \gtrless 0$).

2.2 Symmetry Properties of the Coulomb Field

As is known [2.8,9], the Coulomb field possesses some additional symmetry properties. These properties involve the energy E and the orbital momentum L in the presence of an additional integral of motion, the Runge-Lentz vector A, with A is equal to

$$A = v \times L - err^{-1} \ . \tag{2.23}$$

The physical meaning of vector A in classical mechanics follows from its connection with the coordinate $\{r\}$ averaged over the period of electron motion:

$$A = -\frac{2}{3}\frac{e^2}{a}\langle r \rangle \quad (a = e^2/2|E|) \ . \tag{2.24}$$

The quantum mechanical generalization of the Runge-Lentz vector is the operator

$$A = \frac{1}{2m}(p \times L - L \times p) - err^{-1} , \tag{2.25}$$

where p, L, r are operators of momentum, orbital momentum and electron coordinates, respectively.

All components of the operator A commute with the Hamiltonian H. The remainder of the commutation relations take the form (e_{ijk} is an antisymmetrical tensor)

$$\left. \begin{array}{l} [L_i, L_j] = ie_{ijk}L_k , \quad [L_i, A_j] = ie_{ijk}A_k \\ [A_i, A_j] = -2im^{-1}e_{ijk} - L_kH \end{array} \right\} . \tag{2.26}$$

Operators L and A are mutually perpendicular:

$$LA = AL = 0 \tag{2.27}$$

which satisfy the relation

$$A^2 = \frac{2}{m}\hat{H}(L^2 + 1) = e^2 . \tag{2.28}$$

It is seen from the relations (2.26, 28) that from vectors A and L one may construct new vectors satisfying the commutation relations quite analogous to orbital momenta if one limits one's self to the subspace of atomic states belonging to a fixed value of energy E (isoenergetic surface $H = E$). Introducing new operators

$$J_{1,2} = \frac{1}{2}\left[\hat{L} \pm \left(\frac{m}{a}\right)^{1/2}\hat{A}\right] , \tag{2.29}$$

we can see that they satisfy the usual commutation relations for orbital momentum:

$$[J_{1i}, J_{1j}] = ie_{ijk}J_{1k}; \quad [J_{2i}, J_{2j}] = ie_{ijk}J_{2k} , \tag{2.30}$$

with operators J_1 and J_2 commuting with each other.

Thus, the subspace of atomic states belonging to the given isoenergetic surface $E = \mathrm{const.}$ possesses some transformation symmetry properties which may be described by the rotation of the vectors $J_{1,2}$ in two independent subspaces.

Equation (2.27) results in the equality of the squares of the vectors and their commutivity allows their simultaneous diagonalization to be achieved:

$$\hat{J}_1^2 = \hat{J}_2^2 = j(j+1) \ , \tag{2.31}$$

where the magnitude momentum of the j defined from (2.28) is equal to $(n-1)/2$. Correspondingly, the projections $m_{1,2}$ of the vectors vary within the limits:

$$m_{1,2} = -j, \ -j+1, \ldots, +j; \ j = \frac{n-1}{2} \ . \tag{2.32}$$

The presence of the two commuting vectors J_1 and J_2 enables the construction from wave functions with a given n the different states corresponding to definite magnitudes of the vector projections on different directions. The vectors J_1 and J_2 are independent and the directions of the vector quantization may not coincide. It is also possible, just as in angular momentum theory, to construct states corresponding to the squares of a linear combination of the vectors J_1 and J_2 and one of their projections. The convenience of using this or that representation becomes apparent in the presence of the diagonalization of perturbations possessing different symmetry. Essentially the aforementioned method of constructing states is an extension of a well known method of variable division for a Coulomb field in spherical and parabolic coordinates.

Let us briefly dwell on the connection of parabolic and spherical quantization using the formalism mentioned above.

As is well known [2.8, 9], parabolic wave functions correspond to definite projections on one and the same extracted direction of orbital momentum L defined by the quantum number m and the dipole momentum d, or the same Runge-Lentz vector A defined by the difference of the parabolic quantum numbers $n_2 - n_1$.

Taking into account the definition (2.29) of vectors $J_{1,2}$, it is easy to ascertain that the following relations are correct:

$$\left. \begin{aligned} (J_1)_Z = m_1 = \frac{m+n_2-n_1}{2}; \ (J_2)_Z = m_2 = \frac{(m+n_1-n_2)}{2} \\ L_Z = m, \ \left(\frac{m}{a}\right)^{1/2} A_Z = n_2 - n_1 \ . \end{aligned} \right\} \tag{2.33}$$

Thus, the parabolic state corresponds to a definite z-projection of vectors J_1 and J_2 (along with vectors L and A) in a given direction.

It is also known that the spherical states φ_{nlm} correspond to definite magnitudes of $L^2 = (J_1 + J_2)^2$, and the projection of L_Z equals m. Since the vectors J_1 and J_2 obey the usual rules of momentum composition it is not hard to establish a connection between the wave function in parabolic $(J_1^2, J_{1Z}, \ J_1^2, J_{2Z})$ and spherical $((J_1+J_2)^2 = L^2, J_1^2, J_2^2, L_Z)$ coordinates, or bases. Using the standard results of orbital momentum theory for the calculation of the transformation coefficient

$$\varphi_{n_1 n_2 m} = \sum_{l,m'} \langle n_1 n_2 m \mid nlm' \rangle \varphi_{nlm'} \tag{2.34}$$

we arrive at the relation [2.8, 9]

$$\langle n_1 n_2 m \mid nlm' \rangle = (-1)^{n-1+(2n_2 \pm |m|-m)/2} C \left[\frac{n-1}{2} , \frac{n-1}{2} , l ; \right.$$

$$\left. \frac{m+n_2-n_1}{2} , \frac{m+n_1-n_2}{2} , m' \right] \tag{2.35}$$

where $C[j_1, j_2, j; m_1, m_2, m]$ are the usual Clebsch-Gordan coefficients [2.9].

The result (2.34, 35) is a specific case of the application of Coulomb field symmetry. As has been discussed it is possible to construct the wave functions of a more complicated symmetry, corresponding to the projections of the vectors J_1 and J_2 on different directions. An example of such a construction will be performed in Chaps. 4, 5.

2.3 Nonhydrogenic Atoms. Allowed and Forbidden Transitions. Properties of Multicharged Ion Spectra

Below we briefly state the main properties of nonhydrogenic atoms and multicharged ions which will be used later. The spectral calculation methods for multielectron atoms has been widely discussed in the literature [2.10, 11]. Here, we shall dwell only on some qualitative spectral peculiarities which are used for plasma diagnostics, and we will introduce the main definitions connected with these spectra.

2.3.1 Nonhydrogenic Atomic Spectral Structure. Allowed and Forbidden Transitions

The main difference between the multielectron atomic spectra and the hydrogenic spectra is connected with the difference of the potential $V(r)$ of the atom from the Coulomb potential. As a consequence of this circumstance "accidental" level degeneration is removed over orbital quantum numbers l, so that the transitions from sublevels responding to different values of l become spectroscopically distinguishable. The corresponding selection rules for individual transitions $l \rightarrow l'$ become simple as compared with those for the hydrogen atom with two bases of quantization, namely, parabolic and spherical [2.2]. As expected from the selection rules, the transitions are divided into dipolar allowed and forbidden transitions. For example, in the $4d-2p$, $4f-2p$ transitions in the helium atom, the first one is allowed and the second one is forbidden.

As is known, nonhydrogenic atoms under the influence of an electrical field demonstrate an energy shift proportional to the square of the field strength (the quadratic Stark effect). The theory of the quadratic Stark effect is a typical example of second-order perturbation theory [2.12]. In fact, in the absence of accident degeneration over the orbital quantum number l, the diagonal matrix elements of the perturbation $V = -dF$ turn to zero and the energy shift ΔE arises only in second-order perturbation theory:

$$\Delta E_n^{(2)} = \sum \frac{V_{nm} V_{mn}}{\omega_{nm}} \, . \tag{2.36}$$

The summation in (2.36) covers all intermediate states; however, the main contribution to the sum is made by nearby atomic levels. For example, in the helium atom these levels respond to a change in the orbital momentum quantum number at a value of ± 1. The calculation of the matrix elements $V_{ll'}$ between the states mentioned gives the following results [2.12]:

$$\Delta E_{nlm}^{(2)} = C_4(nlm)F^2, \quad C_4(nlm) = \frac{9}{4} n^2/(Z-1)^2(2l+1)$$

$$\times \{[n^2-(l+1)^2][(l+1)^2-m^2]/(2l+3)(E_{nl}-E_{nl+1})$$

$$+(n^2-l^2)(l^2-m^2)/(2l-1)(E_{nl}-E_{nl-1})\} \, . \tag{2.37}$$

According to (2.37) the energy shift depends on the "spherical" quantum numbers n, l, m, the dependence on n being very strong (which represents the increase of the polarizability of the exited states).

It should be noted that when applying a field F there may be changes not only in energy but also in the wave functions of the states. These changes are connected with the "mixing" of nearby atomic states which leads to the violation of the selection rules for dipolar radiative transitions. Let us reduce the estimation of the ratio R of the intensity I_F $(nl-n_0 l)$ of the forbidden atomic line responding to transitions in helium to the intensity $I_A (nl-n_0 l+1)$ of the allowed atomic line [2.12]

$$R(nlm) = \frac{I_F(nl \rightarrow n_0 l)}{I_A(nl \rightarrow n_0 l+1)} = \frac{9}{4} F^2 [n^2-(l+1)^2] n^2 [(l+1)^2-m^2]/$$

$$\times [4(l+1)^2-1]/(E_{n,l+1}-E_{nl})^2 \, . \tag{2.38}$$

The spectral lines corresponding to the transitions between the exited helium levels lie in the visible and therefore they are convenient for electric field diagnostics in plasmas.

As the electric field strength F increases (or with the transition to hydrogen-like highly exited atomic states), the quadratic Stark effect becomes a linear effect. The transition is often observed in sufficiently dense plasmas. It is a typical situation when for the same transition $n \to n_0$ some atomic states experience a linear Stark effect, whereas other states a quadatric Stark effect.

The energy spectrum of a multi-electron atom is the spectrum of only one of its electrons. The rest of the electrons form an atomic (or ionic) core. The core is essentially an additional (third) body having influence on the interaction of an exited electron with the nucleus. The influence consists of an opportunity of core excitation leading to the appearance of a new type of transition, connected with the excitation of two (or more) atomic electrons. Examples of such transitions are considered below in Sects. 2.3.4, 4.4, and 9.1.

One of the important peculiarities of radiative transitions of an atomic electron in external shells of a neutral atom is the smallness of relativistic effects due to the smallness of the parameter $e^2/\hbar c \ll 1$. Therefore the transitions of higher multipolity (for example, quadrupole transitions as well as magnetic dipole transitions) are strongly depressed (forbidden) as compared to electric dipole transitions. Here lies the difference of the spectra mentioned with spectra of multiply charged ions [2.3.2].

2.3.2 Properties of Multicharged Ions (MCI) Spectra

A vast amount of literature [2.10, 11] is devoted to calculations of multiply charged ion (MCI) spectra. Here we are only interested in the general properties of these spectra for large nuclear charges $Z \gg 1$, which will be used later for the estimation of different effects. If the number of bound electrons is small compared with Z then their motion is determined mostly by their interaction with the nucleus $V_{ie} \sim Ze^2/r$. Taking into account that $r \sim n^2/Z$ for a typical bound energy I_n of the electron on the atomic level n one obtains the hydrogen-like formula: $I_n \sim Z^2/2n^2$. The difference of the transition energy $\Delta E_{nn'}$ is obviously given by the Rydberg formula (atomic units = a.u.):

$$\Delta E_{nn'} = \frac{Z^2}{2} \left(\frac{1}{n^2} - \frac{1}{n'^2} \right) . \tag{2.39}$$

The interaction of electrons with each other $V_{ee} \sim r^{-1}$ leads to the splitting of energy sublevels inside the given n at a value of $\sim Z$ (transitions with $\Delta n \equiv n - n' = 0$). The expansion may be continued and we obtain a series over degrees of Z^{-1} [2.10]. For example, for transitions in Li-like ions, the following relationship is obtained [2.10]:

$$\Delta E_{2s-2p} = 0.0707 Z - 0.120 . \tag{2.40}$$

Analogous expansions may be obtained for transition oscillator strengths $f = 2\omega_{nn'} |d_{nn'}|^2$. For the same transitions with $\Delta n = 0$ in Li-like ions we obtain [2.10] (with account of $\omega \propto Z, d \propto Z^{-1} \dots$):

$$f_{2s-2p} \approx 1.35 Z^{-1} + 2.20 Z^{-2} . \tag{2.41}$$

The dependence on Z of the radiation transition width γ_r follows from the formula $\gamma_r = \frac{2}{3}\omega^3 d_{nn'}^2 c^{-3}$, that leads to

$$\gamma_r = Z^4 \gamma_{r_0} . \qquad (2.42)$$

In the same manner, there are increases in the magnitudes of the atomic level's fine splitting Δ_{fs} and Lamb shift Δ_L. For hydrogenic ions we have [2.12] (j being the orbital momentum):

$$\Delta_{fs} = -\frac{Z^4}{2n^3 c^2} 1/(j+1/2) , \quad \Delta_L(l=0) = \frac{4Z^4}{3\pi n^3 c^3} \ln \frac{1}{Z\alpha} . \qquad (2.43)$$

The detailed calculations of the Lamb shift are performed in [2.13]. The frequencies of the transitions between the fine structure sublevels extend into the visible and even the ultraviolet spectral domains for values $Z \gtrsim 20$.

For nonhydrogenic ions the spin-orbit interaction must be taken into account along with the interaction of the excited electron with a core, and the former may have the same order of magnitude as the latter one. For hydrogen-like ions with $Z \gg 1$, fine splitting leads to the partial removal of the degeneracy inside of the given atomic level, splitting it into separate components with different values of total momentum j. The splitting may be large enough so that when an electric field is applied, some components experience a quadratic Stark effect, whereas the others the linear one. In principle the calculations of the Stark effect for H-like ions does not differ in any way from the corresponding calculations for neutral hydrogen [2.14 – 16]. However, the scale of the electric field strengths causing the analogous splitting picture in a H-like ion is increased Z^4 times as compared with the neutral atom case in accordance with the increase of fine splitting.

The energy-level behavior repeats the corresponding results of *Luders* [2.15] for neutral hydrogen; however, the magnitudes of the field strengths are sharply increased. It is evident that for one and the same magnitude of field strength F some components (for example $2p_{3/2}$ at the level $n = 2$) possess quadratic and others ($2p_{1/2}$, $2s_{1/2}$)-linear splitting over the field. In general, the case for intermediate values of F, the terms' behavior depends on F and it is complicated enough coming to an analytical expression for low levels only.

The interaction V of the atomic dipolar momentum d with a field F in plasmas depends weakly on Z. In fact, the magnitude d decreases with increasing Z ($d \propto Z^{-1}$), whereas the field F created by the ion with charge Z at a distance R increases with Z ($F \sim ZR^{-2}$). On the whole, for a Stark structure observation of multiply charged ions it is necessary to have sufficiently high densities to provide great magnitudes of field strengths F at an average interparticle distance, securing the splitting which may contain fine structure.

The action of the electric field F on H-like ions leads, as in the case of helium (Chap. 1), to removing bans on atomic transitions. In particular, the effects of atomic state "mixing" causes the destruction of the metastable $2s$-level in the field F. Indeed the squared amplitude due to changing of its lifetime [2.9, 12]:

$$|a_s(t)|^2 \sim e^{-I_m E_s(F)t} \quad , \quad I_m E_s(F) = -\frac{\gamma}{2}\left(1 - \frac{\Delta E_{sp}}{\sqrt{\Delta E_{sp}^2 + 4d_{sp}^2 F^2}}\right) \qquad (2.43)$$

where γ is the expansion rate of the $2p_{1/2}$ level, ΔE_{sp} is the Lamb shift dividing the metastable state $(2s_{1/2})$ and radiating state from the $(2p_{1/2})$ atomic levels. Given a large-enough magnitude of the field, the lifetimes of the metastable and radiating levels become equal. The role of these effects in plasmas is examined more in detail in Chap. 9.

The calculations of the Stark spectral splitting of nonhydrogenic ions with regard to fine structure require simultaneous consideration of both the inter-electron and spin-orbital interactions. The latter can be taken into account with the help of perturbation theory in the Pauli approximation up to ion charges $Z \approx 30$ [2.17]. The example of the calculated spectrum of the He-like ion Ar^{+16} is shown in Fig. 2.2. It is evident that the term structure is complex. The same is true of the atomic state wave functions. Therefore, as a rule the spectral line intensity calculations are carried out with the help of numerical methods.

The increase of Z leads to a sharp increase in forbidden transitions. Thus, for example, the probabilities of quadrupole transitions E2 are $(Ze^2/\hbar c)^2$ times smaller than the probabilities of allowed dipole transitions and as a consequence they increase sharply with increasing Z. The width of the two-photon transition $2s-1s$ increases in a multicharged ion as [2.17]

$$\gamma_{2s-1s} = (8.2283 \pm 0.0001)Z^6 (s^{-1}) \qquad (2.44)$$

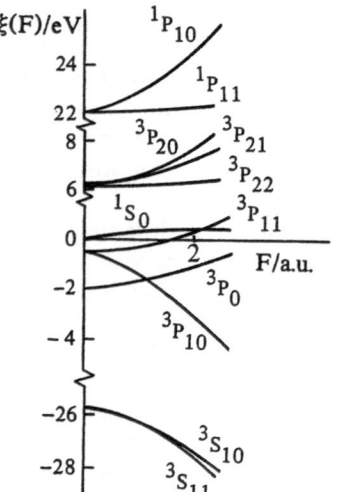

Fig. 2.2. The shift of the energy term of the $1s2l$-level of the He-like ion Ar^{16+} in an electric field F [2.14]; the terms are marked by the value M_z of the total orbital momentum projection on the field direction

and for values as low as $Z = 20$ becomes of the same order of magnitude as the width of the allowed transition $2p - 1s$ in neutral hydrogen. The probabilities of magnetic dipole transitions (M1) increase analogously.

The transition probabilities M1 and E2 between the fine structure sublevels increase especially sharply. Indeed, for such a transition, the transition frequency ω itself increases proportional to Z^4. Therefore, radiation widths proportional to ω^3 increase like Z^{12} or even more sharply. For example, for the transitions $^2P_{3/2} - ^2P_{1/2}$ in the $1s2p$ configuration, we obtain [2.10]:

$$W(\text{M}1) \approx 0.44 \times 10^{-12} Z^{12} \text{s}^{-1} \, , \quad W(\text{E}2) \approx 1.31 \times 10^{-22} Z^{16} \text{s}^{-1} \, . \quad (2.45)$$

The sharp increase of transition probabilities, as expressed in (2.45), leads to the opportunity of observation even under the conditions of laboratory plasmas. The example is the M1 transition in the FeXX ion with a wavelength of $\lambda = 2665$ Å observed in the plasma of the Princeton Tokamak PLT. The corresponding radiation width of the transition is equal to 570 s^{-1} [2.18].

From the above, it is evident that the main peculiarity of MCI spectra is a sharp increase in the number of transitions as compared with the neutral atom's spectra. This is explained as stated above by removing the restrictions on transitions with increasing Z. Under rarified high-temperature plasma conditions of the Sun or of thermonuclear systems, the intensities of forbidden transitions are comparable with the intensities of the allowed transitions. It is explained by the fact that for values as small as $Z \gtrsim 20$ the radiation decay rates γ_r both for the allowed and forbidden transitions begin to dominate over the rates of the states collisional excitation. Under such conditions of "coronal"

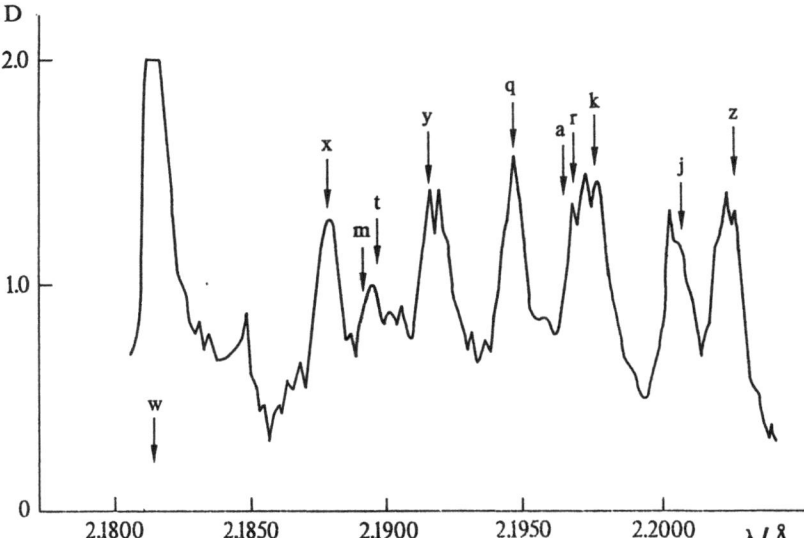

Fig. 2.3. The spectrum of the He-like ion Cr^{+22} near the resonant w-line observed in Tokamak plasmas [2.19]

equilibrium taking place in the Sun's corona, the conditions of the line intensities are obviously determined by atomic states excitation rates (since every act of excitation is certainly accompanied by radiation). The excitation rates of different sublevels, unlike radiation transition's rates, are of comparable values [2.18].

As an example, Fig. 2.3 gives the "corona" spectrum of chromium observed in the Tokamak T-10 [2.19]. Here, the w-line corresponds to the dipole allowed transition $2p-1s$ in the He-like ion Cr XXIII, the rest of the lines (having the standard designations) correspond to the so-called dielectronic satellites as well as to the forbidden transition (x, y, z). It is shown that the intensity of these latter transitions is comparable with the intensity of the w-line.

2.4 Auto-Ionization States. Stationary (Fano) and Time-Dependent (Kompaneets) Descriptions

2.4.1 Auto-ionization States

The excitation of a number of electrons leads to the appearance of new auto-ionization lines in atomic or ionic spectra. The classic examples here are the simplest auto-ionization states of a two-electron helium atom. The peculiarities of these states are connected with the fact that strictly speaking they are not discrete but have a finite width Γ connected with the possibility of the atom's disintegration in the auto-ionization state into an ion and a free electron.

To discern the essence of the matter let us consider an approximate model of the auto-ionization helium state $2s2p\,^1P$ emerging during the excitation of both of its electrons at the atomic level $n = 2$. The usual helium spectrum is the spectrum of excitation of one of its two electrons. This energy spectrum corresponds to the system of rigorously discrete levels, the first excited state, the $n = 2$ level, lies approximately 20 eV above the ground state. These discrete levels condense towards the boundary of the discrete spectrum-continuum, situated approximately 25 eV above the ground state.

Let us assume now that the atom is divided into two parts, namely: the excited electron is in the state $n = 2$ ($\Delta E_{12} \simeq 20$ eV) and the remaining one-electron ion is called the core. The energetic levels of such a core are hydrogen-like and they respond to the nuclear charge $Z = 2$, so that its excitation energy into the state $n = 2$ is approximately equal to $\Delta E_c = Z^2 \text{Ry} (1 - 1/4) \simeq 40$ eV. Hence, the excitation energy of two electrons (the external and the core electrons) at a given time is approximately equal to $\Delta E^{(2)} = \Delta E_{12} + \Delta E_c \simeq 20 + 40 = 60$ eV. Thus, the helium atom has a discrete state with the energy $\Delta E^{(2)} \simeq 60$ eV exceeding the atomic ionization potential $I \approx 25$ eV, that is, it is situated deeply in the continuum (or in the background of the continuum). However, this state is discrete only in the zeroth approximation when we have neglected the interaction of electrons between each other. Taking into account electron interaction leads to the decay of the auto-ionization state. One can imagine the mechanism of such a decay in the following way: the inner core electron makes a transition to the ground state $(1s)$ and the external electron is thrown out into continuum. Such a process is named "core relaxation" and is accompanied by throwing out the external electron (Oge-effect). Though the division of atomic electrons into the "core" and "external" parts is relative, it is at the same time rather visual and we shall use it in future discussion.

2.4.2 The Interaction of Discrete States with a Continuum. Fano and Kompaneets Descriptions

As stated above, the auto-ionized states are nonstationary and experience the decay described by the auto-ionization width Γ. This decay is due to the interaction of the discrete state with a continuous spectrum of free electron states. The description of the auto-ionization decay processes is in many respects similar to the description of radiative decays for which the role of the continuum is played by a number of electromagnetic field harmonics, as discussed in Sect. 9.1. A corresponding theory for auto-ionized states may be developed by analogy with the general theory of states damping. The initial research was performed by *Fano* [2.21] as applied to photoeffect problems.

The essence of *Fano's* approach involves the construction of correct sationary wave functions of a system of a "discrete level-continuum", including the interaction between them. The problem is the generalization to the continuum spectrum from the known problem of the energy spectrum of two interacting, discrete levels. The correct wave functions constructed are used later for the calculation of the transition probabilities between them and another discrete atomic level, or the atomic ground state for photoeffect problems, in particular.

Another approach was developed quite independently and somewhat later by *Kompaneets* [2.22]. In contrast to the *Fano* method [2.21] the approach is based directly on time-dependent approaches being connected by a Fourier-transform and they give an identical result for the final transition probability. Nevertheless, the non-stationary approach far more easily allows further advancement into the domain of strong interactions (Chap. 9). Thus, we shall briefly state both the approaches mentioned.

The existence of auto-ionized states leads to the appearance of sharp peaks for the probabilities of processes with the participation of a continuous spectrum, like photoionization, inelastic scattering and so on.

To calculate effects of this kind, *Fano* [2.21] suggested a method, based as stated above on the determination of a mixed state of the system "discrete level-continuum" by the diagonalization of their interaction. The essence of his methods encompass the following. Consider the Hamiltonian

$$H = H_0 + V \tag{2.46}$$

having the states of discrete ψ_n and continuous ψ_ν spectra

$$\left.\begin{array}{l} \langle \Psi_n | H | \Psi_{n'} \rangle = E_n \delta_{nn'} \\ \langle \Psi_v | H | \Psi_{v'} \rangle = E_v \delta(E_v - E_v') \ . \end{array}\right\} \tag{2.47}$$

Here E_n and E_v are the energies of the discrete and continuous states. The magnitude V depicts the interaction between the discrete and continuous states:

$$\langle \Psi_v | H | \Psi_n \rangle = V_{vn} \ . \tag{2.48}$$

Let us find the eigenfunctions of the Hamiltonian (2.46). For this purpose, let us expand the wave function Ψ over the states of discrete and continuous spectra

$$\Psi = a(E) \Psi_n + \int dE' \frac{dv}{dE'} b_{E'} \Psi_{v'} \ , \tag{2.49}$$

where the indices of the continuous states indicate the integration over the energy E by introducing the density of states dv/dE.

Fano [2.21] determined the coefficients $a(E)$ and $b_{E'}$ from the wave function (2.49) of the discrete level against the background of the continuum. The result for coefficient $a(E)$ takes the form:

$$|a(E)|^2 = \frac{|V|^2}{(E - E_n - \Delta)^2 + \pi^2 |V|^4} \ . \tag{2.50}$$

It is seen from (2.50) that the initial discrete state runs into the resonant curve with a half width of $\pi |V|^2$ determined by the interaction with the continuum. This means that if at some moment in time the system has been in state ψ_n then its average life time determined with respect to the decay into the continuum is equal to $\hbar/2\pi |V|^2$. The further use of the *Fano* method consists in the calculation (with the help of perturbation theory) of a transition probability from some ground state ψ_0 into a determined mixed state ψ (2.49).

Kompaneets [2.22] suggested a non-stationary approach to the description of interactions with a continuum for resonance phenomena in the photoeffect. The basis of such a theory involves taking into account three types of atomic states, namely: the initial state with amplitude c_0, the discrete auto-ionized state c_n, and the continuum state c_E. The matrix elements of the transitions from the initial state to the level n is designated by H_{on}, one to the continuum by H_{0E}, and at least the interaction with the continuum characterizing the auto-ionization decay is designated by $V_{n,E}$. Then, for the amplitudes mentioned, we have the system of equations:

$$i\dot{c}_E = V_{E,n} e^{i\omega_{En}t} c_n + H_{0E} e^{i(\omega_E - \omega)t} c_0 \tag{2.51}$$

$$i\dot{c}_n = \sum_{E'} V_{nE'} e^{i\omega_{E'n}t} c_{E'} + H_{0n} e^{i\Delta\omega t} c_0 \ , \tag{2.52}$$

where $\omega_{En} \equiv \omega_E - \omega_n$, $\Delta\omega \equiv \omega - \omega_n$.

The solution of the system is performed with the help of perturbation theory and assuming $c_0 = 1$. Writing down c_E from (2.51) in quadrature form

$$c_E = -ie^{i(\omega_E - \omega)t} \int_0^t (H_{0E} + V_{E,n} c_n e^{i\Delta\omega t}) e^{i(\omega_E - \omega)t'} dt' , \qquad (2.53)$$

let us substitute (2.53) into (2.52). Making the transition in (2.52) from the sum $\sum_E$ to the integral over energy $\varrho(E) dE$ with the density of states $\varrho(E)$ it is correct for $\omega t \to \infty$ to take out the slowly varying matrix elements at point $E' = \omega$. It results in the following relation:

$$\sum_{E'} V_{n,E'} c_{E'} e^{i(\omega_{E'} - \omega)t} = -\pi i [V_{n,\omega} H_{0\omega}\varrho_\omega + |V_{n\omega}|^2 \varrho_\omega c_n \exp(-i\Delta\omega t)] . \qquad (2.54)$$

Substitution of (2.54) into (2.52) leads to an ordinary differential equation for c_n. Substituting its solution into (2.54), one finds the probability amplitude of the population of the continuum state c_E. Raising the modulus of c_E to the second power, performing the integration over some continuous spectral interval E and performing the limit $t \to \infty$ one obtains the final expression for the transition probability W:

$$W = 2\pi\varrho(\omega) \frac{|H_{0\omega}(\omega_n - \omega) + V_{\omega,n} H_{0n}|^2}{(\omega_n - \omega)^2 + (\pi|V_{\omega,n}|^2 \varrho_\omega)^2} . \qquad (2.55)$$

Equation (2.55) coincides with the *Fano* formula [2.21]. Far from resonance, $\Delta\omega \gg \Gamma = \pi|V_{\omega,n}|^2 \varrho_\omega$, the probability is reduced to the probability of the direct photoeffect:

$$W_{ph} = 2\pi\varrho_\omega |H_{0\omega}|^2 . \qquad (2.56)$$

In the general case, it describes the resonance process of the transition into the continuum with regard to the interference of the direct and auto-ionization channels. The introduction of the dimensionless *Fano* parameters

$$x = \frac{\Delta\omega}{\Gamma} , \quad q = \left| \frac{H_{n0}}{H_{0\omega} 2\pi\varrho_\omega V_{n\omega}} \right| \qquad (2.57)$$

shows that the form of the resonance curve is described by the function

$$f(x) = \frac{(x+q)^2}{x^2 + 1/4} , \qquad (2.58)$$

having a typical asymmetrical form.

Let us discuss the physical sense of the Fano formula (2.55). First of all, it is evident that the photoionization process is of a resonant type with a probability which increases sharply for $\Delta\omega \leq \Gamma$. The first element in the numerator in (2.55) describes the decay of the initial state c_0 into the continuum with amplitude $(W_{ph})^{1/2}$ due to dircet photoionization without any connection with state c_n. The second term describes the decay of the level c_0 into the continuum over the intermediate state c_n, namely, the decay amplitude is proportional to the matrix element H_{on} of the transition from c_0 into c_n, multiplied by the amplitude $\sqrt{\Gamma}$ of the decay of state c_n. Thus, (2.55) allows for two opportunities for the state c_0 to decay into the continuum, namely directly (with probability W_{ph}) and over state c_n. Both states c_0 and c_n "resound" against the background of the continuum in the interaction which leads to the decay of the states with probabilities W_{ph} and Γ as well as to the additional interference connection $(W_{ph}\Gamma)^{1/2}$ of these states. The connection is obviously realized by transitions from level c_0 into the continuum and from the continuum into level c_n. More detailed analysis of the state interaction against the background of a continuum will be presented in Sect. 9.1.

2.5 Rydberg Atomic States in Plasmas

We now pay considerable attention to highly excited (Rydberg) states of atoms and ions. In recent years the investigation of Rydberg states has aroused keen interest [2.23]. The urgency of Rydberg states for plasma applications is connected with the high multiplicity of ionization atom states bringing their spectra to hydrogen-like ones, and with the effective population of highly excited atomic states in a plasma medium.

The usual channel of Rydberg state population for a low temperature plasma is a three-part recombination and cascade excitation, and for high temperature plasmas these are photo and dielectronic recombinations (Chap. 4). An important population channel for MCI excited states is their charge exchange on neutral atoms which exist already or are introduced into plasmas for heating. The typical values of the main quantum numbers n of the excited states total to about $n \sim 20$, for hydrogen-like ions and are $10 \leq n \leq 100$ for multi-charged ions. The special case is atoms in rarefied astrophysical plasmas excited as a rule by photo excitation and leading to very great magnitudes of n up to $n = 400 - 600$ [2.23].

As a rule, the description of the Rydberg states is connected with quasiclassical or pure classical methods. A difficulty arises in that the direct application of quantum mechanical methods with the increase of quantum number magnitudes leads to great calculational difficulties caused by the sharp increase in the number of quantum states ($\propto n^2$). Therefore, the precision of quantum calculations usually diminishes with the increase of n. At the same time it is clear that classical methods should work successfully for large n. We shall now determine the efficiency of the classical methods utilized in the description of the Rydberg states.

3. Radiation Transition Probabilities and Radiation Kinetics in Kramers' Electrodynamics

The quasiclassical calculation methods for radiative transition probabilities and the kinetics of radiative cascades in an atom are presented below. It will be shown that the processes mentioned may be taken into account on the basis of pure classical electrodynamics that is connected with the definite peculiarities of electron motion in attractive potentials (Sect. 2.1).

3.1 Quasiclassical Transition Probabilities

The quantum mechanical transition probabilities $nl \to n'l \pm 1$ in a Coulomb field are determined, as known, by the radial matrix elements of dipole momenta:

$$R_{nl}^{n'l\pm1} = \int_0^\infty R_{nl}(r) R_{n'l\pm1}(r) r^3 dr \ , \tag{3.1}$$

where $R_{nl}(r)$ is the radial wave function in a Coulomb field.

Let us consider following [3.1, 2] the integral (3.1) in the quasiclassical domain $n, n' \gg l \gg 1$. For this purpose, we substitute the quasiclassical wave functions in (3.1)

$$R_{nl}(r) = \frac{\dot{a}_n}{r p_{nl}^{1/2}(r)} \cos \left(\int_{r_{nl}}^r p_{nl} dr - \frac{\pi}{4} \right) \ , \tag{3.2}$$

where the radial momentum and the normalized constant are equal to

$$p_{nl}(r) = \left(-\frac{1}{n^2} + \frac{2}{r} - \frac{(l+1/2)^2}{r^2} \right)^{1/2} , \quad a_n = (2/\pi n^3)^{1/2} \ . \tag{3.3}$$

Substituting (3.2) into (3.1) one may transform the product of cosines under the integral in the sum of cosines corresponding with the sum and difference of the impulses p_{nl} and $p_{n'l} \pm 1$ in their arguments. The member with the sum of momenta is small because of rapid oscillations of the function under the integral. The momentum difference in the argument of the second term may be estimated as

$$p_{nl} - p_{n'l\pm1} = \frac{(n')^{-2} - n^{-2} \pm 2er^{-2}}{2(2r - l^2)^{1/2}} r \ . \tag{3.4}$$

Substituting (3.4) in (3.1) one obtains after some transformations

$$R^{n'l\pm1}_{nl} = \frac{2l^2}{3^{1/2}\pi} \frac{(nn')^{1/2}}{(n^2 - n'^2)} \left[K_{2/3}\left(\frac{\omega l^3}{3}\right) \pm K_{1/3}\left(\frac{\omega l^3}{3}\right) \right] \ , \tag{3.5}$$

where $K_\nu(x)$ are McDonald functions, $\omega = (n'^{-2} - n^{-2})/2$.

Raising the result (3.5) to the second power and summing the probabilities of the transitions with $l' = l \pm 1$ we arrive at

$$w(nl \rightarrow n') = \frac{2l^4\omega}{9c^3\pi^2 nn'} [K^2_{2/3}(\omega l^3/3) + K^2_{1/3}(\omega l^3/3)] \ . \tag{3.6}$$

It is seen that the spectral dependence of the transition probability (3.6) precisely coincides with that determined using classical theory, as in Sect. 2.1. The circumstance is connected as described in Sect. 2.1 with the acceleration of the electron in the attractive potential followed by the exclusion of the electron energy from equations of motion of the radiating section of the trajectory (SIE effect, see Sect. 2.1). It follows from above that the transition probability calculations may be performed with the help of classical mechanics methods using classical intensity distributions obtained in Chap. 2.

3.2 Line Radiation (LR) Probabilities

Let us consider the results of Kramers' electrodynamics the application (KrED, see Sect. 2.1) to the transitions in the discrete energy spectrum. As it follows from the general properties of the KrED, the dependences of the spectral characteristics of the radiation remain the same as in the case of a continuous energy spectrum. The only additional fact to be taken into account is the discrete character of the energy spectrum which corresponds to the following relation between the emitted photon frequency ω and the difference between the initial E_{nl} and final $E_{n'l'}$ atomic state energies:

$$\omega = (E_{nl} - E_{n'l'})/\hbar \ . \tag{3.7}$$

The values of the energies E_{nl} should be taken from the results of the quantum mechanical calculations or from corresponding experimental data.

Equation (3.7) leads to the relationship between the spectral interval $d\omega$ of the emitted photon frequencies and the density dn' of the final states:

$$d\omega/dn' = 2\pi/T_{n'l'} \ . \tag{3.8}$$

Here $T_{n'l'}$ is the period of classical motion (in the general case, of only the radial period) of the electron with the energy equal to the energy $E_{n'l'}$ of the final state. The value T_{nl} is determined by the conventional formulae of classical mechanics [3.3–7] for the case of the central potential $U(r)$.

The general expression for the probability of the transition $\Gamma \rightarrow \Gamma'$ ($\Gamma \equiv \{nl\}$) may be obtained from the classical spectral distribution for the emitted energy whose terms are to be separated with respect to the increase and the decrease of the electron angular momentum:

$$\Delta E_\omega(\varrho)_\pm = \frac{2e^2 m\omega^4}{3\pi c^3} [F_\pm]^2 , \tag{3.9}$$

where the functions $F_\pm$ defined by (2.16) correspond to the processes of radiation emission with the increase (F_+) and decrease (F_-) of electron angular momentum.

In order to obtain the probability $W(\Gamma \rightarrow \Gamma')$ of a radiative transition per unit time we divide the quantity (3.2.3) by the energy $\hbar\omega$ of an emitted photon and by the period T_{nl} of classical motion with an initial energy state and then multiply the result by the final state density (3.8). Thus, we obtain

$$W_{\Gamma \rightarrow \Gamma'} = \frac{\Delta E_\omega(\varrho)}{\hbar\omega} \left| \frac{d\omega}{dn'} \right| \frac{1}{T_{nl}} = \frac{2\pi}{T_{nl} T_{n'l'}} \frac{\Delta E_\omega(\varrho)}{\hbar\omega} . \tag{3.10}$$

Equation (3.10), with account of (3.9) and the relation $M = \hbar(l+1/2) = mv\varrho$, takes the form

$$W(nl \rightarrow n'l\pm 1) = \frac{4}{3} \left(\frac{\omega}{c} \right)^3 \frac{me^2}{\hbar} \frac{1}{T_{nl} T_{n'l\pm 1}} F_\pm^2(\omega,l) . \tag{3.11}$$

This result coincides with the result of the corresponding quasiclassical calculation [3.2, 6] in the limit of $n \gg 1, l \gg 1$. The periods T_{nl} and $T_{n'l'}$ in the latter calculation originate from the normalization constants of the quasiclassical wave functions (for the relations between the functions $F_\pm$ and quasiclassical matrix elements [3.1–6]).

The KrED result for a bound-bound transition corresponds to the high-frequency domain where the emitted frequency ω largely exceeds the frequency of the electron revolution around the field center on its classical trajectory:

$$\omega_{nn'} T_{nl} \gg 1 . \tag{3.12}$$

The KrED method for the description of an electron bound-bound radiative transition in a central potential $U(r)$ may be considered as some alternative to the well-known quantum defect method [3.8–10]. The latter is based on the following relation for the final state density

$$\partial\omega/\partial n' = Z_i^2(\nu_{n'l'})^{-3} , \qquad \nu_{n'l'} = n'-\mu_{l'} , \tag{3.13}$$

where $\mu_{l'}$ is the quantum defect value. Equation (3.13) originates from the corresponding dependence for the Coulomb potential generalized onto the case of a non-integral quantum number n. The essential feature of the quantum defect method is the use of the Coulomb results for the spectral distribution of the transition probabilities with the subsequent replacement of the originally integral quantum number n by the non-integral quantity ν_{nl}. This approach may be ultimately interpreted as a Coulomb-type approximation for the potential $U(r)$.

It should be noted that the KrED approach does not require such an approximation for the potential. Thus, for free-free radiative transitions (bremsstrahlung) in the field of a many-electron atom, the use of the Thomas-Fermi (TF) potential in (2.5) leads to a successful description of the radiation spectral distribution (Sect. 2). The validity of the TF model for the description of bound-bound transitions as well as the comparison of corresponding results of the KrED with the quantum defect method [3.8] (and its classical analog [3.10]) for the case of an arbitrary deviation of the potential from the Coulomb-type form are to be investigated in the future.

The most detailed comparison of quasiclassical results for bound-bound transitions with the corresponding quantum numerical calculations has been carried out for the case of the Coulomb field [3.2]

$$\sum_{l'=l\pm1} W(nl\to n'l') = 2(l+1/2)G_0[\omega(l+1/2)^3/3Z^2]/3\pi^2c^3(nn')^3 , \tag{3.14}$$

where

$$G_0(x)\equiv x[K_{1/3}^2(x)+K_{2/3}^2(x)] = \frac{x}{2}\{[K_{1/3}(x)+K_{2/3}(x)]^2$$
$$+[K_{1/3}(x)-K_{2/3}(x)]^2\}\propto x[F_-^2(x)+F_+^2(x)] . \tag{3.15}$$

The integration of the spectral probability (3.15) over frequencies gives the total probability (per unit time) of the radiative decay of the state $\{nl\}$:

$$W^{\text{tot}}(nl) = 4Z^4/\pi\sqrt{3}\,c^3n^3l^2 . \tag{3.16}$$

The inverse quantity to (3.16) determines the mean lifetime of this state. Equation (3.16) weighted by the factor $(2l+1)$ and averaged over the values of angular momentum l, gives the probability

$$W(n) = 8Z^4\ln n/\pi\sqrt{3}\,c^3n^5 , \tag{3.17}$$

which is close to the results of quantum numerical calculations [3.1, 2, 11].

Using explicit expressions for the functions F_- and F_+ in case of the Coulomb potential, it appears possible to trace the origin of the success of the KrED approach. These functions determine the probabilities of transitions with the decrease and increase of the electron angular momentum, respectively. This fact can be proven in the framework of classical radiation theory by calculating the rate of angular momentum loss $dM/d\omega\,dt$ caused by the classical emission of radiation of a frequency ω (e.g., [3.12, 13] and Sect. 2). Though the net rate of angular momentum change is as it should be, negative, the term containing the function F_+ is positive and therefore corresponds to the increase of the electron angular momentum. Note that the relation between the functions F_+ and F_- indicated is especially transparent within the framework of the quasiclassical approach. In this case these functions correspond to the transitions with a positive and negative change of the electron orbital quantum number ($\Delta l = \pm 1$), respectively (3.14, 15).

In the Coulomb case the values of the function $F_-^2(x)$ largely exceed the values of the function $F_+^2(x)$ in a wide range, $x \geq 10^{-2}$. Figure 3.1 shows the ratio of the transition probability for $\Delta l = +1$ to the total (in Δl) transition probability. The predominance of the transition with $\Delta l = -1$ over a transition with $\Delta l = +1$ and the growth of this predominance with the growth of the transition frequency ($x \propto \omega M^3$) constitutes the essence of the wellknown Bethe empirical rule [3.11] derived originally from the results of quantum numerical calculations in the Coulomb case. However, it follows from our consideration [3.6] that the physical nature of this rule is purely classical. Indeed, this phenomenon can be clearly interpreted classically in terms of the correlation between the angular momentum and the polarization of the classical radiation. The qualitative explanation can be based on the fact that the intensity of the emission of the radiation, circularly polarized along the direction of radiating electron rotation, largely exceeds the intensity corresponding to the case of opposite directions of the electron and radiated electric field rotation (the situation is similar, e.g., to cyclotron radiation emission). The degree of the predominance discussed evidently predetermines the accuracy of the "rotational approximation" (RA) of Sect. 2.2. For the case of a Coulomb field, a

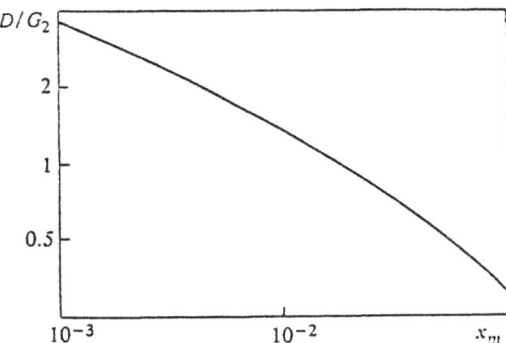

Fig. 3.1. The ratio of the function $D(x)$, which determines the deviation from the Bethe selection rule, to the function $G_2(x)$ which determines the probability of radiative transitions

quantitative estimate of the RA accuracy can be found from a comparison of the corresponding contributions of the $\Delta l = \pm 1$ transitions. Their ratio is equal to [3.2]:

$$\int_0^\infty x F_-^2(x)\,dx \Big/ \int_0^\infty x F_+^2(x)\,dx = \left(1 + \pi\sqrt{3}/6\right) \Big/ \left[\left(1 - \pi\sqrt{3}/6\right)\right] \simeq 20.5 \ . \tag{3.18}$$

Thus, the (integral in ω) accuracy of the RA is of the order of 5% and that agrees with the results of the classical calculations (Sect. 2.2). A detailed numerical comparison of the results of the quantum and quasiclassical calculations for the transition probabilities was carried out in [3.2]. The result of the calculation of the quantum corrections to the classical limit of the transition probabilities in the Kramers domain is presented in [3.5].

The degree of deviation of the transition probability from the *Bethe* rule can be clearly characterized by the function

$$D(x) \equiv x F_+^2(x) = x[K_{2/3}(x) - K_{1/3}(x)]^2 \ . \tag{3.19}$$

It is appropriate to designate this quantity as the "Bethe rule defect" (BRD). An useful analytic approximation of this function is presented in [3.6]. The aforementioned ratio, also shown in Fig. 3.1, can be written according to (3.19) in the form $D(x)/G_2(x)$ (for G_2, (3.26)). Figure 3.1 illustrates the wideness of the domain of the applicability of the Bethe rule.

Thus, the KrED method provides the clues for an universal description of the transitions between those discrete spectral states which dominantly contribute to the integral radiation emission rates. Even for the Coulomb case, in spite of its detailed investigation in the literature [3.3, 11], the KrED approach yields new, simple analytic results. Thus, the replacement of the unwieldy *Gordon* formulae [3.11] for the transition probabilities by the corresponding KrED formulae appreciably reduces the array of variables since these formulae contain a smaller number of independent variables. An application of the KrED method to the non-Coulomb case and a comparison with already existing methods (e.g., the quantum defect method) are now under investigation.

3.3 Photorecombination (PR) Cross Section

The SEI effect lying in the basis of the KrED approach manifests itself most strongly in the process of photorecombination radiation emission. Indeed, this process is surely strongly inelastic since the energy $\hbar\omega$ of an emitted photon is in any case larger than the initial energy E of the recombining electron:

$$\hbar\omega = E + |E_{nl}| \ , \tag{3.20}$$

where E_{nl} is the energy of the electron bound (final) state.

As was demonstrated in Sects. 1, 3, the SEI effect permits the use of the classical approach even for the description of such a strongly inelastic process. We shall use the universal classical formula (3.9) for the spectral distribution of the emitted energy to describe the photorecombination cross section. It may be achieved by means of the continuation of the corresponding results for the bremsstrahlung radiation (BR) cross section onto the domain of negative values of the final electron energy with account of its quantization law (3.8) (Sect. 2).

For the PR, in contrast to the BR, it is of essential interest not only to the cross section, integral over the orbital momentum l, but to the cross section $\sigma_{PR}(E \to n'l')$, differential with respect to l, as well. In order to obtain the latter cross section, one ought to replace the integration over the impact parameters ϱ in the BR formulae by a summation over the final state orbital momentum, $l' = l \pm 1$. Thus, we arrive at the result

$$\sigma_{PR}(E \to n'l') = \frac{\hbar^2 \pi l'}{mE} \frac{\Delta E_\omega(l')}{\hbar \omega} \frac{2\pi}{T_{n'l'}} \tag{3.21}$$

where the relation $\hbar(l' + 1/2) = \varrho m v$ is to be used in $\Delta E_\omega(\varrho)$.

The cross section (3.21) is a function of the atomic potential $U(r)$ which enters the spectral functions $F_-(\omega)$ and $F_+(\omega)$ and the period $T_{n'l'}$. A detailed comparison of the quasiclassical result (3.21) with exact quantum computations for the non-Coulomb potentials have not yet been performed. The Coulomb potential (3.21) has been investigated in [3.14, 15]. In this case, the spectral dependence of the PR cross section is described in terms of the universal spectral function $G_0(x)$ (3.15):

$$\sigma_{PR}(E \to nl) = 8 Z^2 (l + 1/2)^2 G_0 [\omega (l + 1/2)^3 / 3 Z^2] / 3 c^3 n^3 E . \tag{3.22}$$

Note that there is the universal dependence of the cross section on the classical parameter in the argument of the function G_0 similar to the case of the BR.

The total (integral over ϱ or l) PR cross section $\sigma_{PR}(E \to n)$, obtained by the integration of (3.22) for the Coulomb case gives the well known Kramers formula

$$\sigma_{PR}(E \to n) = \frac{8\pi}{3\sqrt{3}} \frac{Z^4}{c^3 n^3} \frac{1}{E\omega} , \tag{3.23}$$

where $\omega = E + Z^2/2n^2$ (in atomic units).

The analytic result (3.22) allows derivation of simple formulae for the photorecombination rate q_{nl} into the state with given quantum numbers n and l for a Maxwellian plasma with temperature T. Multiplying (3.22) by the electron velocity and then averaging over the Maxwellian velocity distribution, we obtain [3.15] (a.u.):

$$q_{PR}^{nl} = 4 \left[\frac{2}{2+x_T} G_2(x_m) + \frac{x_T}{2+x_T} \Psi(x_m, x_T) e^{x_m x_T} \right] \bigg/ \pi^2 n^3 c^3 l^2 \ . \tag{3.24}$$

Here the universal dimensionless parameters are introduced:

$$x_m = E_n M^3/3 = (l+1/2)^3/6n^2 \ , \qquad x_T = 3/TM^3 \ , \tag{3.25}$$

which determine the dependences of the rate q on the level energy $E = 1/2n^2$ and angular momentum $M = \hbar(l+1/2)$. The function G_2 is related to the universal function G_0 by the equation

$$G_2 = \int_x^\infty G_0(x')dx' = xK_{1/3}(x)K_{2/3}(x) \ , \tag{3.26}$$

where the function ψ is expressed in terms of the above defined "Bethe-rule defect" (BRD) $D(x)$ (3.19):

$$\Psi(x_m, x_T) = \int_{x_m}^\infty D(y) \exp(-x_T y)dy \ . \tag{3.27}$$

It follows from (3.24) that the photorecombination rate is determined by two different terms. In the first term described by the G_2 function, the BRD is neglected, whereas the second term is caused exclusively by the BRD and becomes appreciable for small x_m and large x_T. The functions $D(x)$ and $\psi(x_m, x_T)$ may be readily approximated analytically [3.15].

Thus, in the KrED, the PR rate into the states with given n and l is described by an universal function of two parameters. This universal dependence of the PR rate in Kramers domain is in agreement with the exact quantum numerical calculations, as discussed below.

The accuracy of the quasiclassical calculations of the PR rate turns out to be fairly good (within ~20%, according to the results of its comparison with the exact numerical calculations). Detailed tables for the PR rate in the Coulomb case, obtained in quasiclassical approximation, are presented in [3.2].

The applicability of the KrED analytical results for the Coulomb field case to the description of the PR cross sections for an ion with a core was investigated in detail in [3.14]. The authors use the approximation for the potential of such an ion in the form of a modified Coulomb potential with an effective charge Z_{eff}. It has been shown in [3.14] that in a wide range of electron energies and ion charges, this Coulomb type approximation of the potential yields a satisfactory description of the PR cross sections provided the value Z_{eff} is taken equal to the mean value of the charges of the nuclei Z and of the ion Z_i, $Z_{eff} = (Z+Z_i)/2$.

Figure 3.2 from [3.14] shows the dependence of the cross section for the PR into the level with the principal quantum number $n = 5$ and different quantum

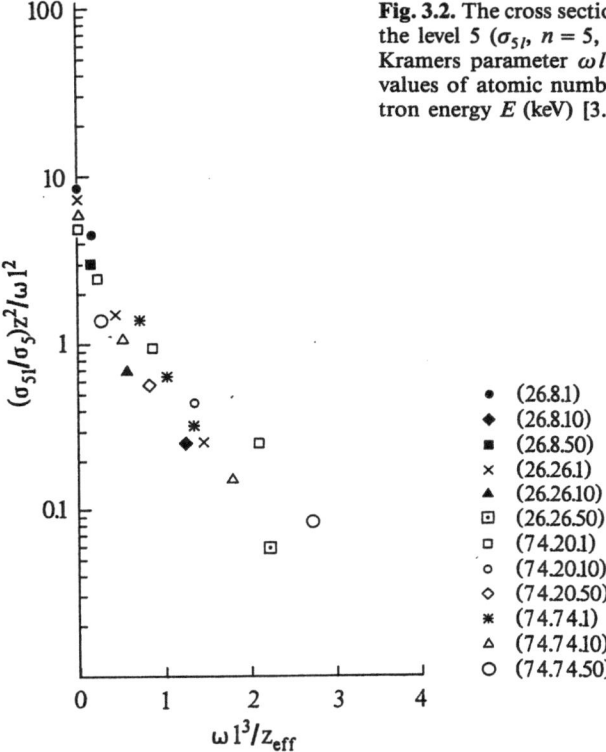

Fig. 3.2. The cross section for the recombination onto the level 5 (σ_{5l}, $n = 5$, l) with a dependence on the Kramers parameter $\omega l^3/Z^2_{\text{eff}}$ for ions with various values of atomic number Z, ion charge Z_i and electron energy E (keV) [3.14], mentioned in brackets

- (26.8.1)
◆ (26.8.10)
■ (26.8.50)
× (26.26.1)
▲ (26.26.10)
⊡ (26.26.50)
□ (74.20.1)
○ (74.20.10)
◇ (74.20.50)
✳ (74.74.1)
△ (74.74.10)
◯ (74.74.50)

numbers l on the universal KrED parameter $\omega l^3/Z^2_{\text{eff}}$ for the ions with different values of Z, Z_i and electron energy E (in keV). The results of quantum numerical calculations for different sets of parameters Z, Z_i, E prove to be satisfactorily described by an universal classical formula with the aforementioned value of Z_{eff}.

The same agreement between classical and quantum results occurs also for the dependences of the cross section on $n (\propto n^{-3})$ which follows from (3.23). The substitution of the Z_{eff} value in this equation gives the following simple analytical approximation for the PR cross section summed over l [3.14]:

$$\sigma^{\text{eff}}_{\text{PR}}(n) = \frac{8\pi}{3\sqrt{3}} Z^4_{\text{eff}}/c^3 n^3 E(E + Z^2_{\text{eff}}/2n^2) \; . \tag{3.28}$$

The total PR cross section is obtained by the summation of (6.22) over all allowed (non-occupied) quantum states according to the following equation:

$$\sigma^{\text{tot}}_{\text{PR}} = W_{n_0}\sigma_{n_0} + \sum_{n \geq n_0+1} \sigma_n \tag{3.29}$$

Here n_0 is the value of principal quantum number of the filled atomic shell, W_n is the statistical weight determined by the ratio of the number n_0 of free

Table 3.1. [3.14]

Z	Z_i	E [keV]	Quantum calculations	KrED, $Z_{eff} = (Z + Z_i)/2$
26	26	1	2.98×10^3	3.3×10^3
Nuclei		5	3.41×10^2	3.6×10^2
		10	1.2×10^2	1.2×10^2
		50	7.21	7.4
26	16	1	3.33×10^2	4×10^2
Ne-like		5	2.35×10	2.1×10
		10	6.64	5.4
26	8	1	0.9×10^2	1.5×10^2
Ar-like		5	4.9	7.3~
		10	1.36	1.9
		50	5.66×10^{-2}	7.6×10^{-2}

places in this shell to their total number. The results for the total PR cross section (3.28) are in good agreement with the results of quantum numerical calculations. This is illustrated in Table 3.1 taken from [3.14].

The agreement between the quasiclassical and quantum results may be improved by means of a proper choice of the lower limit of summation n_{0eff} in (3.29) which depends on the effects of screening and correlations of the electrons in the filled atomic shells. In this case the value n_{0eff} will be an universal parameter for a given isoelectronic sequence. Modifying the Kramers' formula (3.28), it is easy to obtain an analytical approximation for the total PR cross section by replacing the summation over n by an integration [3.14]:

$$\sigma^{tot} \approx \int_{(n_0)_{eff}}^{\infty} \sigma_n \, dn = \frac{8 \pi Z_{eff}^2}{3 \sqrt{3} \, c^3 E} \ln \left(1 + Z_{eff}/2En_{0eff}^2\right) . \tag{3.30}$$

The values of the cut-off parameter $s = 1/2n_{0eff}^2$, obtained from the comparison of (3.30) with the results of quantum mechanical calculations, are as follows: $s = 1.1$ for fully stripped ions (bare nuclei), $s = 0.065$ for Ne-like shells, and $s = 0.045$ for Ar-like shells. Figure 3.3, taken from [3.14], shows the dependence of the function

$$y = \exp \left(3 \sqrt{3} \, \sigma^{tot} E c^3 / 8 \pi Z_{eff}^2 \right) \tag{3.31}$$

on the parameter $Z^2 e^2/E$ for various isoelectronic sequences of Fe, Mo, and W. It is clear that the quantum results are described satisfactorily by a linear dependence of the argument on s, namely, $(1 + s Z_{eff}^2/E)$ for $Z_{eff} = (Z + Z_i)/2$.

The simple approximations (3.28, 30) for the partial and total PR cross sections lead to a simple and reliable analytical result [3.14] for some important

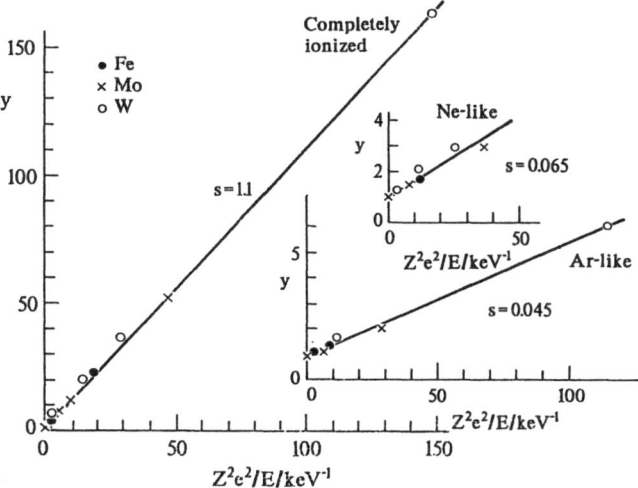

Fig. 3.3. The dependence of the function y in (3.31) on the parameter Z^2e^2/E for various isoelectronic sequences of ions Fe, Mo, and W, illustrating the linear dependence of the results of quantum numerical calculations for each isoelectronic sequence [3.14]

characteristics of a plasma, namely the rate of photorecombination $\alpha = \langle v\sigma^{tot}\rangle$, the rate of electron energy loss $\beta = \langle vE\sigma^{tot}\rangle$, and the total rate of radiation
losses of plasma due to photorecombination $\gamma = \sum_n \gamma_n = \sum \langle v\omega_n\sigma_n^{tot}\rangle$ (here
the brackets denote the averaging over the Maxwellian velocity distribution):

$$\alpha = \frac{16}{3}\sqrt{\frac{2\pi}{3}}\frac{Z^2c^{-3}}{\sqrt{T}}[\exp(b)E_1(b)+C+\ln(b)] , \quad b = \frac{sZ^2}{T} , \quad (3.32)$$

$$\beta = \frac{T}{c^2}\left[1-\frac{16}{3}\sqrt{\frac{2\pi}{3}}\frac{Z^4}{cT^{3/2}}\exp(b)E_1(b)\right] , \quad (3.33)$$

where $C = 0.577$ is the Euler constant, and $E_1(x)$ is the integral exponential function, $Z \equiv Z_{eff}$. The evaluation of γ requires a computation of the sum over n with the aid of approximations for partial cross sections. This gives [3.14]

$$\gamma = \frac{16}{3}\sqrt{\frac{2\pi}{3}}s\frac{Z^4}{c^3T^{1/2}} . \quad (3.34)$$

Table 3.2 shows a comparison of the improved Kramers recombination coefficients α (3.31) with the results of the calculation of the conventional hydrogen-like formulae varied with respect to the value of $Z(Z = Z_i$ and $Z = Z_{eff} \equiv$

Table 3.2. Recombination coefficient $\alpha = \langle v\sigma^{\text{tot}} \rangle$ (in $10^{-12}\,\text{cm}^3\,\text{s}^{-1}$) [3.14]

Z	T [keV]	Nucleus		Ne-like			Ar-like		
		KrED	Coul. Z_i	KrED	Coul. Z_i	Coul. Z_{eff}	KrED	Coul. Z_i	Coul. Z_{eff}
26	1	7.5	8.1	1.4	0.71	1.5	0.57	0.062	0.64
Fe	3	3.1	3.5	0.42	0.2	0.45	0.14	0.016	0.19
	10	1.1	1.2	0.10	0.036	0.11	0.038	0.003	0.039
	30	0.37	0.40	0.027	0.0085	0.025	0.009	0.0007	0.008
42	1	24	26	7.4	5.3	7.8	4.5	2.0	4.9
Mo	3	11	12	2.6	1.8	2.8	1.5	0.62	1.6
	10	4.1	4.5	0.72	0.50	1.1	0.39	0.15	0.42
	30	1.5	1.7	0.19	0.13	0.21	0.097	0.038	0.11
74	1	94	100	40	35	43	31	23	33
W	3	44	48	16	14	17	12	8.2	12
	10	18	20	4.9	4.3	5.2	3.5	2.5	3.9
	30	7.3	8.2	1.4	1.3	1.6	1.0	0.72	1.1

$(Z + Z_i)/2$. The results [3.14] presented in Table 3.2 show that the modification of the value of the effective charge appreciably improves the accuracy of the Coulomb-field approximation for the PR rate.

3.4 Kramers' Electrodynamics and Radiative Cascades Between Rydberg Atomic States

Many physical applications require calculation of the radiative cascade between highly excited atomic states. Examples include calculations of the level populations and line intensities of hydrogen and ionized He (II) in interstellar gas plasmas (nebulae) [3.16 – 18], spectral line calculations for highly stripped ions in hot rarified plasmas whose levels are populated by the processes of charge transfer [3.19], or dielectronic recombination [3.20], level population calculations for atoms excited by stepwise laser transitions [3.21], etc.

Techniques for calculating the parameters of radiative cascade were developed by *Seaton* [3.16] and are discussed in detail in [3.17, 18]. *Beigman* and *Mikhal'chy* [3.22] proposed an analytical method which yields results in close agreement with numerical calculations [3.20, 23]. All of this work deals with one dimensional radiative cascades, in which the populations f_{nl} of atomic states with different orbital quantum numbers l are assumed to be determined by their statistical weights: $f_{nl} = (2l+1)/n^2 f_n$, where the function f_n depends only on the principal quantum number n and corresponds to the total (with respect to l) level population. The radiative transitions in such a consideration thus occur between levels with a definite n, and the corresponding probabilities

$W(n \rightarrow n')$ are obtained by averaging the probabilities $W(nl \rightarrow n'l')$ over l and l' (this is called the n-method).

Pengelly [3.17] and *Summers* [3.18] have carried out numerical calculations for two-dimensional cascades, i.e., dealing with the populations of the individual nl-level (this is called the nl-method). *Summers* [3.18] also considered collisional transitions, making it difficult to trace the role of radiative cascades using his data.

The amount of data that must be handled and the complexity of the numerical calculations in the nl-method clearly increase with the number of the levels considered. Moreover, even in numerical calculations one ought to treat levels with large $n \sim 10^2$ approximately (cf., e.g., [3.20]). The error of such approximations increases with the increase of n and l, and *Pengelly* [3.17] estimates that the error reaches 70% even for a relatively small n and $l (n = 10, l = 5)$. On the other hand, just for $n \gg 1$ and $l \gg 1$ the radiative transition probabilities could be accurately described by quasiclassical methods, and in particular by KrED. This is confirmed by good agreement between quasiclassical results with quantum numerical calculations, as was mentioned in Sects. 3.4.2, 3. We will show that the description of radiative cascade based on the quasiclassical approach leads to manageable analytic solutions which are in good agreement with quantum numerical calculations. These solutions also allow identification of the parameters in terms of which the numerical data can be interpreted in a consistent, unified way without a recourse to laborious numerical methods.

Apart from its practical significance, the study of radiative cascades between Rydberg states is of general physical interest for the light it can shed on the relative importance of direct and cascade populations of atomic levels and on the interrelation between quantum mechanical and classical descriptions of electron motion along the atomic levels. Indeed, the problem can be solved in two extreme cases: 1) The nl-state may be assumed to be populated directly by a source q_{nl}, after which it decays with a probability A_{nl} into all of the lower lying states; the population will then be equal to q_{nl}/A_{nl} (this is the direct population model). 2) One may assume that the electron can reach a certain nl-level only by downward cascading through all of the upper lying states (the cascade population model). The latter approach is closely related to the classical concept of motion in nl-space, in which the electron motion is associated with a gradual loss of energy $(E = -\mathrm{Ry}/n^2)$ and angular momentum $[M = \hbar(l + 1/2)]$ at a rate which is determined by corresponding classical quantities [3.12]. *Belyaev* and *Budker* [3.24] employed this classical description in their treatment of radiative cascades; their method is equivalent to using the equation of continuity in phase space for the population $f(E, M)$. On the other hand, the cascade populations were calculated in [3.22, 23], where it was shown that the classical "flow" description with respect to the energy variable E is invalid – the electron always moves in quantum mechanical jumps. It is therefore of interest to examine the regions of nl-space within which the electron can be considered to move classically or by quantum jumps.

Of particular interest is the cascade population in the case of a photorecombinative source of external population when the free electrons with an equilibrium (Maxwellian) energy distribution fill the bound atomic states, and the radiative transitions determine both the population source and the subsequent radiative cascade. It is noteworthy that the distribution of the atomic electrons with respect to the orbital quantum number l is by no means always proportional to statistical weights, even if the source of electrons populating the levels is in equilibrium (cf. the numerical calculations in [3.17]).

3.4.1 Classical Kinetic Equation

Following [3.24], we will use canonically conjugate action-angle variables to analyze the classical kinetic equation for the electron distribution function (DF) in an atom or ion. These variables are most convenient because the characteristic time of action variables variation for a radiating electron is appreciably larger than the period of electron motion (the latter is the characteristic time of the variation of the angle variables). That is why the DF may be regarded as independent of the angle variables. We shall take the initial kinetic equation to be the continuity equation in six dimensional phase space. After averaging over the angle variables, this equation takes the form

$$\partial f/\partial t + \partial(\dot{I}_k f)/\partial I_k = q \ , \tag{3.35}$$

where I_k are the action variables,

$$I_1 = m\alpha^2/\sqrt{2E} \ , \quad I_2 = M \ , \quad I_3 = M_z \ , \quad \alpha \equiv Ze^2 \ , \tag{3.36}$$

and the $\dot{I}_k$ are the corresponding generalized momenta (averaged over the angle variables):

$$\dot{I}_1 = |\partial I_1/\partial E| \, \dot{E} \ , \quad \dot{I}_1 = (1 - M^2/3I_1^2) m e^{10} Z^4/c^3 M^5 \ , \tag{3.37}$$

$$\dot{I}_2 \equiv \dot{M} = -2 m e^{10} Z^4/c^3 M^2 I_1^3 \ , \quad \dot{I}_3 \equiv \dot{M}_z = M_z \dot{M}/M \ . \tag{3.38}$$

Here and below, $E > 0$ is the modulus of the total energy of the bound electron. Equations (3.37, 38) give the rate at which a classically radiating electron loses energy (I_1), angular momentum (I_2) and its z-component (I_3) [3.12].

We shall consider only the stationary case in what follows. The spherical symmetry of the Coulomb field implies that the DF f must be independent of M_z (we also assume that the source q is independent of M_z). Equation (3.35) thus simplifies to

$$\dot{E}(\partial f^{(3)}/\partial E) + \dot{M}(\partial f^{(3)}/\partial M) = q^{(3)} \ . \tag{3.39}$$

Here the superscript indicates the dimensionality of the space in which f is defined. We note that the variables E, M and M_z satisfy the classical kinematic constraints $M \le M_{max}(E) \equiv (m\alpha^2/2E)^{1/2}$, $|M_z| \le M$. In deriving (3.39) we have used the important property

$$\text{div}_{(3)} \dot{I} = 0 \tag{3.40}$$

of the generalized momentum, which implies that the electron flux in the space E, M, M_z may be uniform ($f^{(3)}$ = const. satisfies (3.35) if $q = 0$).

Solving (3.39) by the method of characteristics, we find

$$f^{(3)}(E, M) = \varphi[M(\tau, E_0)] + \int_{E_0}^{E} dE' q^{(3)}[E', M(\tau, E')]/\dot{E}[E', M(\tau, E')] \ , \tag{3.41}$$

where $\varphi(M)$ is the boundary condition for (3.39) (we take the boundary to be the line $E = E_0$; the generalization to the case of an arbitrary boundary is evident),

$$\tau \equiv \tau(E, M) = M^{-3}(1 - 2EM^2/m\alpha^2) \equiv M^{-3}\varepsilon^2 \ , \tag{3.42}$$

ε is the eccentricity of the electron orbit, and the dependence $M(\tau, E)$ in (3.41) is determined by (3.42). Using (3.41) we can rewrite the Green's function for (3.39) in the form

$$G(E'M' \to EM) = \frac{\eta(E - E')}{\dot{E}(E', M')} \delta[M' - M(\tau, E')]$$

$$\equiv \frac{\eta(M' - M)}{|\dot{M}(E', M')|} \delta[E' - E(\tau, M')] \ , \tag{3.43}$$

where $\eta = 0$ for $x < 0$ and $\eta = 1$ for $x > 0$. The δ-function in (3.43) corresponds to the classical motion of the radiating electron in the two-dimensional $\{E, M\}$-space; the trajectories coincide with the characteristic curves of (3.39) defined by the relation $\tau(E, M)$ = const. Since the energy loss rate exceeds the angular momentum loss, ε decreases during the radiation emission process so that the orbits eventually become "rounder".

3.4.2 Quantum Kinetic Equation in the Quasiclassical Approximation

We will consider the quantum mechanical kinetic equation for the distribution function $f^{(2)}$ in the two-dimensional space $\{I_1, I_2\}$ and use the formulae

$$I_1 = \hbar n \ , \quad I_2 = \hbar(l + 1/2) \ , \quad I_3 = \hbar m_z \ , \tag{3.44}$$

which relate the action variables to the quantum numbers n, l and m_z. Because $f^{(3)}$ is independent of M_z, $f^{(2)}$ and $f^{(3)}$ obey the simple relation

$$f^{(2)}(I_1, I_2) = 2M f^{(3)}(I_1, I_2) \equiv (2l + 1) f^{(3)}(I_1, I_2) \ . \tag{3.45}$$

The kinetic equation has the standard form ($\Gamma = \{n\,l\}$)

$$\sum_{n'=n+1}^{\infty} \sum_{l'=l\mp 1} f^{(2)}(\Gamma')\,W(\Gamma'\to\Gamma)+q(\Gamma) = A(\Gamma)f^{(2)}(\Gamma) \; , \tag{3.46}$$

where we have allowed for cascades from all higher-lying states; W is the probability per unit time for a radiative transition $\Gamma\to\Gamma'$, q is the external population source, and A is the total rate of radiative decay from the Γ level:

$$A(\Gamma) = \sum_{n'=l+1}^{n-1} \sum_{l'=l\pm 1} W(\Gamma\to\Gamma') \; . \tag{3.47}$$

For $n\gg 1$ we can replace the sum in (3.46) by an integral, and for $l\gg 1$ $f(\Gamma')$ can be expanded in l near the state Γ. This leads to an integral-differential equation (we will henceforth write f in place of $f^{(2)}$ where no confusion may arise)

$$q+ \int_{n+1}^{\infty} \left[f(n',l)\,W(n'\to nl)+\frac{\partial f(n',l)}{\partial l} \sum_{\Delta l = \pm 1} (l'-l)\,W(\Gamma'\to\Gamma) \right] dn'$$
$$= A(\Gamma)f(\Gamma) \; , \tag{3.48}$$

where

$$W(n'\to nl) = \sum_{\Delta l = \pm 1} W(\Gamma'\to\Gamma) \; . \tag{3.49}$$

The quasiclassical kinetic equation (3.48) reduces to a simpler one-dimensional integral or two-dimensional differential equation, depending on the specific region in $n\,l$-space, and the solutions can be pieced together uniquely because of the corresponding regions overlap.

Indeed, consider (3.48) for the region $l\ll n$, for which the Kramers approximation is valid for the radiative transition probabilities W. The radiative angular momentum loss ($\Delta l = \pm 1$) for $l\ll n$ is slower than the energy loss, because transitions with $\Delta n\gg 1$ (including those with $\Delta n\approx 1$) are more likely to occur (Sect. 7.5). If the DF is smooth enough we can therefore discard the differential term in (3.48), so that l appears to be merely a parameter of the resulting integral equation ($x_m \equiv (l+1/2)^3/6n^2$)

$$\int_0^{x_m} G_0(x)f\left[E\left(1-\frac{x}{x_m}\right),M \right] dx-f(E,M) \int_0^{\infty} G_0(x)\,dx = Q \equiv \frac{\pi q(\Gamma)}{\sqrt{3}\,A(\Gamma)} \; . \tag{3.50}$$

Here $E = 1/2n^2$ (in atomic units, a. u.), $M = \hbar(l+1/2)$, and, as before, f is normalized in Γ space. The function G_0 is related to the leading term in the expansion of the transition probability $W(n' \to nl)$ (3.49) with respect to $\hbar$ for $l \ll n$ and is given by (3.6). The function $A(\Gamma)$ is the total radiative decay rate for the level $\Gamma = \{nl\}$ (3.16, 17):

$$A(\Gamma)_{\text{a.u.}} = 4 \left[\sqrt{3}\, \pi c^3 n^3 (l+1/2)^2 \right]^{-1} . \tag{3.51}$$

The first (cascade) integral in (3.50) is negligible for small x_m, so that the population of level Γ is determined by the external source q,

$$f(\Gamma) = q(\Gamma)/A(\Gamma) . \tag{3.52}$$

The cascade term becomes important as x_m increases.

Since the Kramers' probability W depends only on the difference between the energies of the initial and final states, the integral equation (3.50) can be solved by taking Laplace transforms. The latter satisfy the equation

$$\bar{f}(s) = \bar{Q}(s)/s\bar{G}_2(s) , \tag{3.53}$$

where s is the Laplace variable conjugate to x_m,

$$G_2(x) = \int\limits_x^\infty G_0(x')\,dx' = xK_{1/3}(x)K_{2/3}(x) , \quad s\bar{G}_2(s) = \bar{G}_0(0) - \bar{G}_0(s) . \tag{3.54}$$

We can approximate G_2 to within 10% by the expressions

$$G_2(x) \simeq \alpha \exp(-2x) , \quad \bar{G}_2(s) = \frac{\alpha(s)}{s+2} , \tag{3.55}$$

where the function $\alpha(s)$ is slowly varying, $\alpha(s=0) = \pi^2/6 = 1.64$, $\alpha(s=\infty) = \pi/\sqrt{3} = 1.81$. If we set $\alpha = 1.7$, ensuring at most a 10% error in (3.55), we obtain the approximate analytic expression

$$f(\Gamma) = q(\Gamma)/A(\Gamma) + \int\limits_{n+1}^\infty dn'\, q(n',l)/|\dot{n}(n',l)| \tag{3.56}$$

for an arbitrary source q; here the quantity $\hbar\dot{n} \equiv \dot{I}_1$ is the rate of energy loss (see E in (3.37)) in Kramers' domain $l \ll n$.

To make the essence of the approximation (3.55) clear it should be pointed out that the exact relation between G_2 and G_0 takes into account (3.54) and the form

$$G_0(x) - 2G_2(x) = x[K_{1/3}(x) - K_{2/3}(x)]^2 \equiv D(x) \equiv \text{BRD} . \tag{3.57}$$

The correction $D(x)$ is the "Bethe-rule defect" (3.19) and is proportional to the Kramers' transition probability for a transition with $\Delta l = -\operatorname{sgn}(\Delta n)$. Such transitions, as was pointed out (Sect. 2.2), are suppressed (relative to the transitions with $\Delta l = \operatorname{sgn}(\Delta n)$) with stronger and larger Δn. In the Kramers' domain this leads to an approximate coincidence of the averaged Δl transition probability with one corresponding to $\Delta l = \operatorname{sgn}(\Delta n)$ transitions only. The transition to the limit of a classical trajectory (in Γ space, see below) corresponds to the motion with averaged Δl probabilities. That is why the transitions with $\Delta l = -\operatorname{sgn}(\Delta n)$, in spite of their existence as an elementary, one-step transition, can, within the framework of the KrED, be neglected in multistep transitions.

The DF (3.56) satisfies (3.48) for $x_m \leq 1$ (including $x_m \ll 1$), where both integrals in (3.50) are of the same order of magnitude. The integrals cancel each other for $x_m \gg 1$ that corresponds to the classical limit in (3.46). We can follow this limit by expanding $f(\Gamma')$ in the integrand with respect to both n and l (not only with respect to l as in the derivation of (3.48)). This expansion, which is valid for $x_m \gg 1$, leads to the two-dimensional differential equation

$$\dot{E}\partial f^{(2)}/\partial E + \dot{M}\partial f^{(2)}/\partial M - \dot{M}f^{(2)}/M = q^{(2)} \tag{3.58}$$

for $f^{(2)}(\Gamma)$. Recalling (3.45), we see that (3.58) is equivalent to (3.39).

We note that since the classical limit is consistent with the inequality $l \ll n$, it can be described in terms of Kramers transition probabilities. The contribution of the leading term in the $\hbar$-expansion for the transition probability, which is proportional to $\hbar^{-1}$, vanishes due to the aforementioned cancellation between the contribution of cascades from all upper levels to the nl-level under consideration and the contribution of cascades from the nl-level to all lower levels. This cancellation takes place (in the two-dimensional consideration) only for the leading terms of the $\hbar$-expansion for the contributions mentioned. The calculation of these contributions, with account of the quantum corrections to the leading term of the $\hbar$-expansion for W, gives the third term on the left-hand side of (3.58). As $\hbar \to 0$, a continuous classical flow of electrons described by (3.58) thus replaces the discrete quantum mechanical "jumps" specified by the non-local coupling in the integral equation (3.50).

We will now consider how the quasiclassical and classical distributions (3.41, 56) are to be matched. Comparison in the Kramers' domain $l \ll n$ shows that the first term in (3.41) (the contribution from the boundary condition for a classical differential equation) must be replaced by the contribution from the direct population. The resulting distribution function is valid for the entire quasiclassical domain of n and l, including the non-Kramers region $n \sim l$:

$$f(\Gamma) = q(\Gamma)/A(\Gamma) + M \int\limits_{n+1}^{\infty} \frac{q[n', l(\tau, n')]dn'}{|\dot{n}[n', l(\tau, n')]|M(\tau, n')} \equiv q/A + \hat{C}[q] \ , \tag{3.59}$$

where $l(\tau, n)$ is given by (3.42). Indeed, the boundary condition contributes to the classical distribution function (3.41) mostly for large n and, respectively, small x_m, for which the purely classical description breaks down. In Sect. 3.4 we will carry out calculations for a specific (photorecombination) source and explicitly piece the solutions together. The results will prove the correctness of the quasiclassical expression (3.59).

3.4.3 Relationship of the Quasiclassical Solution to the Quantum Cascade Matrix. The Solution in the General Quantum Case

We will interpret the above result (3.59) by using the quantum cascade matrix formalism, in which the cascade matrix $C(\Gamma' \to \Gamma)$ plays the role of the Green's function for the quantum mechanical (3.46). The DF obeying (3.46) can be expressed in the form [3.16, 17, 20]:

$$f(\Gamma) = A^{-1}(\Gamma) \sum_{n'=n}^{\infty} \sum_{l'=0}^{n-1} C(\Gamma' \to \Gamma) q(\Gamma') \equiv \frac{q(\Gamma)}{A(\Gamma)}$$

$$+ A^{-1}(\Gamma) \sum_{n'=n+1}^{\infty} \sum_{l'=0}^{n-1} C(\Gamma' \to \Gamma) q(\Gamma') . \qquad (3.60)$$

The matrix C can be regarded as the probability of a $\Gamma' \to \Gamma$ transition via all possible cascades ($C(\Gamma \to \Gamma) = 1$) and obeys the two equivalent recursion formulae:

$$[C](\Gamma' \to \Gamma) = \sum_{n''=n}^{n'-1} \sum_{l''=l\pm1} \frac{W(\Gamma' \to \Gamma'')}{A(\Gamma')} [C](\Gamma'' \to \Gamma)$$

$$\equiv \sum_{n''=n+1}^{n'} \sum_{l''=l\pm1} [C](\Gamma' \to \Gamma'') \frac{W(\Gamma'' \to \Gamma)}{A(\Gamma'')} . \qquad (3.61)$$

Comparison of (3.60) with the quasiclassical function (3.59) shows that the cascade population will be purely classical if f is smooth enough (so that $f(\Gamma')$ can be expanded in (3.46) as a Taylor series near the point $\Gamma' = \Gamma$). In the classical limit, matrix C takes the form

$$C(\Gamma' \to \Gamma) \propto M A(\Gamma) \delta(\tau - \tau') ,$$

where the δ-function of the argument τ (cf. (3.42)) describes the classical trajectory. A similar expression for $[C]$ also follows directly from (3.61) in the classical limit. If we let $\hbar \to 0$ as in the derivation of (3.58), we find that $[C](\Gamma' \to \Gamma) \propto M A F(\tau, \tau')$, where the function F is arbitrary.

We will now estimate the error in the classical description of cascades for an arbitrary source q (including a selective population source $q \propto \delta(\Gamma - \Gamma_0)$) by substituting the approximate solution (3.56) for the Kramers domain $l \ll n$ into the corresponding (3.50). The non-canceling term can be transformed into

$$\int_0^{x_m} dx q(E',M) \left[G_0(x) - \frac{4\sqrt{3}}{\pi} G_2(x) \right] \Big/ A(E',M) \ ,$$

$$x \equiv [(E'-E)M^3/3]_{\text{a.u.}} \ . \tag{3.62}$$

The expression in square brackets in the integrand coincides with the above defined "Bethe-rule defect" (BRD) (3.19) to within 10%. Equation (3.62) implies that the terms in square brackets cancel only for those x for which the BRD can be neglected. The distribution function f given by (3.59) cannot be used for sources q whose main contribution to the integral in (3.61) comes from small x, for which the terms in square brackets do not cancel.

Let us analyze the case of a δ-function source. Equation (3.59) is clearly not applicable if direct transitions from the level Γ' populated by the source to the level Γ under consideration are important (this corresponds to the leading term in the BRD as $x \to 0$). In any case, such direct transitions will be important for levels Γ close to Γ', as well as for more remote levels that are populated solely by BRD transitions, i.e., by electrons lying far from the classical trajectory. Classical cascade may occur between the levels which lie close to the classical trajectory provided they are sufficiently far from the levels Γ' populated directly by the source ($\Delta x_m \geq 1$).

The situation depicted (i.e., the transition from the quantum direct population, in the domain close to an externally populated level, to the classical cascade population) can be described in terms of a modified classical cascade. For example, in Kramers' domain this gives [here x is the same as in (3.62)]:

$$f(\Gamma) = q(\Gamma)/A(\Gamma) + \int_{n+1}^{\infty} \frac{q(n',l)G_0(x)dn'}{|\dot{n}(n',l)|2G_2(x)} \ . \tag{3.63}$$

However, there is an alternative, more systematic method for treating "the quantum mechanical properties" of the external source of population. This method exploits the fact that the form of the quantum mechanical kinetic equation remains unchanged if we subtract an arbitrary number of the leading terms in the expansion of the distribution function in powers of the number of the transitions in a cascade from the externally populated level Γ' to the investigated level Γ. Indeed, (3.46) continues to hold for the function $(f-q/A)$ if we replace q by

$$\langle q \rangle \equiv \sum_{n'=n+1}^{\infty} \sum_{\Delta l = \pm 1} q(\Gamma') W(\Gamma' \to \Gamma)/A(\Gamma') \ . \tag{3.64}$$

Proceeding as in Sect. 3.2, we thus arrive at the distribution function

$$f = \langle f \rangle \equiv q/A + \langle q \rangle/A + \hat{C}[\langle q \rangle] \ . \tag{3.65}$$

[Compare with (3.59)]. The generalization of the result (3.65) in the case of an arbitrary number of averaging procedures for the source q gives the result

$$f = \langle f \rangle_N \equiv q/A + A^{-1} \sum_{i=1}^{N} \langle q \rangle_i + \hat{C}[\langle q \rangle_N] \ , \tag{3.66}$$

where the effective source $\langle q \rangle_N$ describes the population of the level Γ by all possible N-step (i.e., N-photon) cascade transitions from all points of the source,

$$\langle q \rangle_N = \sum_{n_1 = n+N}^{\infty} \sum_{n_2 = n+N-1} \cdots \sum_{n_N = n+1}^{\infty} \sum_{l_1, \ldots, l_N = 0}^{n_i - 1} q(\Gamma_1)$$

$$\times \frac{W(\Gamma_1 \to \Gamma_2)}{A(\Gamma_1)} \cdots \frac{W(\Gamma_N \to \Gamma_{N+1})}{A(\Gamma_N)} \ , \tag{3.67}$$

and the appropriate selection rules for the radiative transition probabilities must be used in calculating (3.67). Each additional summation in (3.67) further smoothens the effective source and thus decreases the error caused by summing the remainder terms in the series (3.60) "classically" to $\cong 10\%$ (Sect. 3.4.4). The error in the final result depends both on the specific form of the source q and on the values of quantum numbers n and l. The error will be small if the relative change of $f(\Gamma)$ is small due to subtracting one more term (corresponding to $(N+1)$-step transitions) out of the classical cascade.

The above algorithm can be used to calculate the distribution function f for radiative electron cascades between Rydberg atomic or ionic states for arbitrary sources and quantum numbers (in particular, n and l may be of the order of unity).

Note that the extent to which the population source q is of essentially "quantum mechanical" character depends partly on the sharpness of its form in its distribution in Γ-space (3.61) and partly on the range of values of n, l within which the source is concentrated. For example, the distribution function (DF) is of essentially quantum character if a smoothly distributed (i.e., "classical") source is concentrated in the "quantum" region $x_m \ll 1$ (cf. Sect. 3.4.4). On the other hand, the cascade population can be described quasiclassically even for a selective source if the latter is concentrated in the "classical" region $l \sim n$. Thus, if the levels with $l = n-1$ are selectively populated by the external source, the population of the lower levels by cascades can be described purely classically and the result agrees with the exact quantum calculation. Specifically, if we use the quantum cascade matrix and recall the relations for this case ($l = n-1$),

$$W(n, l \to n-1, l-1) = A(n, l) = |\dot{n}| = |\dot{l}| = 2/3 n^{-5} \ , \tag{3.68}$$

we find from (3.60) that

$$f(n, l) = q(n_0, l_0) A^{-1}(n, l) \delta(n, n_0 - k) \delta(l, l_0 - k) \ , \tag{3.69}$$

where δ is the Kronecker symbol and $k \geq 0$. The calculation using (3.59) leads to the same result.

3.4.4 Atomic-Level Populations for a Photorecombinative Source. Quasiclassical Scaling Laws

The general results of Sects. 3.2, 3 are now applied to the calculation of level population for a Rydberg atom externally populated by a photorecombination source. Since the latter involves the same radiative transitions as the cascade between the atomic levels, the above approximations for the cascade can be also applied to the recombination source (see approximation (3.55) to the error in the quasiclassical DF (3.59)) in a specific case.

The calculation of the DF (3.56) for the source (3.24) corresponds to including the BRD contribution to the source q but neglecting it in the Green's function for (3.50). It is worthwhile to express the result for the DF in terms of the equilibrium DF [as the corresponding ratio $b(\Gamma)$]

$$f(\Gamma) \equiv 2 A_m M \exp(E/T) b(\Gamma) , \quad A_m = (2\pi m_e T)^{-3/2} \tag{3.70}$$

(reminder, we use $E > 0$). The result for the Kramers domain is

$$b_{nl} = \frac{2}{2+x_T} \exp(-E/T) + \psi(x_m, x_T)/\alpha , \tag{3.71}$$

where ψ and α are given by (3.27) and (3.55), respectively.

For $n, l \gg 1$, the second term in b_{nl} is of importance only for $x_m \ll 1$, $x_T \gg 1$; the DF is therefore independent of the energy E at the edge ($l \sim n$, or $x_m \geq 1$) of the Kramers domain. This implies that the solution outside of the Kramers domain could be found from the first term in (3.71), regarded simply as a classical boundary condition. Because this term is independent of E, the resulting DF will be the same regardless of which line in nl-space is chosen as the boundary. If we then use (3.41, 42) to continue the DF (3.71) along the characteristic curves, we obtain the final result

$$b_{nl} = \frac{2}{2+x_T \varepsilon^2} \exp(-E/T) + \psi(x_m, x_T)/\alpha , \tag{3.72}$$

which is valid for all quasiclassical values of n and l. It is legitimate to continue the solution in this way because the source (3.27) is concentrated in the Kramers domain, so that there is no need to evaluate (3.59) directly (recall that A and $\dot{n}$ in (3.59) are the transition probabilities for an arbitrary ratio l/n). Indeed, a calculation using (3.59) for $(l/n - 1) \ll 1$ reveals that these states are populated solely by classical cascades; moreover, most of the contribution comes from the transitions whose initial state is far from the curve $l \sim n$. The latter result corresponds precisely to the classical behavior, in which the states

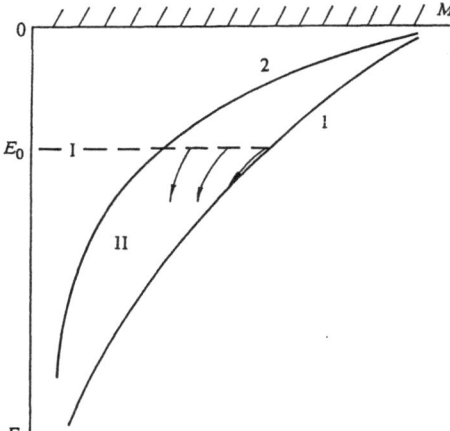

Fig. 3.4. Regions in the *EM* plane corresponding to quasiclassical (*I*) and classical (*II*) electron motion. Curve *1* demarcates the region of classically allowed motion: $2EM^2 = 1$ (or, equivalently, $M = M_{\max}(E)$]; curve *2* separates regions *I* and *II*: $EM^3 \sim 1$ (or $x_m \sim 1$). The arrows indicate the classical trajectories (the characteristics (3.37) of (3.42); the classical boundary conditions are imposed on the line $E_0 = \mathrm{const.}$)

near the boundary $M = M_{\max}(E)$ (Fig. 3.4) can be populated by a source concentrated within a region with an eccentricity $\varepsilon \to 1$ (Sect. 3.1). For a recombination source, the Kramers domain shrinks along the *n* axis as *l* increases (as the edge of the continuum is approached) and thus is effectively transformed into a boundary condition.

We will now show that the use of the algorithm discussed in Sect. 3.3 leads to the incorporation of some additional BRD contributions to the DF. For a singly averaged source ($N = 1$), (3.64) gives

$$\langle Q \rangle \equiv \langle q \rangle / A = 3 \pi^{-2} \int_0^{x_m} G_0(x) Q(x_m - x) dx \tag{3.73}$$

to within a 10% error in approximating the coefficient α in (3.55). This gives

$$\langle f \rangle = f + 3 \pi^{-2} \int_0^{x_m} D(x) Q(x_m - x) dx \tag{3.74}$$

for $\langle f \rangle$ in (3.65); here f is defined by (3.59). A calculation of $\langle f \rangle$ for our photorecombination source reveals that the corrections due to the BRD contribution are smaller than the 10% error arising from the approximation (3.55). Thus, if we include the linear correction to (3.71), in accordance with (3.74), we find that

$$\langle b \rangle = \frac{2}{2 + x_T} \left[1 + e^{-2x_m} - \sqrt{3}\, G_2(x_m)/\pi \right] e^{-E/T}$$

$$+ \frac{\psi}{\alpha} \left[1 + \frac{x_T}{2 + x_T} \int_0^{x_m} D(x) dx / \alpha \right]. \tag{3.75}$$

Since the factor multiplying ψ is significant only for $x_m \ll 1$, $x_T \gg 1$, we conclude, as in Sect. 7.3, that the accuracy of the DF (3.71) is the same as for (3.55).

The quasiclassical DF (3.72) derived above reveals approximate scaling laws for the exact quantum level populations, these laws being unknown from the results of complex quantum numerical calculations. These laws are a consequence of the fact that the quasiclassical DF (3.72) depends on a fewer number of variables than in the case for the quantum DF. Indeed, f_{nl} depends only on x_m and x_T for $x_m \ll 1$, $x_T \gg 1$. If one of the parameters x_m or $1/x_T$ becomes ~ 1, the second term in (3.72) becomes much less than the first term and f depends only on x_T. Elsewhere in nl-space, f depends on the parameter $x_T \varepsilon^2$. We thus have a smooth transition between the three scaling laws for $n, l \gg 1$. Comparison of the quasiclassical DF (3.72) with the results of numerical quantum calculations [3.17] reveals that the quasiclassical DF (3.72) can also be used for relatively small values of n and l.

The validity of the scaling laws derived by the quasiclassical method can be verified by means of a corresponding transformation of quantum numerical data. Thus, replotting of quantum results [3.17] in terms of the parameter x_T (for a fixed x_m) reveals that they form a single curve dependent on x_T (Fig. 3.5). Figure 3.6 gives an illustration of the accuracy both of the quasiclassical scaling laws and the DF. Here the data from [3.17] corresponding to two different states $n = 6, l = 2$ and $n = 10, l = 3$, for which x_m ($\simeq 0.072$) and ε^2 ($\simeq 0.83$ and 0.87) are close, are indicated as a function of x_T. We see that although these values correspond to different n, l, they lie on a single curve which coincides to within 20% of the quasiclassical result (3.72).

The population distribution with respect to l is also of interest. Figure 3.7 gives the relative populations b as functions of l for several different temperatures. We see that, as noted in Sect. 3, the dependence $b(l)$ differs, in general,

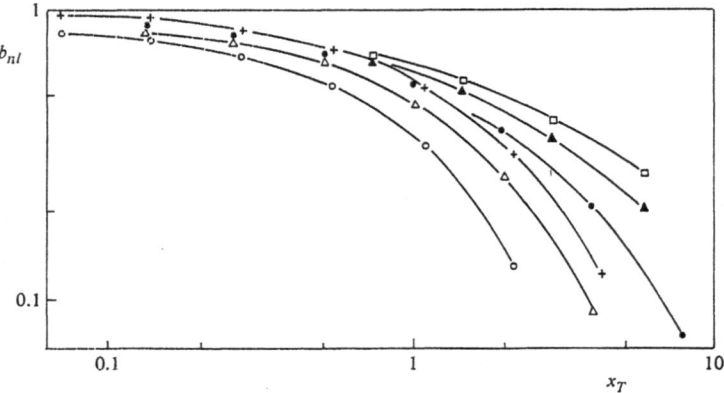

Fig. 3.5. Atomic level population b_{nl} from [3.17] replotted as a function of the parameter $x_T = 3/TM^3$ for several values of $x_m = EM^3/3$: circles, $x_m = 0.77$; triangles, $x_m = 0.42$; crosses, $x_m = 0.28$; points, $x_m = 0.15$; filled triangles, $x_m = 0.072$; squares, $x_m = 0.026$

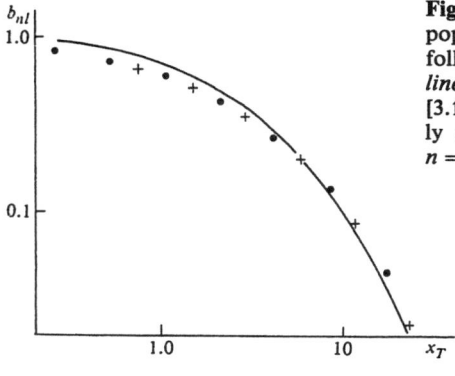

Fig. 3.6. Universal dependence of the level population b_{nl} on the parameter $x_T = 3/TM^3$ following from the quasiclassical (3.72) (*solid line*) and the results of quantum calculations [3.17] for various states with an approximately equal value of parameter x_m ($\simeq 0.072$): $n = 10, l = 3$ (*points*) and $n = 6, l = 2$ (*crosses*)

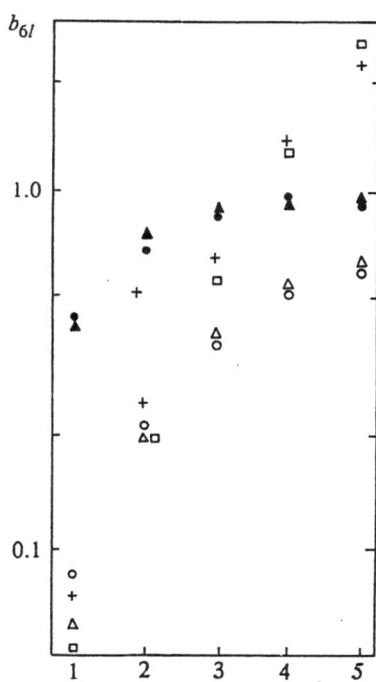

Fig. 3.7. The population b_{6l} for $n = 6$ hydrogen level as a function of the orbital momentum l for several temperatures. $T = 8 \times 10^4$ K: points, numerical calculations [3.17], filled triangles (3.72); $T = 10^4$ K: circles [3.17], triangles (3.72); for $T = 1/8 \times 10^4$ K the values $100 \times b_{6l}$ are given: crosses [3.17], squares (3.72)

from the statistical weight distribution corresponding to $b(l) = $ const. The latter is the case only for large T (small x_T), whereas for low temperatures (large x_T) $b(l)$ increases sharply with l. These results clearly agree with the quasiclassical formula (3.72).

Finally, the scaling law with respect to x_m is illustrated in Fig. 3.8, which plots b_{nl} for $l = 1$ (the p-state) as a function of x_m for a fixed n. We see that

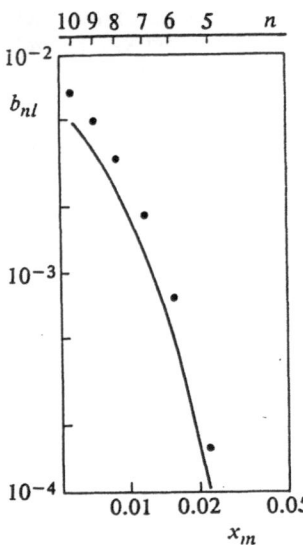

Fig. 3.8. The level population b_{nl} as a function of the parameter $x_m = EM^3/3$ for a fixed $x_T = 214$ ($T = 1/8 \times 10^4$ K, $l = 1$); x_m has been varied by changing the principal quantum number n (*top scale*). The points show numerical values from [3.17]; the curve was calculated using the quasiclassical (3.72)

even for this small value of l, the quasiclassical result (3.72) agrees to within 35% of the quantum numerical calculations in [3.17].

It is worthwhile to illustrate the relative importance of direct and cascade population in the Kramers domain $l \ll n$ for a photorecombination source. We find from (3.56)

$$f_C \propto \frac{2}{2+x_T}(1-e^{-2x_m}) + \frac{2}{2+x_T}\,\psi\,e^{E/T}/\alpha \;,$$

$$f_D \propto \frac{2}{2+x_T}\,e^{-2x_m} + \frac{x_T}{2+x_T}\,\psi\,e^{E/T}/\alpha \tag{3.76}$$

for the cascade (f_C) and direct (f_D) populations. The contribution from f_D clearly decreases as x_m increases while the sum ($f_C + f_D$) coincides with (3.71). The numerical values of the ratio $f_D/(f_D + f_C)$ agree reasonably well with the data in [3.17], e.g., for $n = 6$ and $T = 10^4$ K, (3.76) implies that this ratio is equal to 96% for $l = 1$ and 87% for $l = 2$; the corresponding values from [3.17] are 81% and 73%, respectively.

It is important to note that the dependence of the integral over the l population differs from the one found by the n-method [3.22, 23]. The major contribution to $f(n)$ comes from the first term in (3.72), which gives the following dependence on the parameter Th^3:

$$f(n)n^{-2} \propto \int_0^1 x \left(1+\frac{3(1-x^2)}{2Tn^3x^3}\right)^{-l} dx \simeq \begin{cases} 1/2 \;, & Th^3 \gg 1 \;, \\[2mm] \dfrac{Th^3}{3}\ln\dfrac{3}{Th^3} \;, & Th^3 \ll 1 \;, \end{cases} \tag{3.77}$$

whereas in the n-method the universal parameter is Th^2.

The quasiclassical method for an analytic description of radiative cascades developed in this section gives the possibility of an approximate (to within 10%) calculation of the contribution of multistep cascade transitions to the atomic level populations. This calculation is known to be the most difficult part of the corresponding numerical calculations. Indeed, the δ-function properties of the cascade matrix that correspond to radiating electrons moving (in nl-space) along the characteristics in the classical domain [cf. (3.36−41)] are difficult to reveal from the results of quantum numerical calculations. For example, the calculations in [3.17] detected only the boundary characteristics corresponding to $l = l_{max} = n-1$.

The algorithm in Sect. 3.3 for calculating populations in the general quantum case and for arbitrary sources can thus be used to correctly treat cascades through an arbitrary large number of Rydberg states. The number of quantum mechanical cascade transitions which cannot be described classically may be quite small in practice, particularly for the case of distributed sources. For example, the cascade population is purely classical (to within 10%) for a photorecombination source (Sects. 3.4.1, 2).

4. Fermi Method of Equivalent Photons and the Probabilities of Radiative-Collisional Transitions in Atoms

4.1 Applicability of the Fermi Method

The radiative-collisional processes appear to be a wide domain of the Kramers electrodynamics (KrED) application. These are the processes in which the electron participates while moving along the classical highly curved quasiparabolic orbits [4.1]. We shall demonstrate the applicability of the KrED method to the description of collisional (e.g., excitation of ions by electron impact) and radiative-collisional (e.g., dielectronic recombination (DR) and polarizational radiation (PLR)) processes in plasmas. The most natural domain of the KrED application is the physics of multicharged ions (MCI). The description of these processes is obtained within the framework of the following assumptions:

1) According to the Fermi concept [4.2] of equivalent photons (EPh), the electromagnetic field produced by an external particle (e.g., an electron) at the MCI location may be interpreted as a flux of equivalent photons incident on the MCI. It may be shown that such a description is applicable provided the dipole approximation for the interaction is valid between the bound electron of the MCI and the incident electron of the plasma. The latter approximation universally treats all the processes of energy loss by the incident electron (either due to radiation emission during a collision with an ion or due to an inelastic non-radiative collision with an ion) as the processes of the emission of (real or equivalent, respectively) photons. The probability of both processes is determined by the dipole matrix element for the corresponding inelastic (radiative or non-radiative) transition of the incident electron.

2) The spectral intensity distribution of the EPh may be described on the basis of classical radiation theory (for a detailed discussion of the applicability of the classical approach for real photons, see Sect. 2.3.4). In this case the intensity of the EPh flux is simply determined by the Fourier transforms of the electron coordinates determined in turn by its classical trajectory.

Such an approach makes it possible to treat some radiative-collisional processes as follows:

- the excitation of an ion by electron impact as an absorption of the EPh by this ion;
- the same excitation with a subsequent re-emission of a real photon as a resonance fluorescence of the EPh;
- the dielectronic recombination as a resonance fluorescence of the EPh, which results in a recombination of incident electron onto the ion.

An essential advantage of this method comes from the extension of available results for purely radiative processes to the description of non-radiative processes both collisional and radiative-collisional. The processes discussed are of a resonant character with respect to the absorption of the EPh by the ion. Except for these processes, there are non-resonant processes as well, for which the intermediate state of a two-step "absorption-re-emission" process is not real and consequently is not obeying the energy conservation law (this state is formed by the process of virtual excitation with the energy $E \neq \hbar \omega_0$, where ω_0 is the frequency of the resonant transition). These non-resonant processes are known as polarization radiation [4.3] and can be treated as the non-resonant scattering of the EPh by the ion. The polarization radiation is determined by the dynamical polarizability of an ion in a non-resonant frequency domain.

For the Fermi method to be applicable to processes involving MCI, the effective distances r_{eff} which are responsible for the main contribution to the cross section of the inelastic collision of an incident electron with MCI should be much greater than the characteristic size of the bound electron orbit in MCI. This requirement is fulfilled especially well for MCI. Let us illustrate this for the process of excitation of $\Delta n = 0$ transitions. The electron orbit size is of the order $1/Z$ (in atomic units), transition energy ΔE for $\Delta n = 0$ transitions in MCI is typically of the order Z, and the values of r_{eff} for the corresponding cross section can be estimated as

$$r_{eff} \sim r_\omega = \Delta E/\hbar \sim (Z/\Delta E^2)^{1/3} \sim Z^{-1/3} \gg Z^{-1} \; . \tag{4.1}$$

This inequality justifies the use of the dipole approximation for the potential of interaction between bound and incident (with space coordinate vectors r_1 and r_2, respectively) electrons, $V = e^2 r_1 \cdot r_2 / r_2^3$. In this framework the static Coulomb interaction between the bound and incident electrons transforms to the processes of emission and absorption of the EPh by electrons, and the corresponding probabilities are determined by the conventional dipole matrix elements.

The electric field produced by the incident electron at the location $r = 0$ of the ion is equal to

$$F(0, t) = r_2(t)/r_2^3(t) \; , \tag{4.2}$$

where the dependence $r_2(t)$ describes the classical trajectory of the incident electron. Using the equation of the motion of the incident electron in the field of the MCI ($\ddot{r}_2 = Z r_2/r_2^3$), it is convenient to transform (4.2) into the form

$$F(t) = \ddot{r}_2(t)/Z \; . \tag{4.3}$$

The spectral distribution for the EPh flux, $I(\omega) = dE/d\omega \, dS_\perp$, produced by the incident electron's electric field, can be expressed in terms of the Fourier transforms of the field

$$I(\omega) = c[\,|F_x(\omega)|^2 + |F_y(\omega)|^2]/4\pi^2 = c\omega^4 Z^{-2}[x_\omega^2 + y_\omega^2]/4\pi^2\ , \qquad (4.4)$$

where x and y are the coordinates of the incident electron in the plane of its motion. The Fourier transforms of the electron space coordinates in the Coulomb field are well known [4.4–6]. Thus we obtain

$$I(\omega) = c\omega v^{-4}\{[H_{i\nu}^{(1)\prime}(i\nu\varepsilon)]^2 - \frac{\varepsilon^2-1}{\varepsilon^2}\,[H_{i\nu}^{(1)}(i\nu\varepsilon)]^2\}/4\ , \qquad (4.5)$$

where v is the electron's initial velocity.

In the limit of low EPh frequencies, $v \ll 1$, the main contribution to the spectral distribution of the EPh flux integrated over the electron impact parameters ϱ is due to the distance from the field center trajectories, $\varrho \gg a \equiv Z/2E$, which are nearly rectilinear, with eccentricity $\varepsilon \gg 1$. In this case, (4.5) is transformed to

$$I(\omega) = c\omega v^{-4}[K_0^2(\omega\varrho/v) + K_1^2(\omega\varrho/v)]/\pi^2\ , \qquad (4.6)$$

where $K_0(x)$ and $K_1(x)$ are the *McDonald* functions. *Fermi* used (4.6) for the description of atomic excitation by a rectilinearly moving particle [4.2].

For the description of the processes resulting in a loss of a considerable part of the incident electron energy, it is necessary to consider the EPh with high frequencies, namely $v \gg 1$. The main contribution to the emission of such EPh comes from the strongly curved electron trajectories, $\varepsilon - 1 \ll 1$, which are close to the field center, $\varrho \ll a$. In this Kramers' domain we arrive at the result

$$I(\omega) = \pi^{-2} Z^{-2} cM G_0(\omega M^3/3Z^2)\ . \qquad (4.7)$$

4.2 Excitation by Electron Impact as Absorption of Equivalent Photons by an Ion

The equivalent photon method makes it possible to obtain a simple analytical description of the collisional processes and treat them as purely radiative. Within this framework the excitation of MCI by electron impact may be clearly considered as the absorption of the EPh with a resonant frequency $\omega = \Delta E_{if}/\hbar$ by MCI. The relationship between the collisional cross section σ_{exc} and the cross section σ_{abs} for the absorption of the EPh can be obtained by means of equating the number of excitation events, during the time interval dt, caused by the collisions of the MCI with the electron flux with a space density n_e and a particle velocity v_e

$$dN_{exc} = n_e v_e \sigma_{exc} dt$$

to the corresponding number of transitions caused by the absorption of the EPh produced by a single electron. This is multiplied by the total number of electrons in the volume dV corresponding to the time interval dt, $dV = 2\pi\varrho\,d\varrho\,v_e dt$,

$$dN_{abs} = \int 2\pi\varrho\,d\varrho\,n_e v_e dt \int d\omega\,(cF_\omega^2/4\pi^2\hbar\omega)\sigma_{abs}(\omega)\ , \tag{4.8}$$

where the expression in curly brackets corresponds to the spectral distribution of the EPh flux (4.4) produced by a single electron with a fixed value of the impact parameter ϱ. Assuming the following relation for the total and partial (with respect to the quantum orbital number l) cross sections

$$\sigma_{exc} = \int \sigma_{exc}^l\,dl$$

we arrive at the result

$$\sigma_{exc}^l = 2\pi(\hbar/mv_e)^2(l+1/2)\int \sigma_{abs}(\omega)(cF_\omega^2/4\pi^2\hbar\omega)d\omega\ . \tag{4.9}$$

Furthermore, the expression for the EPh flux can be taken out of the integral sign at the frequency ω_0 of the radiative transition under consideration in the MCI core because of its slow frequency dependence in comparison with that of the absorption cross section. The resulting integral over ω gives the well known expression [4.6]

$$\int \sigma_{abs}(\omega)d\omega = \pi^2(c/\omega)^2 g_f 4\omega_0^2 |d_{if}|^2/3\hbar c^3\ ,$$

where d_{if} is the dipole moment matrix element of the transition considered and g_f is the statistical weight of the upper level.

Substituting the spectral distribution (4.5) into (4.9) for the EPh flux produced by the electron in the Coulomb field of the MCI, we finally obtain

$$\sigma_{exc}^l = \frac{8\pi^3}{3}(\hbar/mv)^2\omega_0^2|d_{if}|^2 g_f v^{-4}(l+1/2)$$

$$\times\{[H_{iv}^{(1)'}(iv\varepsilon)]^2 - (\varepsilon^2-1)\varepsilon^{-2}[H_{iv}^{(1)}(iv\varepsilon)]^2\}\ . \tag{4.10}$$

The transition in (4.10) to the KrED domain ($v \gg 1$) corresponds to the transition from (4.5) to (4.7). Thus we obtain the result in the KrED domain:

$$\sigma_{exc}^l = (8\pi/3)(\hbar/mv_e)^2(g_f/g_i)f_{if}Z^{-2}(l+1/2)^2 G_0[\omega(l+1/2)^3/3Z^2] \tag{4.11}$$

where f_{if} is the oscillator strength for the transition considered, and g_i is the statistical weight of lower level.

Equation (4.11) explicitly manifests the interrelation between the SEI effect and the well known fact of the finiteness of the excitation cross section at the threshold. Thus, we once more face the phenomenon, inherent to KrED, of the independence of the spectral distribution on energy, which leads to a smooth transition between the discrete and continuous energy spectra for the processes with both real (from the BR to the PR) and equivalent photons (the transition from the Born approximation domain for the excitation to its threshold and further to the DR, Sect. 4.3).

The total excitation cross section is obtained by summing the partial cross section (4.9) over l, yielding the expression in terms of the well known spectral distribution for the Coulomb bremsstrahlung Gaunt-factor $g(v)$ [4.1].

$$\sigma_{exc}^{if} = \frac{8\pi^2}{\sqrt{3}} |d_{if}|^2 g_i^{-1} v^{-2} g[Z\omega_0/(2E)^{3/2}] \ . \tag{4.12}$$

Remember that the function $g(v)$ has a simple analytic approximation [4.1].

The result (4.12) was derived earlier [4.7] in a somewhat different way. It should be noted that (4.12) is valid up to the excitation threshold where Kramers EPh spectrum (4.7) does not depend on the incident electron energy at all. In the opposite limit of a fast incident particle, the cross section (4.12) exhibits a logarithmic (Born-type) structure. It is this result that was derived by E. Fermi for atomic excitation and ionization by fast particles. Equation (4.12) is in good agreement with quantum numerical calculations as well as with experimental data. Figure 4.1 from the *Bazylev* and *Chibisov's* survey [4.7] shows the comparison of the excitation cross section (4.12) with the results of quantum numerical calculations by the strong coupling method [4.8].

It should be also noted that for the first time the interrelation between the excitation cross section for allowed dipole transitions and the Gaunt-factor for bremsstrahlung in a Coulomb field for the general quantum case (Sommerfeld formula in [4.6]) was investigated by *Gailitis* [4.9]. In this section we have

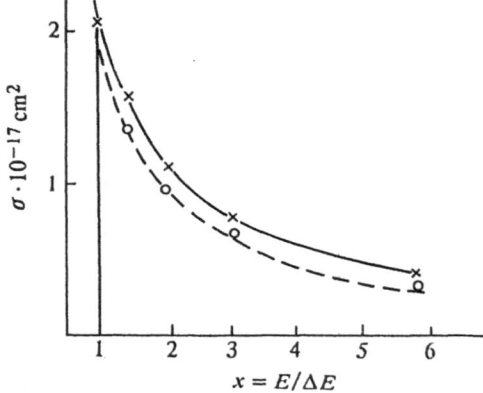

Fig. 4.1. Cross section of the electron impact excitation of the transition $2s-2p$ of energy ΔE in the Li-like ion Ar^{15+} as a function of electron energy E expressed in terms of the quantity $E/\Delta E$: stars (4.12); circles, calculation by the strong coupling method [4.8]

restricted ourselves to the case of quasiclassical incident electron motion, however, keeping in mind the applicability of the KrED approach to a calculation of excitation cross sections for an arbitrary (not necessarily purely Coulombic) ionic potential (e.g., of a Thomas-Fermi ion). The description of those excitation cross sections may be achieved by the replacement of the Coulomb EPh intensity by the corresponding EPh spectral intensity in (4.5).

4.3 Dielectronic Recombination as the Resonance Fluorescence of Equivalent Photons

Firstly recall the essence of the dielectronic recombination (DR) process. An incident electron with an energy E_i excites an ion core with an excitation energy $\Delta E = \hbar \omega_0$. In this case, if the energy E_i is smaller than ΔE, then the electron is finally captured by the ion into a state with the energy $E_f = -\text{Ry}/n_f^2$ obeying the condition

$$E_i - E_f \equiv \Delta E = \hbar \omega_0 \ .$$

This capture results in a double excited state of the ion, namely, the ion core electron is excited with the energy ΔE while the captured electron occupies a highly excited level of the ion. This state of the ion can decay in two possible ways: (i) by the relaxation of the ion core electron into the initial ground state with the simultaneous rejection of the captured electron from the ion, this process is known as auto-ionization; (ii) by the radiative decay of the ion-core electron, resulting in its return to the initial state after the emission of a photon of energy $\hbar \omega \cong \hbar \omega_0 = \Delta E$, whereas the captured electron remains in the ion.

It is the latter channel of reaction that is known as dielectronic recombination (DR). Thus, the DR process as well as the photo recombination (PR) process results in the capture of an incident electron and its simultaneous photon emission. The difference is that the photon is emitted by the ion core electron in the DR process rather than by the incident electron as in the PR process. The relationship between the PR and the DR is analogous to the interrelation between conventional and polarizational bremsstrahlung (Sect. 4.4).

As a rule, the DR rate is large for ions with a complex core which possess transitions between the levels with the same quantum number n (the transitions with $\Delta n = 0$, e.g., $2s-2p$ transitions in lithium-like and more complex ions). The transition energy $\Delta E = \hbar \omega_0$ for $\Delta n = 0$ and $Z \gg 1$ is of the order of $Z\text{Ry}$, while the ionization energy is of the order of $Z^2 \text{Ry} \gg \Delta E$. Since the energy E of the incident recombining electron is in any case smaller than the excitation energy, this implies the following inequality

$$(Z^2 \text{Ry}/E)^{1/2} \sim Z e^2 / \hbar v \gg 1 \ , \tag{4.13}$$

which justifies the application of a quasiclassical approach to the description of dielectronic recombination processes.

An application of the proposed approach to the DR description implies the treatment of a DR process as a resonance fluorescence with a complicated intermediate state which appears after the capture of the incident electron by the ion and possesses an additional channel of decay via the auto-ionization process. The resonance fluorescence thus involves three types of quantum states, as follows:

- an initial state with total energy E_1, it includes a non-excited ion and an initial spectral distribution I_0 of equivalent photons;
- an intermediate state with total energy E_2, it includes an excited ion with a captured incident electron on a highly excited ionic level (double excited ion with the ion charge diminished by unity) and an EPh distribution I_0 without one EPh of energy ω_{eq};
- a final state with total energy E_3, it includes an once excited ion with charge, decreased by unity, the EPh of energy ω and the EPh distribution I_0. The state energies are connected by the conservation laws:

$$E_3 - E_1 = \omega - \omega_{eq} \; , \quad E_3 - E_2 = \omega - \omega_0 \; .$$

The resonance fluorescence probability has the form [4.10]

$$W_{RF} = |V_{21}|^2 |V_{32}|^2 [(\omega - \omega_{eq})^2 + \Gamma^2/4]^{-1} [(\omega - \omega_0)^2 + \gamma^2/4]^{-1} \; , \qquad (4.14)$$

where V_{21} and V_{32} are the matrix elements which correspond to the absorption of a EPh of a frequency ω_{eq} and the emission of a real photon of frequency ω, respectively, by the ion; Γ and γ are the total probabilities (per unit time) for the photon absorption and emission by the ion, respectively:

$$\gamma(E) = 2\pi \sum_k |V_{32}|^2 \delta(E - E_3) \; , \qquad (4.15)$$

$$\Gamma(E) = \gamma \sum_{k_{eq}} |V_{21}|^2 [(E - E_2)^2 + \gamma^2/4]^{-1} \; . \qquad (4.16)$$

The quantities $\Gamma(E)$ and $\gamma(E)$ should be taken in (4.16) at the energy $E = E_3$, but in fact they depend weakly on energy.

It is noteworthy to recall an implication of (4.14): for the case of the elementary process of the absorption-emission, the energies of the absorbed and emitted photons are equal (within the practically very small width Γ). This conservation of "memory" about the absorbed photon by the ion (atom) manifests itself in the probability (4.14) which does not reduce to the product of absorption and emission probabilities. Indeed, the first factor in the denominator in (4.14) binds (approximately via the corresponding δ-function) not the energies of the initial (ω_{eq}) and exactly resonant (ω_0) photons, but the energies of the initial and final (ω) photons. It is the continuity of the incident photon spectrum that reduces the resonance fluorescence process to the two independent processes of absorption and subsequent emission. Thus, the DR width γ_{DR} is formed by the width γ (4.15) and by the probability (per unit time) of the auto-ionization process in intermediate state:

$$\gamma_{DR} = \gamma + \Gamma_A . \tag{4.17}$$

This relation implies the possibility of the return of the recombined electron to the continuum energy state with the simultaneous equivalent photon re-emission by the ion core.

The matrix element V_{21} is determined by the oscillator strength of the radiative transition in an ion ($V_{21} \propto d_{21}$, where d_{21} is the matrix element of the dipole moment of a bound radiating electron in an ion) and is proportional to the flux density of the EPh incident on an ion. For the case of the continuum EPh spectrum, the total probability of absorption Γ_{RF} is expressed in terms of I_0:

$$\Gamma_{RF} = 2\pi I_{0\omega} \overline{|V_{21}|^2} , \tag{4.18}$$

where the overbar denotes the averaging over the angles of the absorbed photon's wave vector. As applied to the specific conditions of the model for the DR considered, the summation over k_{eq} in (4.16) should be supplemented by a summation over the final states of the captured electron. This procedure combined with the conservation law for the incident electron energy leads to the result:

$$\gamma_{DR} = \Gamma_{RF} Z^2 n^{-3} . \tag{4.19}$$

Then, using the expression for the DR total probability summed over the EPh frequencies

$$\sum_{\omega} W_{DR} = \gamma \Gamma_A / (\gamma + \Gamma_A) \tag{4.20}$$

we obtain the rate of the auto-ionization process

$$\Gamma_A = \pi^{-1} f_{12} n^{-3} l G_0(\omega_0 M^3 / 3 Z^2) , \tag{4.21}$$

where f_{12} is the oscillator strength of the excited radiative transition in the ion core, and the function G_0 is given by (4.7). The result (4.21) coincides with the result which may be obtained from the exact quantum calculation [4.11] of the auto-ionization rate in Kramer's domain (quasiclassical motion of the incident electron, $Ze^2/\hbar v \gg 1$, along a quasiparabolic orbit, $\varrho \ll a$).

The total DR rate corresponding to the capture of an electron into an ionic nl-state is given by the expression

$$\alpha_{DR} = (2\pi)^{3/2} g(2)(2l+1)\gamma\Gamma_A \exp[-\omega/T + Z^2/2n^2 T]/g(1)T^{3/2}(\gamma+\Gamma_A) , \tag{4.22}$$

where T is the electron temperature, and $g(1)$ and $g(2)$ are the statistical weights for the ground and excited ion levels, respectively.

The result (4.22) may also be derived on the basis of the relation between the auto-ionization probability and the cross section for the ion excitation by electron impact near the ionization threshold. This relation follows from the detailed balance equation for the mutually inverse processes of auto-ionization and electron capture into an ion nl-level with the excitation of the $1\rightarrow2$ transition in an ion core [4.12]

$$(2l+1)g_2\Gamma_A(nl) = Z^2 n^{-3}\omega g_1 \sigma_{ex}(l)/\pi^2 a_0^2 . \tag{4.23}$$

Substituting the KrED result (4.11) for the excitation cross section σ_{exc} we arrive at (4.21).

All of the methods of the derivation of auto-ionization and DR rates are equivalent in the sense that they are in any case based on the dipole approximation for the interaction between an incident and a bound electron. It is this approximation that allows us to treat all of the processes of incident electron energy loss as the processes of the effective radiation of real (bremsstrahlung and photorecombination radiation) or equivalent (excitation, dielectronic recombination, polarizational bremsstrahlung and polarizational recombination) photons.

In order to illustrate the potential of the KrED method, let us apply the analytic result for the dielectronic recombination (DR) rate (4.22) to the description of the multicharged ion (MCI) level populations f caused by radiative cascades from the Rydberg levels populated, in turn, by the DR process. In Kramers' domain, $l \leq n$, the population contains the contributions of the direct and cascade transitions (Chap. 3)

$$f_{nl} = (f_{nl}^D + f_{nl}^C) \propto l^6 \left(\frac{\pi\sqrt{3}}{4}\frac{(n/l)^3}{1+(n/n_l)^3} + n_l \int_{n/n_l}^{\infty}\frac{dx}{1+x^3}\right) ,$$

$$n_l = 137\left[\frac{l}{2\pi\omega^2}G_0(\omega l^3/3Z^2)\right]^{1/3} . \tag{4.24}$$

The ratio $R_{C/D}$ for the cascade and direct contributions, which follows from (4.24), is equal to

$$R_{C/D} \sim (l/n)^3 \max\{n, n_l\} . \tag{4.25}$$

Thus, for the Li-like ion of iron ($Z = 26$, $\omega \equiv \omega_0 = 0.07\,Z$), the cascade mechanism of level population turns out to be appreciable ($R_{C/D} \sim 1$), in particular, for the following sets of parameters: $l \sim 10$ (respectively, $n_l \sim 80$) and $n \sim 40$; $l \sim 5$ ($n_l \sim 60$) and $n \sim 20$.

4.4 Polarization Radiation as Non-Resonant Scattering of Equivalent Photons

Polarizational radiation (PlR), which has been thoroughly investigated in recent years [4.13 – 16], belongs to that wide class of phenomena which can be treated as the radiation emission by a compound system "atom + incident particle". Such a general approach was formulated for the first time by *Born* [4.17]. References [4.13 – 16] reveal the importance of polarization radiation for the electron bremsstrahlung in the atomic frequencies domain. The effects considered are close to the more general *Born* concept [4.17] for radiative processes in a many-body system. This concept has been used by *Percival* and *Seaton* [4.18] for the description of polarization effects in the resonance frequencies domain for collisionally induced radiation, and by *Jablonsky* [4.19] for the description of atomic spectral line broadening. The importance of polarization effects outside of the resonance domain (i.e., in the spectral line wings) was demonstrated for free-free electron transitions in [4.13 – 16]. A detailed discussion of polarization radiation is given in *Tsytovich* and *Oiringel's* monograph [4.3].

    M. *Born* [see 4.17] has used the general Hamiltonian of the total system which includes the incident particle, atomic electrons and their interaction with an electromagnetic field. However, in subsequent calculations a static screening potential was used (in contrast to [4.13 – 16, 20]). Contrary to the conventional treatment of bremsstrahlung radiation as induced by the dipole moment d_e of the incident electron only, in the framework of M. *Born*, a more general approach to the radiation emission is due to the total dipole moment of the system presented by the sum of atomic and incident electron dipole moments, $d_e + d_a$. Though the PlR is determined by the atomic electrons dipole moment (and therefore it has the same origin as the radiation emission of the atom excited by electron impact) it has been introduced in the literature [4.13 – 16] for free-free incident electron transitions as bremsstrahlung PlR. The EPh approach to the description of the PlR provides the possibility both to obtain a simple analytical description for the PlR and to trace the common roots of the PlR and line radiation emission of the atom excited by electron impact.

    We shall start with the description of the PlR in the Kramers domain ($\tilde{\omega} \ll \omega_{eq} \ll Z_i^2 \text{Ry}$, Z_i is the ion charge, Sects. 4.1 – 3) which corresponds, according to Sect. 4.3, to a strongly inelastic transition of a quasiclassical incident electron. Just in the Kramers domain it can be seen, due to the independence of the process on the specific type of electron trajectory, an universal character of polarizational radiation emission in the sense that the PlR can occur for all three types of a transition of the electron emitting the equivalent photons, namely, not only for free-free but for free-bound and bound-bound transitions, as well. In the latter case we are dealing with a new process that was previously not considered in the literature, which may be designated as polarizational recombination and polarizational bound-bound transitions (below and Table 4.1).

Table 4.1.

Type of transition	Processes with real photons	Processes with equivalent photons	
		Line core	Line wing
free-free	Bremsstrahlung radiation	Impact excitation	Polarization bremsstrahlung radiation
free-bound	Photorecombination radiation	Dielectronic recombination	Polarization recombination
bound-bound	Line radiation	Dielectronic bound-bound transitions	Polarization bound-bound transitions

For free-bound transitions of incident electron and $\Delta n = 0$ transitions of a bound electron in an ion core, we have already derived the formulae for dielectronic recombination in the previous section. The DR process corresponds to the case when the intermediate state of the compound system (incident electron + excited ion core) is real since it is a double excited state of an ion with energy ω_{eq}, transition energy $\hbar\omega_0$. The process of the PIR (in the case of the free-bound incident electron transition, it is the process of polarization recombination radiation) corresponds to an appreciable deviation of ω_{eq} from the resonance. Here the resonance approximation for the description of the process is not applicable. An adequate description may be achieved by a simple replacement of the resonance denominator $(\omega_0 - \omega - i\gamma/2)$ in the resonance fluorescence amplitude by the sum

$$(\omega_0 - \omega - i\gamma/2)^{-1} + (\omega_0 + \omega - i\gamma/2)^{-1} \tag{4.26}$$

(cf. e.g., § 59, 63 in [4.6]). This procedure leads to the expression of the PIR spectral intensity in terms of the cross section σ_{sc} for elastic scattering of nonresonant (with respect to ω_0) EPh by the ion core:

$$dE_{PIR}/d\omega \equiv I_{PIR}(\omega) = I(\omega)\sigma_{sc}(\omega) , \tag{4.27}$$

where $I(\omega)$ is the spectral distribution (4.4) for the EPh flux. Equation (4.27) reflects the origin of the PIR which is determined by the processes of atomic (ionic) dynamic polarization induced by an incident electron electric field. Indeed, the cross section σ_{sc} is expressed in terms of the dynamic polarizability $\alpha(\omega)$ of an ion at the frequency ω

$$\sigma_{sc}(\omega) = 8\pi\omega^4\alpha^2(\omega)/3c^4 ,$$

$$\alpha(\omega) = \hbar^{-1} \sum_n |d_{1n}|^2 [(\omega_{n1} - \omega - i0)^{-1} + (\omega_{n1} + \omega - i0)^{-1}] , \tag{4.28}$$

where, due to the non-resonant character of the process, the virtual transitions of the ion core electron to all of the other ion levels, denoted by index n, should be taken into account.

Equation (4.28) allows us to find a quantity of practical interest, namely the ratio of the intensities of conventional bremsstrahlung radiation and the PIR since both of these processes are the channels of the process of electron-ion inelastic collision. This ratio, which is valid in the framework of the dipole approximation for an electron-ion interaction discussed above, is dependent not on the parameters of the incident electron trajectory but on the dynamic polarizability of the ion only:

$$-\frac{I_{\mathrm{PIR}}}{I_{\mathrm{BR}}} = \left[\frac{m_e \omega^2 \alpha(\omega)}{Z_c e^2}\right]^2 \equiv R(\omega) \ . \tag{4.29}$$

It is worthwhile to point out that (4.29) can be also derived with the help of the following simple consideration. The Fourier component of the induced dipole moment D_ω of the core is determined by the dynamic polarizability $\alpha(\omega)$ of the core and by the Fourier component of the electric field F_ω of the excited electron at the point of the core location. This electric field can be found from the (4.3) for the excited electron motion in the Coulomb field of the core:

$$F = er/r^3 = -m_e \ddot{r}/Z_c e \ ; \quad F_\omega = -m_e \omega^2 d_\omega/Z_c e^2 \ . \tag{4.30}$$

Here r is the space coordinate of excited electron, Z_c is the charge of the core, which is located at the point $r = 0$. Allowing for the ratio $I_{\mathrm{PIR}}/I_{\mathrm{BR}} = |D_\omega|^2/|d_\omega|^2$, we arrive at (4.29).

The last derivation of (4.29) suggests that it does not depend on the character of the external electron motion (either quantum or classical), i.e., it is valid for an arbitrary value of the parameter $Ze^2/\hbar v$, not only $Ze^2/\hbar v \gg 1$ as in the KrED. This is the case. Thus, (4.29) is also valid for the bremsstrahlung of electrons on an ion in the alternative (with respect to the character of the electron motion) case of the Born approximation for the electron motion [4.3, 20].

It follows from (4.29) that the PIR being the processes which can be interpreted in terms of the equivalent photons approach, are interrelated in the same way as the corresponding processes with real photons, i.e., conventional bremsstrahlung, photorecombination and line radiation. For the latter processes there exist the well known fact that the universal analytical description can be attained simply by an analytic continuation, with respect to the sign of the incident initial (E) and final (E') electron energies, and results in a single formula. This phenomenon pertains first of all to the transition matrix elements, the subsequent derivation of the formulae for the spectral intensities being different only in the quantization law for the densities of the initial and/or final states of the electron. The possibility of the description of the PIR in terms of the EPh method leads us to the conclusion that the afore mentioned property of the processes with real photons can be directly extended to the processes with the EPh. In particular, in the KrED domain this interrelation between, e.g., free-free and free-bound transitions of an incident electron, reaches its closest point (the point $\hbar\omega = E$ plays no role for the spectral distribution functions).

The interrelation between the processes with real and equivalent photons is illustrated in Table 4.1. The processes with equivalent photons are divided into two groups in this table: resonant (whose intermediate states are real ones), corresponding to the emission in the core of spectral line (or in its nearest wings); and non-resonant (whose intermediate states are virtual), corresponding to transitions with frequencies which are far from atomic transition frequencies ω_0 (wings of spectral lines). The transition from resonant to non-resonant processes turns out to be of a continuous nature, as discussed above (4.26).

In addition to the well known processes discussed in Sects. 4.2, 3, Table 4.1 points out three new types of transitions: polarizational recombination, dielectronic bound-bound, and polarizational bound-bound transitions. The two latter processes are constructed in a way analogous to the processes for a continuous energy spectrum. They correspond to bound-bound non-radiative transitions of excited ionic electrons resulting in a real (or virtual) excitation of the ion core, respectively. An estimate for the intensity of these processes can be obtained from the general formula (4.29), since this universal ratio for the intensities of processes with real and equivalent photons is applicable, according to the KrED concept, both to transitions in the continuous and discrete energy spectrum.

Let us consider polarizational bound-bound transitions in an ion with the core (in its ground state $n_0 l_0$) and a highly excited electron on the level $n'l'(n' \gg 1)$. The electron may transit to a lower lying level nl with a simultaneous (virtual or real) excitation of the core onto the level $n_1 l_1$. It results in the formation of a double excited state $(n_1 l_1, nl)$, the probability of its appearance being dependent on the closeness of its energy to that of the initial state $(n_0 l_0, n'l')$. If the difference between the energies of these states is less than their widths, there will be a real excitation of the $(n_1 l_1, nl)$ state, otherwise the excitation will be virtual. It is clear that the real excitation is possible only in the case of some occasional resonance and therefore it is much less probable. Moreover, in this case there will be an appreciable mixing of corresponding one- and two-electron configurations. Therefore we restrict ourselves below to the investigation of a widely encountered situation, namely, the virtual excitation of the core without a change of the principal quantum number (i.e., $n_0 = n_1$), for which the energy of the $(n_1 l_1, nl)$-state lies within the discrete spectrum.

For bound-bound transitions of an external (with respect to the ion core) electron with the transition frequency ω_0, the dynamic polarizability $\alpha(\omega)$ in (4.29) is determined by the oscillator strength f of the closest radiative transition in the core. Assuming $\alpha(\omega) \sim f/2\omega_0\Delta\omega$, where $\Delta\omega = \omega - \omega_0$, we arrive at the following estimate for the value of $R(\omega)$:

$$R(\omega) \cong (\omega_0 f/2 Z_c \Delta \omega)^2 . \tag{4.31}$$

The dielectronic bound-bound transitions considered above can play a significant role for the transitions of an external electron just from the highly excited states. Indeed, for these states the value of $\Delta \omega$ in any case does not exceed the energy difference between the neighbouring levels, $\Delta \omega \leq Z_c^2/n^3$, so that (4.31) gives:

$$R \cong (\omega_0 f/2)^2 (n/Z_c)^6 . \tag{4.32}$$

The typical value of the factor $\omega_0 f/2$ for the transitions without a change of principal quantum number in the core of Li-, Be-like, and more complex ions varies within the range $0.1-1.0$ a.u. (e.g., it approximately equals 0.2 for transitions $2s^2-2s2p$ in ions N^{3+}, O^{4+}, Ne^{6+}, etc.). Therefore, for the states with $n \geq 2Z_c$, the non-radiative dielectronic bound-bound transitions dominate over the radiative transitions with the same frequency.

The importance of polarizational bound-bound transitions is illustrated here for the ion Ne IV (Fig. 4.2 from [4.21]). The level which corresponds to the double excited state $2s2p^3 3l$ is located in the region of the Rydberg spectrum of the one-electron excited states $2s^2 2p^2 n'l$, namely, between levels $n' = 9$ and $n' = 10$. In this case $\Delta \omega/\omega \sim 10^{-2}$, and, finally, $R \geq 30$. It follows

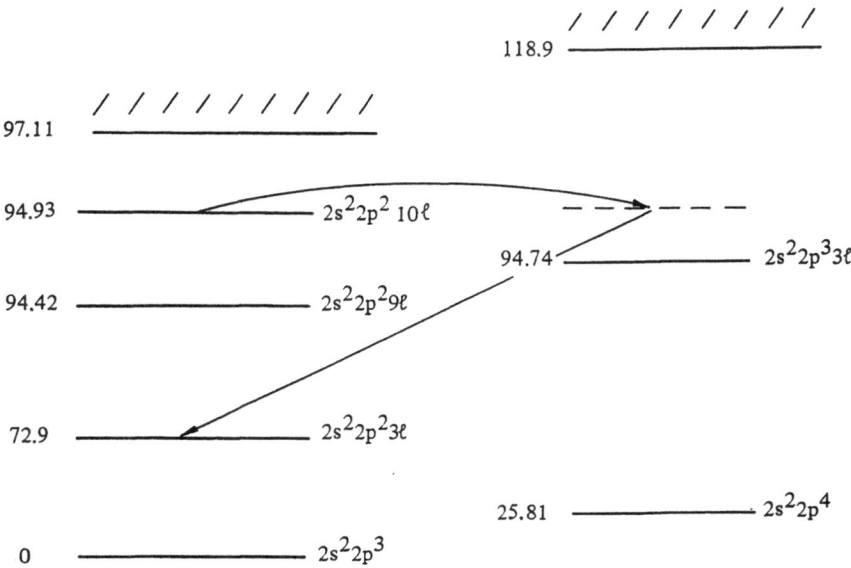

Fig. 4.2. The scheme of ion Ne IV levels corresponding to one- and two-electron excitation (excitation energies in keV are indicated). The radiative and polarization channels of the decay of the state $2s^2 2p^2 10l$ are shown

that the population of the level $2s2p^3 3l$ by the transitions from those upper lying levels which are close to the double excited level $2s2p^3 3l$ (in this case, these are the levels $2s^2 2p^2 10l$ and $2s^2 2p^2 9l$) will be produced effectively by the polarizational transitions (i.e., transitions via the virtual double-excited state, for the level $2s^2 2p^2 10l$ as indicated on Fig. 4.2) rather than by a direct radiative transition. This fact, in turn, can crucially influence the kinetics of the $2s2p^3 3l$ level population, especially in the case when this level is predominantly populated by transitions from highly excited states (in the case considered, from the level $2s^2 2p^2 9l$ and higher ones). The latter situation is typical, e.g., recombining non-equilibrium plasmas with multicharged ions.

5. Hydrogenic Atom in an Electric Field. Quasiclassical Consideration

5.1 Quasiclassical Results for Transition Probabilities and Lifetimes in Parabolic Coordinates

5.1.1 Introductory Comments

Investigations of the splitting of the energy levels of atoms in an electric field F (Stark effect, 1913) and in a magnetic field B (Zeeman effect, 1896) have provided important confirmations of the validity of the main assumptions of quantum theory. Fundamentals of the quantum theory of the Stark and Zeeman effects are treated in detail in the well known monograph of *Bethe* and *Salpeter* [5.1], and also in textbooks on quantum mechanics [5.2] and theoretical spectroscopy [5.3].

In recent years the interest in these effects has shifted to applications. It has been found that many theoretical results are not very suitable for specific calculations or are still incomplete. Further development of the theory has mainly involved investigations of atomic spectra in strong B and F fields and of highly excited states characterized by $n \gg 1$. The interest in these problems arises due to a wide range of applications such as the ionization in an electric field of Rydberg atomic states selectively populated by laser radiation [5.4, 5], absorption spectra of excitons in a magnetic field [5.6, 7], structure of atoms in very strong magnetic fields on the surfaces of neutron stars [5.8 – 10], splitting and broadening of atomic spectral lines by electric and magnetic fields in a plasma [5.11, 12], structure of radio frequency (*rf*) lines emitted by highly excited atoms in the interstellar medium [5.13, 14].

Recent experiments on Rydberg atoms have been carried out in connection with a wide range of theoretical problems including the quantum defect method [5.15], the dynamic Stark and Zeeman effects in alternating F and B fields [5.16], and other topics (discussed in a monograph by *Stebbings* and *Dunning* [5.17]). An important place among these problems is occupied by the simplest atomic system, the hydrogen atom, to which the present review is devoted.

Selective population of hydrogen states with $n \sim 10 - 50$ has been detected under laboratory conditions (for a review: chapter by *Koch*, in [5.17], p. 473). These experiments provide, in principle, a technique for precision measurements of fundamental atomic constants. The hydrogen atom is also of great interest in astrophysics because it can be excited to states with $n \sim 100 - 400$ in space (for a review: chapter by *Dalgarno*, in [5.17], p. 1). In the interpretation of both laboratory and astrophysical data, one requires detailed information on the dependences of atomic parameters on the intensities of external fields and on the individual quantum numbers of atomic states. Therefore, the account given below is designed not only to introduce the general theoretical principles, but also to provide specific analytic and numerical results suitable for use in many applications. At the same time an attempt will be made avoid to omitting problems of a fundamental nature. They include, above all, the existence of an additional integral of motion of an electron moving in Coulomb and magnetic fields, and the possibility of stochastization of the electron motion in this case. These problems are closely related to the fundamental principle of quantization of a system with inseparable variables.

It is surprising that in spite of the traditional nature of the topics related to the Stark and Zeeman effects, a whole series of new results have been obtained and many new observations have been made. This has revealed the major capabilities of the semiclassical and purely classical solution methods. The success of the classical approach is clearly due to the incompleteness of the classical descriptions of the atom which have been "prematurely" discarded by the establishment of the theoretical apparatus of quantum theory. Quantum mechanical calculations based on perturbation theory, asymptotic approach, and numerical solutions of the Schrödinger equation are treated fully in the monograph of *Stebbings* and *Dunning* [5.17]. Therefore, we shall concentrate our attention on semiclassical methods.

We shall frequently use the atomic system of units (a.u.) without mentioning the fact explicitly. However, in some cases it is convenient to retain dimensional units. We shall therefore immediately mention the characteristic ranges of the F and B fields. The interatomic electric field F_A is

$$F_A = e a_0^{-2} \approx 5.1 \times 10^9 \text{ V/em} \approx 1.7 \times 10^7 \text{ SGS} , \qquad (5.1)$$

where e and m are the charge and mass of an electron, and $a_0 = \hbar^2/me^2$ is the Bohr radius.

In the case of magnetic fields it is convenient to introduce a field B_0 such that the magnetic interaction $\mu_B B_0$ ($\mu_B = e\hbar/2mc$ is the Bohr magneton) is comparable with the scale of the atomic energy $\text{Ry} = me^4/2\hbar^2$:

$$B_0 = \text{Ry}/\mu_B = 2.35 \times 10^9 \text{ G} = 2.35 \times 10^5 \text{ T} . \qquad (5.2)$$

The magnetic field B_A inside an atom is in fact less by a factor $(\hbar c/e^2) \approx 137$ because of the nonrelativistic nature of the motion of electrons

$$B_A \approx B_0/137 \approx 1.7 \times 10^7 \text{ G} . \qquad (5.3)$$

The values (5.1−3) of the fields F_A and B_0 are very high and they are attainable only under fairly exotic conditions. However, we must bear in mind that these fields decrease rapidly with the increase in the principal quantum number n of the atom. For example, the critical electric field F_c which suppresses the potential barrier of an atomic electron is

$$F_c = F_A/16n^4 \; , \tag{5.4}$$

which is five orders of magnitude less than the atomic field F_A if $n \simeq 10$.

Similarly, hydrogen excitations in a solid (excitons) correspond to effective values $B_0 \sim 10-10^2$ T because of a reduction in the effective mass of an electron in a solid and also because of the high permittivity. Therefore, many of the effects considered below occur in near-critical F and B fields, but can be observed in practice.

5.1.2 General Relationships

The fundamentals of the theory of a hydrogenic atomic in an electric field F are well known and have been presented in detail in, for example, the monographs mentioned earlier [5.1 − 3]. However, from the point of view of practical applications many topics of this theory have not been finally resolved until very recently. This particularly applies to the interpretation of the spectra of highly excited atoms when the high degree of degeneracy of the hydrogen levels enormously complicated the calculations based on the direct application of the general formulas for the intensities of transitions, etc. We shall present a number of new theoretical results on the spectra of the hydrogen atom in an electric field and these provide simple and reliable analytical results suitable for the application to cases of practical interest.

A fundamental feature of the theory of the Stark effect is the ability to separate the variables in the equation for the hydrogen atom expressed in parabolic coordinates ξ and η [5.2]:

$$\frac{d^2\chi_1}{d\xi^2} + \frac{1}{4}\left(2E + \frac{4\beta_1}{\xi} - \frac{m^2-1}{4\xi} - F\xi\right)\chi_1 = 0 \; , \tag{5.5}$$

$$\frac{d^2\chi_2}{d\eta^2} + \frac{1}{4}\left(2E + \frac{4\beta_2}{\eta} - \frac{m^2-1}{\eta^2} + F\eta\right)\chi_2 = 0 \; . \tag{5.6}$$

Here $\chi_1(\xi) = f_1/\sqrt{\xi}$ and $\chi_2(\eta) = f_2/\sqrt{\eta}$ are the reduced wave functions [5.2], E is the energy, and $\beta_{1,2}$ are the constants of the process of separation of the variables $\beta_1 + \beta_2 = 1$, m is the magnetic quantum number.

Equations (5.5,6) are one dimensional and they are characterized by the effective potentials

$$V_\xi = \frac{m^2-1}{8\xi^2} - \frac{\beta_1}{\xi} + F\xi \; ; \quad V_\eta = \frac{m^2-1}{8\eta^2} - \frac{\beta_2}{\eta} - F\eta \tag{5.7}$$

governed at short distances by the Coulomb and centrifugal terms, and at large distances by the term containing the field F (Fig. 5.1). We can see from Fig. 5.1

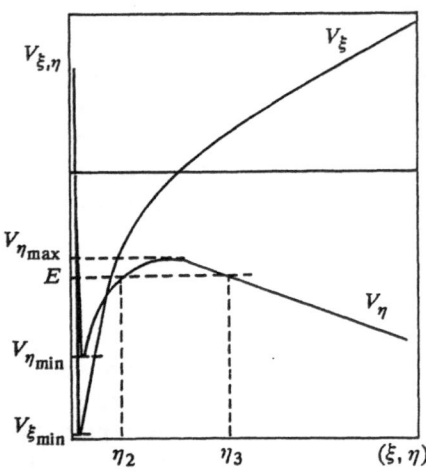

Fig. 5.1.   Potentials $V_\xi = V_1(\xi)$ and $V_\eta = V_2(\eta)$ for the motion of an electron of energy E in an electric field plotted using the parabolic coordinates ξ and η

that the potential barrier expressed in terms of the variable η has a finite penetrability, so that an atomic electron can escape to the continuous spectrum (i.e., an atom can decay).

It is convenient to rewrite (5.1,6) in terms of dimensionless variables $x = \nu^{-1}\xi$ and $y = \nu^{-1}\eta$, where ν is the effective principal quantum number of the level, which allows for the shift of the level in an electric field F:

$$\nu = (-2E)^{1/2} \; . \tag{5.8}$$

We now introduce the reduced field intensity $w = F\nu^3$ and the new separation constants α_1 and α_2 $(\alpha_1 + \alpha_2 = \nu)$, yielding the following equations for the reduced wave functions φ_1 and φ_2 [5.18]:

$$\frac{d^2\varphi_1}{dx^2} + \frac{1}{4}\left(-1 + \frac{4\alpha_1}{x} - \frac{m^2-1}{x^2} - wx \right)\varphi_1 = 0 \; , \tag{5.9}$$

$$\frac{d^2\varphi_2}{dy^2} + \frac{1}{4}\left(-1 + \frac{4\alpha_2}{y} - \frac{m^2-1}{y^2} + wy \right)\varphi_2 = 0 \; . \tag{5.10}$$

The states of the hydrogen atom in the field F are described, as is well known [5.1,2], by the parabolic quantum numbers n_1 and n_2 and by a magnetic quantum number m, which are related by

$$n_1 + n_2 + |m| = n - 1 \; . \tag{5.11}$$

It is convenient to introduce an "electric" quantum number $k = n_2 - n_1$, which defines the projection of the dipole moment of an atom $e\mathbf{r}$ along the direction of an electric field F:

$$\langle n_1 n_2 m | e\mathbf{r} \cdot \mathbf{F} | n_1 n_2 m \rangle = eF \langle n_1 n_2 m | z | n_1 n_2 m \rangle = -\frac{3}{2} nkea_0 F \ . \tag{5.12}$$

The energy levels can be described, up to the second order in respect to the field intensity F, by the expression (in atomic units)

$$E = -\frac{1}{2n^2} - \frac{3}{2} Fnk - \frac{1}{16} F^2 n^4 (17 n^2 - 3 k^2 - 9 m^2 + 19) \ . \tag{5.13}$$

The parabolic wave functions $n_1 n_2 m$ correspond to specific projections, along the direction of the electric field F, of the vectors representing the dipole d and orbital angular momenta l of an atom. The specific symmetry properties of the Coulomb field make it possible to simply express these parabolic functions in terms of spherical functions ψ_{nlm} corresponding to specific values of l^2 and m^2 (§ 37 in [5.2]):

$$\psi_{n i_1 i_2} = \sum_{l=0}^{n-1} C(l, i_1 + i_2 | i_1, i_2) \psi_{nlm} \ , \tag{5.14}$$

where $i_{1,2} = [m \pm (n_1 - n_2)]/2$, $C(l, m | l_1, l_2)$ are the Clebsch-Gordan coefficients.

The intensities I of the Stark components are described by the matrix elements of the coordinate r of an atomic electron:

$$I(n n_1 n_2 m \Rightarrow n' n'_1 n'_2 m') = \frac{4 e^4 \omega_0^4}{3 c^3} |\langle n n_1 n_2 m | r | n' n'_1 n'_2 m' \rangle|^2 \ , \tag{5.15}$$

where ω_0 is the unperturbed frequency of the $n \rightarrow n'$ transition.

Depending on the polarization (linear, characterized by $m = 0$, or circular when $m = \pm 1$), the Stark component can be divided into π and σ components. The intensities of the π components are governed by the matrix elements of the z component of r and the intensities of the σ components are determined by the x or y components of r. The general formulas for the matrix elements have been obtained by *Gordon* (§ 65 in [5.1]) and can be expressed in terms of the hypergeometric function

$$\psi_m(n_i, n'_i) = F\left(-n_i, -n'_i, m+1 \ , \quad -\frac{4 n n'}{(n - n')^2} \right) \ . \tag{5.16}$$

In the case of the parabolic quantum numbers n_1 and n_2 (in contrast to the spherical quantum number l) there are no rigid selection rules for dipole radiation. Nevertheless, the distributions of the intensities of the Stark π and σ components obey certain relationships which we shall discuss later (Sect. 5.1). The *Gordon* formulas are very cumbersome and suitable only for calculations in a few special cases [5.1]. The formulas for the intensities expressed in terms of spherical coordinates can be considerably simplified in the semiclassical range [5.14]. Since the relationship between the parabolic and spherical bases is determined by the Clebsch-Gordan coefficients (5.14) whose properties are known, we can expect to obtain satisfactory semiclassical expressions for the matrix elements also in terms of parabolic coordinates. However, such expressions have not yet been derived.

5.1.3 Radiative Lifetimes of States

We shall now consider the simplest radiation parameter of the sublevels which is their lifetime governed by all possible radiative transitions to lower states.

The probability $A_{nn'}^l$ of radiative transitions and the lifetime T_{nl} of excited atomic states are usually analyzed using the spherical quantum number of atoms nl [5.1]. *Bureeva* [5.19] obtained general semiclassical formulas for the calculation of the probabilities $A_{nn'}^l$ of radiative transitions. *Goreslavskii* et al. [5.20] derive simple analytic expression for $A_{nn'}^l$, which are highly accurate even when the numbers n and l are not too large. The structure of these expressions is closely related to the familiar formulas in the classical intensity of radiation in a Coulomb field (§ 70 in [5.21]). The quantum corrections to these probabilities can be found in [5.22, 23].

Simple dependences of the probabilities A_{nl} on the orbital momentum l in the range $3 < n < 25$ are obtained in [5.24]

$$A^0(l) \simeq A(1) \frac{1}{[1+\alpha(l-1)]^2} , \tag{5.17}$$

where the numerical coefficient α is selected so that it agrees with the exact result for $l = n-1$:

$$\alpha = 0.7148 \pm 0.0004 . \tag{5.18}$$

If we use the exact probability of a transition in the case when $l = n-1$ [5.1], we obtain the following expression for $l > 0$:

$$A^0(l>0) = 2.6759 \times 10^9 \, \text{s}^{-1} \left[\frac{1+\alpha(n-2)}{1+\alpha(l-1)} \right]^2$$

$$\times \frac{(2n-1)(2n^2-n-1)}{n^6(n-1)^2} . \tag{5.19}$$

Equation (5.19) ensures a high degree of accuracy so that the maximum deviations amount to a few percent in the range $l < n/2$. If $l = 0$, the numerical results can be approximated satisfactorily by the formula [5.24]

$$A^0(0) \simeq 5.97 \times 10^8 \, \text{s}^{-1} \frac{1}{n^2(n+1.46)} \,. \tag{5.20}$$

Following *Herrick* [5.24], we shall describe the probability $B(nkm, n')$ of a transition from a parabolic state $|nkm\rangle$ to all the states of the level n':

$$B(nkm, n') = w(n, n') \sum_{k', m'} |\langle nkm |r| n'k'm'\rangle|^2 \,, \tag{5.21}$$

where

$$w(n, n') = \frac{4e^2 a_0^2}{3 \hbar c^3} \left[\frac{1}{n^2} - \frac{1}{n'^2} \right]^3 \,. \tag{5.22}$$

Then, the total probability $B(k, m)$ of a transition from a given Stark sublevel to all of the lower levels is

$$B(k, m) = \sum_{n' = |m|+1}^{n-1} B(nkm, n') \,. \tag{5.23}$$

The corresponding lifetime is given by

$$\tau(nkm) = B^{-1}(k, m) \,. \tag{5.24}$$

The formulas (5.23, 24) are "parabolic" analogs of the corresponding spherical quantities A_{nl} and $T_{nl} = A_{nl}^{-1}$. Using the familiar relationship between the parabolic and spherical functions given by (5.14), we obtain an expression relating the two types of probabilities:

$$B(k, m) = \sum_{l = |m|}^{n-1} A(n, l) [C(n, k \,|\, lm)]^2 \,. \tag{5.25}$$

It follows from the properties of the Clebsch-Gordan coefficients $C(n, k \,|\, lm)$ that the symmetry properties of the probabilities are

$$B(k, m) = B(-k, m) = B(k, -m) \,. \tag{5.26}$$

Summation over all the values of k and m clearly gives the total lifetime $A(n)$, which is independent of the summation bases

$$A(n) = \sum_{k, m} B(k, m) = \sum_{l = 0}^{n-1} (2l+1) A(n, l) \,. \tag{5.27}$$

An important sum rule is obtained from (5.25) by adding all the values of k and m in such a way that either $k+m$ or $k-m$ remains constant [5.24]

$$n^{-1}A(n) = B(k,0) + \sum_{m=1}^{(n-1-k)/2} B(k+m,m) + \sum_{m=1}^{(n-1+k)/2} B(k-m,m) . \quad (5.28)$$

In (5.28) we sum the values of the number $k = n-1, n-3, \ldots -(n-1)$ over all n, and we again obtain (5.27).

It follows from (5.28) that the distribution of the transition probabilities in the $m = 0$ case is determined uniquely by the distribution $B(k,m)$ when $m \neq 0$. The latter are found to depend weakly on the "electric" quantum number k. It is therefore convenient to introduce average (in terms of "k") values of $B(k,m)$ described as

$$\bar{B}(m) = \frac{1}{n-|m|} \sum_k B(k,m) = \frac{1}{n-|m|} \sum_{l=|m|}^{n-1} A(n,l) . \quad (5.29)$$

Since in the $m \neq 0$ case we can accurately assume that $B(k,m) = \bar{B}(m)$, we can obtain $B(k,0) \approx B^0(k,0)$ from (5.28), which gives

$$B^0(k,0) = n^{-1}A(n) - \sum_{m=1}^{(n-1-k)/2} \bar{B}(m) - \sum_{m=1}^{(n-1+k)/2} \bar{B}(m) . \quad (5.30)$$

If $k = n-1$, this relationship reduces to

$$B^0(n-1,0) = n^{-1}A(n) - \sum_{m=1}^{n-1} \bar{B}(m) . \quad (5.31)$$

In the range of lower values $k < n-1$ the quantities $B^0(k,0)$ are found from the recurrence relationship

$$B^0(k-2,0) = B^0(k,0) - \bar{B}\left(\frac{n+1-k}{2}\right) + \bar{B}\left(\frac{n-1+k}{2}\right) . \quad (5.32)$$

The quantities $\bar{B}(m)$ can be found from (5.29) by means of the approximate expressions given in (5.17) for $A(n,l)$

$$\bar{B}^0(m) = \frac{A^0(m)}{(n-m)} + \frac{A^0(1)}{(n-m)} \frac{(n-1-m)}{[1+\alpha(m-1/2)][1+\alpha(n-1/2)]} , \quad m > 0 . \quad (5.33)$$

A comparison of the results for $\bar{B}(m)$ and $B(k,0)$ based on the use of the approximation represented by (5.32, 33) with the results of exact calculations is made in Table 5.1. It is clear from Table 5.1 that these approximations are quite accurate. Therefore, the method described above makes it possible to determine the lifetime $B(k,m)$ of the Stark sublevels of the hydrogen atom using simple analytic formulas and thus avoid the direct summation of the series of (5.25) containing the Clebsch-Gordan coefficients.

Table 5.1. Comparison of probabilities of radiative $\bar{B}(i)$ and $B(i,0)$ transitions obtained using approximation formulae (5.32, 33) (columns denoted by A) with exact results [5.25] (columns denoted by B) for the $n = 10$ level of hydrogen [5.24]

i	$\bar{B}(i)$		$B(i,0)$	
	A	B	A	B
1	1.0905	1.0871	0.4753	0.4806
2	0.5472	0.5523		
3	0.3623	0.3658	0.5666	0.5685
4	0.2703	0.2722		
5	0.2155	0.2163	0.7757	0.7706
6	0.1791	0.1794		
7	0.1532	0.1532	1.1891	1.1733
8	0.1338	0.1338		
9	0.1186	0.1188	2.1610	2.1473

5.2 Intensities of the Stark Components

The intensities of the Stark components are described by the general *Gordon* formulas [5.1] for the matrix elements of the components of the radius vector of an atomic electron expressed in parabolic coordinates. However, the application of these formulas involves very time consuming numerical calculations, particularly in the case of highly excited levels. The situation is also complicated by the absence of rigorous selection rules for the parabolic quantum numbers.

In the case of large values $n \gg 1$ we can establish simple relationships governing the distributions of the intensities of the components in the case of transitions characterized by a small change in the quantum number $n - n' \equiv \Delta n \ll n$ which are of practical interest. Following *Gulyaev* [5.14] we shall consider these relationships in the case of highly excited lines $H_{n\alpha}$ ($\Delta n = 1$) and $H_{n\beta}(\Delta n = 2)$ observed under astrophysical conditions.

The change $\Delta \omega$ in the frequency of a transition in an electric field F, corresponding to the Stark component $n_1 n_2 m \rightarrow n'_1 n'_2 m'$, is according (5.13) given by

$$\frac{\Delta \omega}{\omega_F} = n(n_1 - n_2) - n'(n'_1 - n'_2) \ , \quad \omega_F \equiv \frac{3}{2} \frac{e a_0}{\hbar} F \ . \tag{5.34}$$

The intensities of the components are sensitive functions of the combinations of quantum numbers $K = (n_1 - n_2) - (n'_1 - n'_2) = k - k'$ and $i \equiv n'_1 - n'_2 \equiv k'$, where $\Delta \omega$ becomes

$$\frac{\Delta \omega}{\omega_F} = Kn + \Delta ni \ . \tag{5.35}$$

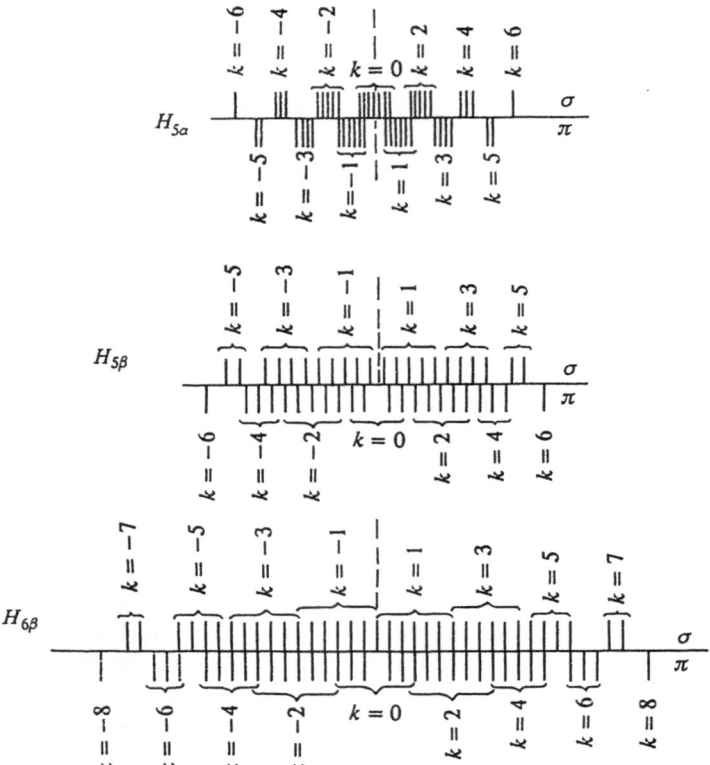

Fig. 5.2. Stark splitting of the $H_{5\alpha}$ lines, belonging to the Brackett series, and of $H_{5\beta}$ and $H_{6\beta}$ lines [5.14, 26]. The upper rows of lines are the σ-components and the lower rows are the π-components

The parameter K, which ranges from $-(2n-2-\Delta n)$ to $(2n-2-\Delta n)$, groups the components of $2(2n-2-\Delta n)$ series, each of which contains $2(n-\Delta n)-(K+1)$ terms labeled by the parameter i. Figure 5.2 shows the grouping of the components in the case of $H_{5\alpha}$ and $H_{5\beta}$ and $H_{6\beta}$ lines. We can see that the parity (or nonparity) of the number of K corresponds to the σ (or π) polarization of the components of the lines H_α and H_β. The separation between the components within the series $\Delta \omega_i$ and the separation $\Delta \omega_K$ between the centers of the series are:

$$\Delta \omega_i = \omega_F \Delta n \,, \quad \Delta \omega_K = \omega_F n = \frac{\Delta \omega_i n}{\Delta n} \,. \tag{5.36}$$

We shall now consider the nature of changes in the intensity on increase in the number K. We shall utilize the fact that arguments of the hypergeometric functions in (5.16), which occur in the Gordon formulas [5.1], are large if

$1 \sim \Delta n \ll n \sim n'$ so that these functions can be replaced by the last (largest) terms:

$$\psi_m(n_i, n_i') \approx \frac{n_i'! \, m!}{(n_i' - n_i)! \, (m + n_i)!} \, b^{2n_i} (-1)^{n_i} \, , \tag{5.37}$$

where $b = 4nn'/(n - n')^2 \gg 1$ (to be specific, we shall assume that $n_i' > n_i$ and instead of $|m|$ we shall write simply m).

The approximation of (5.37) allows us to obtain simple analytical expressions for the matrix elements of the coordinate governing the intensities of the π and σ components. We shall consider specific transitions characterized by $\Delta n = 1$ ($H_{n\alpha}$-lines) and $\Delta n = 2$ ($H_{n\beta}$-lines).

$H_{n\alpha}$-Lines (Transitions with $\Delta n = 1$). The central series ($K = 0$) is formed by the σ components and the transitions in this case are described by the relationships

$$n_1 = n_1' \, , \quad n_2 = n_2' \quad \text{for} \quad m \Rightarrow m - 1 \, ,$$
$$\tag{5.38}$$
$$n_1 = n_1' + 1 \, , \quad n_2 = n_2' + 1 \quad \text{for} \quad m \Rightarrow m + 1 \, .$$

The next series ($K = \pm 1$) represents the π components for which the similar conditions are

$$n_1 = n_1' + 1, n_2 = n_2' \quad \text{for } K = +1 \, ,$$
$$\tag{5.39}$$
$$n_1 = n_1', n_2 = n_2' + 1 \quad \text{for } K = -1 \, .$$

Using the approximation of (5.37) and the relationships given by (5.38, 39), we find from the *Gordon* formulas the following simple expressions for the matrix elements:

$$X_m^{m-1} = \frac{a_0}{4} b \sqrt{(n_1 + m)(n_2 + m)} \left(1 - \frac{(n_1 + 1)(n_2 + 1)}{b^2} \right) , \tag{5.40}$$

$$X_m^{m+1} = \frac{a_0}{4} b \sqrt{n_1 n_2} \left(1 - \frac{(n_1 + m)(n_2 + m)}{b^2} \right) , \tag{5.41}$$

$$Z_m^m = \frac{a_0}{4} b \left[\sqrt{n_1(n_1 + m)} \, \delta_{K, +1} + \sqrt{n_2(n_2 + m)} \, \delta_{K, -1} \right] . \tag{5.42}$$

The decrease of intensity with increasing number K is described by the ratio

$$\frac{[(X_m^{m+1})^2+(X_m^{m-1})^2]_{K+2}}{[(X_m^{m+1})^2+(X_m^{m-1})^2]_K} \approx \frac{n_1(n_1+m)n_2(n_2+m)}{b^4} , \qquad (5.43)$$

which cannot exceed $2^{-8} \approx 4 \times 10^{-3}$ (a similar estimate applies also to z_m^m). Therefore, there is practical interest only in the nearest σ and π series, corresponding to the first few values of K.

We can calculate the intensity of a line shifted by $(Kn+i)$ by summing the squares of the matrix elements following for the remaining degeneracy with respect to m. In the case of the central σ series $(K=0)$ and a given value of i, such summation yields

$$I_i^\sigma \sim \sum_m [(X_m^{m-1})^2+(X_m^{m+1})^2] \approx \frac{a_0^2}{24} n^2(2n^3-3n^2i+i^3) . \qquad (5.44)$$

Similarly in the case of the nearest π series $(K=1)$, we find that

$$I_i^\pi \sim \sum_m (Z_m^m)^2 \approx \frac{a_0^2}{24} n^2(n^3+3n^2i-3ni|i|-i^2|i|) . \qquad (5.45)$$

Figure 5.3 schematically shows the splitting of the $H_{n\alpha}$ line in the case when $n \gg 1$. The central $\sigma(K=0)$ series, the nearest π series $(K=\pm1)$, and the distribution of the total (summed over K) intensity are shown in the figure. The intensity of the series decreases by half when the width $\Delta\omega_{1/2} = (n/3)\omega_F$ is reached. The intensity minimum in the π series corresponds to $4\Delta\omega_{1/2}$ and represents approximately 40% of the maximum intensity of the central σ series.

$H_{n\beta}$-Lines (Transitions with $\Delta n = 2$). In the case of the odd n the components of the π series of these lines never coincide with the components of the σ series (Fig. 5.2). Beginning from the $H_{6\beta}$ line, the terms of the series with $K=+1$ and -1 corresponding to the σ polarization begin to overlap. This overlap is the reason for the nonzero intensity at the line center $(\Delta\omega = 0)$

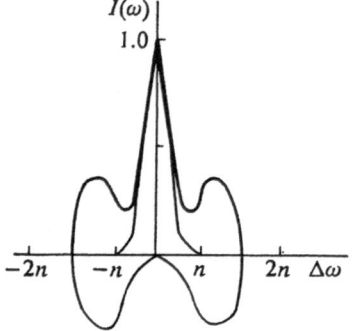

Fig. 5.3. Distribution of the intensity $I(\omega)$ plotted as a function of the frequency shift $\Delta\omega$ (in units of $(3/2)ea_0F/\hbar$) for $H_{n\alpha}$ lines characterized by $n \gg 1$ [5.14, 26]. The lines are the envelopes of the σ (*top part*) and π (*lower part*) components and the thick line is the combined line profile

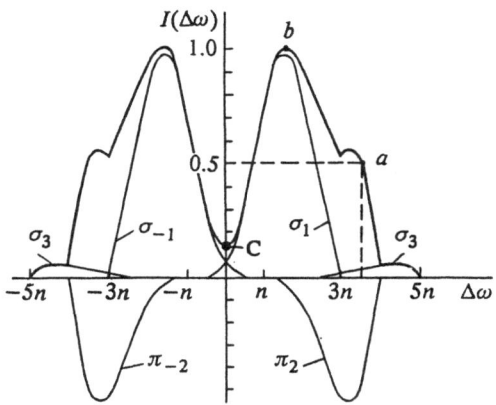

Fig. 5.4. Structure of the $H_{n\beta}$ lines in the $n \gg 1$ case [5.14, 26]. The notation is the same as in Fig. 5.3. Points: *a* level corresponding to half intensity $I_{max}/2$; *b* maximum intensity; *c* central dip

because the intensity of the components of the central ($K = 0$) series vanishes for $\Delta\omega = 0$. If we follow the procedure used above for the $H_{n\alpha}$ line, we can consecutively calculate the intensities of the series $I_{\pi 0}$ ($K = 0, \Delta m = 0$), $I_{\sigma 1}(K = \pm 1, \Delta m = \pm 1)$, $I_{\pi 2}(K = \pm 2)$, etc. The structure of the $H_{n\beta}$ lines is shown in Fig. 5.4, which also gives the separate contributions of the π and σ series [5.26].

The positions of the line maxima correspond to frequencies obeying $\Delta\omega_n = \pm 1.6n$. Their width at mid-amplitude amounts to $2.85 n\omega_F$. The intensity at the center is 13% of I_{max} irrespective of the value of n. The line halfwidth is $\Delta\omega_{1/2} = 3.6\,\omega_F$.

5.3 Weak Fields. Asymptotic Theory of the Decay of an Atom

The behavior of an atom in an electric field F depends on the ratio of F to the critical intensity $F_c \sim 1/16n^4$ (5.4) at which the barrier along the coordinate η disappears for a given level n and the classical above-barrier motion of an electron becomes possible (Fig. 5.1). If $F \ll F_c$, the barrier width is fairly wide and the energy levels of an electron are well localized, i.e., the level width Γ is exponentially small. This case was treated by *Smirnov* and *Chibisov* [5.27] who developed an asymptotic method for the calculation of the atomic parameter. *Damburg* and *Kolosov* [5.28, 29] suggested a method based on the similarity of decay to the scattering of an electron by a quasidiscrete level. In both cases we are dealing with a resonance energy level lying against the background of a continuous spectrum (continuum). When the electron energy E approaches the energy of a discrete level E_0, the phase φ of the wave function changes abruptly [5.29, 30]

$$\varphi = \varphi_0 + \arctan\left[\Gamma/2\left(E - E_0\right)\right] . \tag{5.46}$$

Having determined the wave function, we can then find the parameter Γ and the value of E_0 from (5.46). Determination of the wave function of an atomic electron in the limit of weak fields F is based on [5.29, 30] and on the matching of solutions at low and high values of the coordinate η (along which an electron can reach the continuum). The details of the method are fully described in the review of *Damburg* and *Kolosov* ([5.17], p. 31). The results

$$\Gamma = (4R)^{2n_2+m+1} \exp(-2R/3) [n^3 n_2! (n_2+m)!]^{-1} , \tag{5.47}$$

where $R = (-2E_0)^{3/2}/F.$

Equation (5.47) was first derived in [5.27], but there E was assumed to be the unperturbed value. In fact, terms of the order of F in the expansion for the energy are important in the argument of the exponential function, whereas terms of higher order occur in the correction terms. The results of an expansion of Γ right up to terms of the order of F^2 are given in [5.17, 27, 28]. The results of this asymptotic theory are in good agreement with the numerical calculations for fields of intensities $F \ll F_c$ [5.17, 29].

5.4 Classical Theory of the Decay of an Atom in an Electric Field

An increase in the electric field F reduces the effective potential barrier V_η along the coordinate η. In the case of the highly excited states of an atom, we can have a situation when the energy level coincides with the maximum of the potential barrier, i.e. when $E_c = V_{\eta \, max}$ (Fig. 5.1). This critical energy corresponds to a critical electric field F_c in which two points of intersection of the straight line $V_\eta = E$ with the potential V_η coincide (roots η_2 and η_3 in Fig. 5.1 merge). In this case the above-barrier emission of an electron from an atom is clearly possible and it is allowed by the laws of classical mechanics. Hence, obviously the critical values E_c and F_c can be found by purely classical calculations. This was done by *Banks* and *Leopold* [5.32] and we shall now follow their treatment.

The separation of variables in terms of the parabolic coordinates and of an electron in a Coulomb field e^2/r and in an external field F makes it possible to treat the motion of an electron in the effective potentials of (5.7), as already shown in Sect. 5.1.2. It is convenient to introduce the moment p_η and p_ξ for each variable η and ξ, defined by

$$\frac{p_\xi^2}{2m} + V_\xi(\xi) = E , \quad \frac{p_\eta^2}{2m} + V_\eta(\eta) = E , \tag{5.48}$$

where E is the total energy of an electron in an atom. In the case of bound states, the value of E lies within the range (Fig. 5.1)

$$\max\left(V_{\xi\,\min}, V_{\eta\,\min}\right) \le E \le V_{\eta\,\max} \le 0 \ . \tag{5.49}$$

In the case of this value of E the points of intersection of the straight line $E = \text{const.}$ with the curves representing the effective potential V_{ξ} and V_{η} are described by cubic equations with three roots each: $\xi_1, \xi_2, -\xi_3, \eta_1, \eta_2,$ and η_3.

Our task is to find the dependences of the critical parameters E_c and F_c on the classical action variables I_{φ}, I_{ξ} and I_{η}, representing the state of an electron in an atom. It is convenient to derive this dependence in a parametric form by expressing all of the dependences in terms of the roots of the equations $V_{\xi} = E$ and $V_{\eta} = E$ and then requiring that the roots $\eta_2 = \eta_3$ coincide for the critical values of the parameters E_c and F_c. The expressions for the classical action variables I_{ξ} and I_{η}, expressed in terms of the roots ξ_i and η_i, are

$$I_{\xi} = \frac{1}{2\pi}\oint p_{\xi}d_{\xi} = \frac{(2meF)^{1/2}}{\pi}\int_{\xi_1}^{\xi_2}\frac{d\xi}{\xi}[(\xi_2 - \xi)(\xi - \xi_1)(\xi + \xi_3)]^{1/2} \ . \tag{5.50}$$

$$I_{\eta} = \frac{1}{2\pi}\oint p_{\eta}d\eta = \frac{(2meF)^{1/2}}{\pi}\int_{\eta_1}^{\eta_2}\frac{d\eta}{\eta}[(\eta_2 - \eta)(\eta - \eta_1)(\eta_3 - \eta)]^{1/2} \ , \tag{5.51}$$

$$I_{\varphi} = \text{const.} \ . \tag{5.52}$$

The formulas (5.50–52) together with the equations for the roots and the condition $\eta_2 = \eta_3$ yield a parametric relationship between the critical parameters E_c and F_c and the action variables. This relationship can be written in the form

$$F_c = \frac{m^2 e^5}{I^4}\,\Phi_c(u,v) \ ,$$

$$\tag{5.53}$$

$$E_c = -\frac{me^4}{2I^2}\,\varepsilon_c(u,v) \ ,$$

where we have introduced the total action $I = I_{\xi} + I_{\eta} + I_{\varphi}$ and the parameters $u = I_{\eta}/I$ and $v = I_{\xi}/I$, the values of which lie in a triangular region defined by

$$\Delta\{u \ge 0, v \ge 0, u + v \le 1\} \ .$$

The functions $\phi_c(u,v)$ and $\varepsilon_c(u,v)$ are generally found by numerical solution of the above equations. In the most interesting limiting cases the classical values of the critical parameters are

$$n^4 F_c = 2^{10}/3^4\pi^4 = 0.13; \ |E_c|\,n^2 = 2^6/3^2\pi^2 = 0.72 \text{ for } n_1 = n \ ;$$

$$n^4 F_c = 0.3834, \ E_c = 0 \text{ for } n_2 = n \ ; \tag{5.54}$$

$$n^4 F_c = 2^{12}/3^9 = 0.208; \ |E_c|\,n^2 = 2^7/3^5 = 0.527 \text{ for } m = n \ .$$

This classical method for the calculation is effective when estimates are being obtained of the ionization of an atom from highly excited states $n \gg 1$ and $l \gg 1$, and the general quantum mechanical theory meets with considerable computational difficulties. The critical values of the parameters given in (5.54) are in good agreement with the values found by quantum calculations in the relevant range of the parameters (Sect. 5.6).

5.5 Decay of States Near the Critical Value of an Electric Field

The classical results obtained in Sect. 5.4 for the characteristics of an atom in an electric field F can be generalized allowing for the quantum (tunneling) effects in the semiclassical approximation. According to the Bohr-Sommerfeld rules, the values of the action variables (5.50, 51) are related to the parabolic quantum numbers by

$$I_\xi = \pi(n_1 + \tfrac{1}{2}) , \quad I_\eta = \pi(n_2 + \tfrac{1}{2}) . \tag{5.55}$$

The conditions of (5.55) have been frequently discussed in the literature [5.17]. For example, *Zaretskii* and *Krainov* [5.33] used the relationships in (5.55) to find the behavior of an atom in a low frequency electric field. *Kadomtsev* and *Smirnov* [5.34] investigated the atomic parameters near the critical field F_c.

We shall find, following [5.34], the field F_c which suppresses the barrier. This is accomplished by employing the semiclassical quantization conditions given by (5.55) combined with (5.50), and the additional condition

$$\frac{dp_\eta}{d\eta} \Big|_{\eta = \eta_2} = 0 , \tag{5.56}$$

where p_η is the momentum of an electron in η space and η_2 is the right-hand turning point which coincides (when $F = F_c$) with the maximum of the effective potential energy. These equations establish a unique relationship between the separation constants β_1 and the electron energy E, and the critical field intensity F_c. The solution of this system of equations is simplest in the case $m = 0$. For example, in the limit $n_2 \to 0$, the solution gives [5.34]:

$$F_c n^4 = \frac{2^{10}}{3^4 \pi^4}\left(1 + 0.4\frac{n_2}{n}\right), \quad |E_c| n^2 = \frac{2^7}{3^2 \pi^2}\left(1 - 0.4\frac{n_2}{n}\right) . \tag{5.57}$$

We can see that the zeroth-order terms of the expansion exactly coincide with the results of a classical analysis of (5.54). In the limit $n_2 \to 0$, we find that

$$F_c n^4 \simeq 0.383 \left[1 - 1.75 \left(n_2/n\right)^{2/3}\right]^6 , \quad n_2/n \ll 1 , \tag{5.58}$$

$$|E_c| n^2 \simeq 1.48 \left(\frac{n_1}{n}\right)^{2/3} \left[1 - 1.1 \left(\frac{n_2}{n}\right)^{2/3}\right]^2 . \tag{5.59}$$

In the limiting case these results also agree with (5.54). The correction factors given in the square brackets were obtained by *Drukarev* [5.18] and Sect. 5.6.

The solution of the system of semiclassical equations is fairly cumbersome in the general case when $m \neq 0$ and it was obtained numerically in [5.34]. In the case when $m = n$ the results of the semiclassical analysis reduce to those given by the classical formulas (5.54).

We can now plot the critical fields F_c and the corresponding energies E_c throughout the plane of the variables n_1 and n_2 (Fig. 5.5). The corresponding classical results of (5.54) are located at the corners of the triangles in Fig. 5.5 in the region of $n_1 \sim n$ (if $n_2 \sim n$). This method allows us to calculate the rate of decay of an atom near the critical field $F - F_c \ll F_c$. The calculations reported in [5.34] are based on the approximation of a barrier near its maximum by a parabola, followed by the determination of the above-barrier transmission coefficient. We shall not consider the details, but simply give the rate of decay Γ for $F = F_c$ and $n_2 = n$:

$$\Gamma \simeq \frac{1}{3.27 \, n^3 \ln \left(3 \pi n/4\right)} \ll 1 . \tag{5.60}$$

We can see that the rate of decay is not exponentially small (in contrast to the case when the fields are weak $F \ll F_c$), and moreover, it is comparable with the period of motion of an electron along an orbit. It is interesting to estimate the ratio of the width of a level to its energy E_c at the critical point. According to (5.59, 60), we have

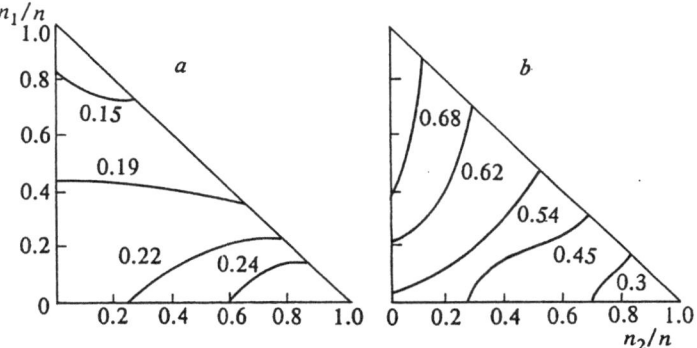

Fig. 5.5. Reduced values of the critical electric field $F_c n^4$ (a) and of the level energy En^2 (b) plotted in the plane of the quantum numbers n_1 and n_2 [5.34]

$$\Gamma/E_c \simeq \frac{1}{2.35\,n\,\ln{(3\pi n/4)}} \ll 1 \ . \tag{5.61}$$

Therefore, the ratio of the indeterminacy of the energy of a level to the energy itself amounts to 1.8×10^{-3} for $n = 50$ and 5.5×10^{-3} for $n = 20$.

5.6 General Theory of Atomic States in an Electric Field

5.6.1 Basis of the Semiclassical Approach

We shall consider a more general semiclassical theory of the decay of levels in an electric field, which makes it possible to follow the smooth transition from the case of weak fields $F \ll F_c$ to fields which are comparable with the critical value. It should be pointed out that although the fundamentals of the semiclassical theory were provided by *Lanczos* [5.35] in the 1930's, specific calculations have remained incomplete. Such calculations have been recently carried out, as mentioned above, using more rigorous methods [5.36–39]. We shall follow the results of *Drukarev* [5.18] who performed a consistent calculation of the energies and widths of levels by the semiclassical method.

The semiclassical theory is based on the quantization rules given by (5.50–52) and (5.55). If the relevant numbers are sufficiently small so that $|m|/n \ll 1$, the integrals of (5.48–51) can be represented in the form

$$L_1(x_1) = \frac{1}{2}\int_0^{x_1}\left(-1+\frac{4\alpha_1}{x}-wx\right)^{1/2} dx = \left(n_1+\frac{|m|+1}{2}\right)\pi \ , \tag{5.62}$$

$$L_2(y_1) = \frac{1}{2}\int_0^{y_1}\left(-1+\frac{4\alpha_2}{y}+wx\right)^{1/2} dy = \left(n_2+\frac{|m|+1}{2}\right)\pi \ , \tag{5.63}$$

where we shall use the notation of (5.9, 10) for the separation constants α_1 and α_2 and for the field. In this approximation the energy E (or the effective principal quantum number) are functions of two parameters

$$S = (n_1 - n_2)/n \quad \text{and} \quad T = 4n^4 F \ . \tag{5.64}$$

We shall find by noting that the integrals of (5.62, 63) can be expressed in terms of the hypergeometric function $F(-1/2, 1/2, 2, +z) \equiv F(+z)$, so that

$$L_1 = \pi\alpha_1(1-Z_1)F(-Z_1) \ , \quad L_2 = \pi\alpha_2(1+Z_2)F(Z_2) \ , \tag{5.65}$$

where

$$Z_1 = \frac{(1+16\alpha_1 w)^{1/2}-1}{(1+16\alpha_1 w)^{1/2}+1} \; , \quad Z_2 = \frac{1-(1-16\alpha_2 w)^{1/2}}{1+(1-16\alpha_2 w)^{1/2}} \; . \tag{5.66}$$

Using the quantization conditions of (5.62, 63), we obtain

$$\alpha_1 = \frac{n_1 + \frac{1}{2}(|m|+1)}{(1-Z_1)^{1/2} F(-Z_1)} \; , \quad \alpha_2 = \frac{n_2 + \frac{1}{2}(|m|+1)}{(1+Z_2)^{1/2} F(Z_2)} \; . \tag{5.67}$$

Hence, it is clear that

$$\alpha_1 \geq n_1 + \frac{|m|+1}{2} \text{ and } \alpha_2 \leq n_2 + \frac{|m|+1}{2} \; . \tag{5.68}$$

We shall next express $\alpha_1 w$ and $\alpha_2 w$ in terms of Z_1 and Z_2 using (5.66) and substitute expressions from (5.67) for α_1 and α_2, which yields

$$\left(n_1 + \frac{|m|+1}{2} \right) w = \frac{Z_1}{4(1-Z_1)^{3/2}} F(-Z_1) \; , \tag{5.69}$$

$$\left(n_2 + \frac{|m|+1}{2} \right) w = \frac{Z_2}{4(1+Z_2)^{3/2}} F(Z_2) \; . \tag{5.70}$$

It follows from the last relationship that the maximum value is attained at $Z_2 = 1$ and it is

$$w_c = \left[3\pi\sqrt{2} \left(n_2 + \frac{|m|+1}{2} \right) \right]^{-1} \; . \tag{5.71}$$

The value clearly determines the critical field F_c and we can find this field if we know the energy E (parameter ν). If we use (5.70), we can reduce (5.70) to

$$\frac{w}{w_c} = \frac{3\pi\sqrt{2}\,Z_2}{4(1+Z_2)^{3/2}} F(Z_2) \; . \tag{5.72}$$

5.6.2 Energy Levels

We shall not find the equations for the determination of the effective principal quantum number $\nu = (-2E)^{1/2}$. We shall do this using the relationship $\alpha_1 + \alpha_2 = \nu$ between the separation constants in (5.67), which yields the relationship between ν/n and the parameter S (5.64) and the variables Z_1 and Z_2 (5.66). The relationships (5.69, 70) then give two other equations relating the parameters T, S, and ν to the combinations of the functions $F(-Z_1)$ and $F(Z_2)$. These three equations and simple transformations yield the parametric relationship between ν/n and the parameters S and T [5.18]:

Table 5.2. Comparison of quantum-mechanical [5.28] and semiclassical [5.18] calculations

n, F	v		$\Gamma[10^{-6}\,s^{-1}]$	
	[5.28]	[5.18]	[5.28]	[5.18]
$n = 5, \quad F = 1.8 \times 10^{-4}$	4.9240	4.929	2.282	2.55
$n = 11, \, F = 10^{-5}$	10.6882	10.722	2.815	3.3
$n = 15, \, F = 3 \times 10^{-6}$	14.5771	14.619	1.338	1.74

$$v/n = v/n(Z_1, Z_2) \; ,$$

$$S = S(Z_1, Z_2) \; , \quad T = T(Z_1, Z_2) \; . \tag{5.73}$$

The procedure of finding v with the aid of the system of (5.73) reduces to the following: given S and T, we use (5.66–70) to find Z_1 and Z_2; substituting them in the expression for v/n, we then find the required quantity. In general, this procedure is carried out numerically. In [5.18] the dependences of Z_1 and Z_2 on T were reported for different values of S. Some of the numerical data on the parameter v/n can be found in Table 5.2.

We shall now compare the results of the semiclassical theory (5.73) with those obtained by classical calculations (5.54) in the case when $m = 0$, $n_2 = n$, and $F = F_c$ [5.18]. We shall do this by assuming that the parameter Z_2 in (5.73) is unity, which corresponds (according to (5.66)) to the critical value $F = F_c(w = w_c)$. Bearing in mind that $F(Z_2 = 1) = 8/3\pi$ is given by (5.72), we obtain $v/n = 3\pi \times 2^{-7/2}$ and hence the energy $E = -v^2/2$ is exactly equal to the classical value of (5.54). Using then (5.71), we find the critical field $F_c = w_c v^{-3}$, we can see that it also agrees with the classical estimate of (5.54).

5.6.3 Decay Rates

A calculation of the rate of decay in [5.18] is based on finding the asymptotic form of the wave function based on the semiclassical method [5.36]. Determination of the asymptote of the semiclassical function is related to the problem of the determination of the penetrability of the potential barrier $V(y)$ in y-space. This problem can be solved exactly either for a barrier of parabolic shape or in the limiting case of low penetrability (large width) of the barrier. In our case the barrier shape is nearly parabolic near its top, and the penetrability is weak far from the top. Consequently, we can derive a single analytic expression which is approximately valid for any barrier penetrability.

If K and Φ are the parameters governing the penetrability and the phase of the wave function

$$K = \int_{y_1}^{y_2} p(y)dy \; , \quad \Phi = \int_{y_{min}}^{y_1} p(y)dy + \delta(K) \; , \tag{5.74}$$

where $y_{1,2}$ are the turning points to the left and right of the barrier ($p(y)$ is the momentum in the y-space), and the approximate expression for Γ becomes

$$\Gamma \approx e^{-2K}/2(d\Phi/dE)_{E_n} \quad . \tag{5.75}$$

The physical meaning of (5.75) is clear: the rate of decay Γ is proportional to the frequency of motion of an electron in a potential well $(d\Phi/dE)_{E_n}^{-1}$ multiplied by the decay probability e^{-2K} on approach to the barrier. Both these parameters can be expressed, by analogy to Sect. 5.6.1, using the analytic functions $h(Z_2)$ and $g(Z_2)$ related to the hypergeometric functions

$$\Gamma = \frac{1}{v^3 h(Z_2)} \exp\left(-\frac{g(Z_2)}{w}\right) \quad . \tag{5.76}$$

The general form of the functions h and g can be found in [5.18]. In the case of weak fields $F \ll F_c$ we can use the relationship (5.66) between the parameter Z_2 and the field w, which readily yields an asymptotic expression for the decay parameter Γ which is identical, as expected, with the result of the asymptotic theory.

When the field F is close to the critical value $F_c (F_c - F \ll F_c)$, the parameter Z_2 is close to unity: $1 - Z_2 \ll 1$. We then have

$$g(Z_2) \approx \frac{\pi}{2^{1/6} 16}(1 - Z_2)^2 \quad . \tag{5.77}$$

The limiting value of the function $h(1)$ is

$$h(1) = \sqrt{2}\,[6 - \ln(32\sqrt{2}\,w_c) - \psi(1/2)] \tag{5.78}$$

(ψ is the logarithmic derivative of Γ at the point $F = F_c$). Substituting (5.77, 78) into (5.76), we find that the two results for the linewidth diverge by a factor of approximately 2.5. This divergence may be entirely due to the difference between the analytic approximations [5.18].

5.7 Results of Numerical Calculations

Numerical calculation methods have been developed in several papers [5.37 – 39] and their results generally agree quite well with one another, however, they differ considerably from the results of calculations carried out using the earlier method of *Lanczos* [5.17, 18]. The numerical results can be conveniently represented (for many of their applications) in a semi-analytic form based on perturbation theory with respect of the field [5.37]. The energies

$E(n_1, n_2, m, \lambda)$ and the widths $\Gamma(n_1, n_2, m, \lambda)$ of the Stark sublevels can be written in the form of a power series in terms of the parameter $\lambda = n^3 F/4$:

$$E(n_1, n_2, m, \lambda) = -\frac{1}{2n^2}(1 + 4\sum_p C_p(n_1, n_2, m)\lambda^p) \,, \tag{5.79}$$

where the coefficients C_p are found by perturbation methods [5.37]. The first three coefficients are described by

$$C_1 = -3(n_1 - n_2) \,, \quad C_2 = \tfrac{1}{2}[17n^2 - 3(n_1 - n_2)^2 - 9m^2 + 19] \,,$$

$$C_3 = -3(n_1 - n_2)[23n^2 - (n_1 - n_2)^2 + 11\,m^2 + 39] \,. \tag{5.80}$$

The first two coefficients clearly correspond to the familiar linear and quadratic Stark effects. The coefficients C_p are given right up to the ninth order [5.37]. Analytic expressions for the widths Γ of levels are closely related to the frequency of motion of an electron in a potential well governed by the derivative $\partial E/\partial n_2$:

$$\Gamma(n_1, n_2, m, \lambda) = (2\pi)^{-1}\left(\frac{\partial E(n_1, n_2, m, \lambda)}{\partial n_2}\right) \exp[-K(n_1, n_2, m, \lambda)] \,. \tag{5.81}$$

The function $K(n_1, n_2, m, \lambda)$ clearly describes the barrier penetrability. The derivative $\partial E/\partial n_2$ is found by differentiating the series of (5.79). The function K describing the barrier penetrability can be found using an asymptotic series for Γ of the type (Sect. 6.4)

$$\Gamma(n_1, n_2, m, \lambda) = \frac{\exp[3(n_1 - n_2) - 1/6\lambda]}{n^3 n_2!(n_2 + m)!\lambda^{2n_2 + m + 1}}\sum_{k \geq 0} a_k(n_1, n_2, m)\lambda^k \,, \tag{5.82}$$

where the coefficients $a_k(n_1, n_2, m)$ are expressed in terms of the coefficients of (5.79) using the dispersion relationship

$$C_p(n_1, n_2, m) = \frac{n^2}{4\pi}\int_0^{+\infty}\frac{d\lambda}{\lambda^{p+1}}[\Gamma(n_1, n_2, m, \lambda) + (-1)^p\Gamma(n_1, n_2, m, \lambda)]. \tag{5.83}$$

The relationship (5.82) is deduced from the condition of analyticity of the energy in the plane of complex values of the field F (for details, [5.38], p. 328). A comparison of the coefficients a_k with the coefficients C_p from (5.82) makes it possible to obtain the following relationship for the function

$$K(n_1, n_2, m, \lambda) = (6\lambda)^{-1} + \ln[\lambda^{2n_2 + m + 1}n_2!(n_2 + m)!/2\pi] + 3(n_2 - n_1)$$

$$-\frac{1}{6}\int_0^\lambda\left[n^3\frac{\partial E(n_1, n_2, m, \lambda')}{\partial n_2} - 1 + 6(2n_2 + m + 1)\lambda'\right]\frac{d\lambda'}{\lambda'^2} \,. \tag{5.84}$$

Table 5.3. Rates Γ of ionization of states corresponding to the $n = 10$ level in an electric field [5.38]

n_1, n_2, m	$\Gamma[\text{s}^{-1}]$			
	F [10^4 V/cm]	[5.37]	[5.28]	[5.38]
090	4.058	8.236 (2)	–	6.889 (3)
	4.603	8.015 (5)	7 (6)	7.090 (6)
	4.178	1.425 (8)	1.310 (9)	1.344 (9)
	5.814	7.560 (9)	6.707 (10)	7.611 (10)
900	8.082	1.539 (7)	6 (6)	5.802 (6)
	9.134	8.525 (8)	3.170 (8)	3.272 (8)
	10.81	1.802 (10)	6.585 (9)	7.675 (9)

It therefore follows that the coefficients C_p tabulated in [5.38] allow us to find the energy levels of (5.79), the frequencies of motion inside a barrier, and also the decay half-widths from (5.81, 83).

Table 5.3 is based on [5.38] and it compares the results of the numerical calculations of [5.28, 38] for the rates of ionization of a level with $n = 10$ and also the data of [5.37] based on the *Lanczos* theory. We can see that the data of [5.28, 38] agree well but they differ considerably from the earlier results [5.37] based on the Lanczos theory.

Figure 5.6 shows the Stark structure of a highly excited lithium atom [5.39]. It illustrates a set of Stark components corresponding to the projection of the

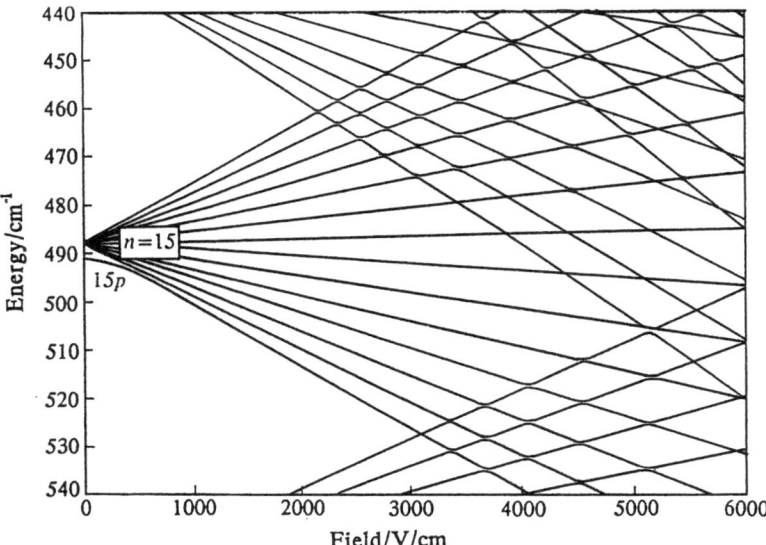

Fig. 5.6. Stark splitting of a highly excited lithium atom [5.39] (the small repulsion of the terms in the crossing points is shown)

quantum number $|m| = 1$. This set (because of the smallness of the quantum defects of the p states) is very close to the pattern of the Stark splitting in hydrogen. A discrepancy is observed only in the case of the weakest fields when the quantum defect of the p levels is important. We can clearly see the pattern of the Stark components crossing at various levels. The repulsion between the crossing terms is due to the nonzero value of the quantum defect, and on the whole it rises with the increase of this defect [5.39]. In contrast to non-hydrogen atoms, a nonrelativistic theory of the hydrogen atom admits exact crossing of the levels. This is possible because of an additional degeneracy: the crossing levels have not only the parabolic quantum numbers n_1 and m, but they can have different values of the additional integral of motion which is the constant of separation of the variables $\alpha_{1,2}$ (5.56). On the whole, the calculated pattern of Stark splitting agrees very well with the experimental results [5.39].

6. Atom in a Magnetic Field and Crossed $F-B$ Fields

6.1 Introductory Remarks.
Energy Spectrum of Low Lying Atomic States

One of the effects of a magnetic field B on an atom is the well known Zeeman splitting of atomic levels into separate components corresponding to specific values of the projection of the orbital momentum m of the atom along the direction of the field B. The characteristics of the Zeeman effect which are in particular due to the spin-orbit interaction have been described in detail in a number of textbooks and monographs [6.1−2]. Therefore we shall consider only the less well known aspects of the effect observed at high values of the field B. The application of a magnetic field to an atom imposes an additional constraint on the motion of an electron across the field. When the magnetic field is still further increased, there is a strong reduction in the transverse motion of an electron and consequent transformation of a three dimensional potential well of an atom into a one dimensional well. A considerable change in the energy spectrum of an atom may follow.

We shall initially consider the first order Zeeman effect in the case of the simplest zero-spin one-electron (hydrogen) atom. The Hamiltonian of a perturbation V due to the interaction of the orbital momentum l of an electron with the field B is

$$V = -\mu_B l \cdot B , \tag{6.1}$$

where $\mu_B = e\hbar/2mc$ is the Bohr magneton. The eigenvalues of the energy can be found (i.e., the perturbation of (6.1) can be diagonalized) simply by selecting the wave functions corresponding to a specific projection of l along the direction of B. Usually these functions are spherical wave functions ψ_{nlm} corresponding to specific values of the total angular momentum $l^2 = l(l+1)$ and on its projection $l_z = m$.

In the case of the hydrogen atom, the Zeeman sublevels corresponding to a specific value of m are degenerate with respect to the quantum number l. This special feature of the Coulomb degeneracy is manifested also by the fact that the wave functions of (6.1) which become diagonalized can be parabolic wave functions $\psi_{n_1 n_2}$ with the $0z$ axis along the field B. In view of the relationship $n_1 + n_2 + m = n - 1$ corresponding to a given value of m these states remain degenerate with the values of n_1 and n_2 corresponding to the constant

sum $n_1 + n_2$. Therefore, the Zeeman component of a hydrogen level is characterized not by one but usually by several wave functions. The intensity of the
component is governed by the sum over the degenerate states and this sum
should no longer depend on the selection of the basis (spherical or parabolic
quantization).

For the simplest case of the L_α line (representing the $2 \to 1$ transition), the
state with $m = 0$ corresponds to two parabolic functions with $n_1 + n_2 = 1$ and
two values of $n_1 - n_2$, which amount to $+1$ and -1 and which represent two
different projections of the dipole moment of the atom along the field B.
Clearly, the sum of the intensities of the transitions from these two states is
equal to the intensity of the transition from one "spherical" p state characterized by $l = 1$ and $m = 0$. In general, the transition from the parabolic to the
spherical basis is described by the formulas in (5.14).

6.1.1 Energy Spectrum of Lower States

We shall briefly consider the evolution of the lower states of the energy spectrum of a hydrogenic atom when the magnetic field B is increased to values
comparable with (or exceeding) the interatomic electric field. This evolution
has become important, as already mentioned, in connection with the absorption spectrum of excitons, which are characterized by a very low electron-hole
binding energy (because the permittivity of the medium is high and the effective mass of an electron in the medium is small): this energy is comparable
with the energy of an electron in a magnetic field of moderate intensity
$(10-20^2 \, \text{T})$. We shall introduce the following parameter as a measure of the
field intensity

$$\gamma = \frac{\mu_B B}{\text{Ry}} \simeq 4.26 \times 10^{-10} \, B(\text{G}) \,, \tag{6.2}$$

where $\text{Ry} \simeq 13.6 \, \text{eV}$ is the Rydberg constant.

The Hamiltonian of such an atom in a field B is [6.3]

$$\hat{H} = \frac{1}{2m} \left(p + \frac{1}{2} \frac{e}{c} B \times r \right)^2 - \frac{e^2}{r} \,. \tag{6.3}$$

In view of the invariance of H relative to orientation about the $0z$ axis,
parallel to the field B and passing through the nucleus of the atom, the z component of the orbital angular momentum $L_z = -\hbar M$ is conserved. Introduc-
ing a cylindrical coordinate system $0z \| B$ and bearing in mind that the
dependence of the wave function ψ on the angle of rotation about φ the z axis
is trivial $\psi \propto e^{iM\varphi}$, we can write the Schrödinger equation in the form

$$\left[\frac{\partial^2}{\partial \varrho^2} + \frac{1}{\varrho} \frac{\partial}{\partial \varrho} + \frac{\partial^2}{\partial z^2} - \frac{M^2}{\varrho^2} - 4\gamma^2 \varrho^2 + \frac{4}{r} + \left(\frac{E}{\text{Ry}} - \gamma M \right) \right] \psi(\varrho, z) = 0 \,. \tag{6.4}$$

The two dimensional equation (6.4) cannot be solved analytically in its general form because the Coulomb interaction term containing $r = \sqrt{\varrho^2 + z^2}$ prevents separation of the variables. We shall therefore demonstrate the nature of the solution for $\gamma \ll 1$ and $\gamma \gg 1$ and obtain some approximation formulas for the transition range $\gamma \sim 1$.

If $\gamma \ll 1$, allowance for the terms containing γ can be made by using perturbation theory. In the case of the ground state of the hydrogen atom, this gives [6.5]

$$E_{100} = -1 + \frac{\gamma^2}{2} - \frac{53}{96} \gamma^4 + \frac{5581}{2304} \gamma^6 - \frac{21\,577\,397}{1\,105\,920} \gamma^8 + O(\gamma^{10}) . \tag{6.5}$$

The expression (6.5) agrees well with the results of numerical calculations up to $\gamma \sim 0.1$ (i.e., up to $B \sim 10^4$ T). Similar results were reported in [6.6] for the case when $n = 2$.

If $\gamma \gg 1$, the motion of an electron across the applied magnetic field is governed by the size of its cyclotron orbit $\lambda = (\hbar c/em)^{1/2}$ whereas along the field it is determined by the Coulomb interaction. The potential of this "longitudinal" Coulomb interaction can be obtained by averaging the total Coulomb potential $e^2(\varrho^2 + z^2)^{-1/2}$ over the small parameter of transverse motion. Therefore, bearing in mind that on the average we have $\langle \varrho^2 \rangle \sim \lambda \ll \langle z^2 \rangle$, we may assume that the longitudinal motion of an electron occurs in a one-dimensional Coulomb potential. Successive separation of the transverse and longitudinal types of motion in (6.4) when $\gamma \gg 1$ can be achieved if the wave function $\psi(\varrho, z)$ is represented as the product of the wave function $\psi_{NM}(\varrho)$ of the transverse motion of an electron in the applied magnetic field and the function $\chi_{NM}(z)$ of its longitudinal motion in a one-dimensional "longitudinal" potential.

The transverse motion of an electron in a magnetic field is equivalent, as demonstrated by (6.4), to the motion of an oscillator whose wave functions are well known [6.2, 7].

$$\psi_{NM}(\varrho) = \begin{cases} \left[\dfrac{(N-M)!}{2\pi\lambda^2(N!)^3} \right]^{1/2} (-\mathrm{i})^N \sigma^{M/2} e^{-\sigma/2} L_N^M(\sigma) , & M \geq 0 \\[3mm] \left[\dfrac{(N-M)!}{2\pi\lambda^2((N-M)!)^3} \right]^{1/2} (-\mathrm{i})^N \sigma^{-M/2} e^{-\sigma/2} L_{N-M}^{-M}(\sigma) , & M < 0 ; \end{cases} \tag{6.6}$$

where $N = 0, 1, 2 \ldots$ are integers governing the number of energy (Landau) levels in a magnetic field; $|M| \leq N$; $L_N^M(\sigma)$ are the Laguerre polynomials; $\sigma = \varrho^2/2\lambda^2$. Consequently, the energy levels of the transverse (oscillator) motion of an electron are

$$E_N = \mathrm{Ry}\,\gamma(N + \tfrac{1}{2}) . \tag{6.7}$$

The equation for the wave functions $\chi^i_{NM}(z)$ of the longitudinal motion is obtained from (6.4) after averaging over the transverse motion functions of (6.6) [6.3, 4]:

$$\left[-\frac{\hbar^2 d^2}{2m\,dz^2}+V_{NM}(z)\right]\chi^i_{NM}=E^z_{NM_i}\chi^i_{NM}\;,\tag{6.8}$$

where the energy of the longitudinal motion E^z_{NM} should be added to the energy of the transverse motion of (6.7) and the average potential is given by

$$V_{NM}(z)=\int\psi^*_{NM}(\varrho)\frac{e^2}{\sqrt{\varrho^2+z^2}}\psi_{NM}(\varrho)\varrho\,d\varrho\;.\tag{6.9}$$

The explicit form of the potential (6.9) is not too complex and (6.8) can be solved analytically. However, it can be satisfactorily approximated by a function of the type [6.4]

$$V(z)=-\frac{e^2}{a+|z|}+\frac{Aae^2}{(a+|z|)^2}\;,\tag{6.10}$$

where the size a and the coefficients A are selected for each NM so as to best approximate the true potential of (6.9). If the parameter a is sufficiently small, the potential $V(z)$ is close to a one-dimensional Coulomb potential $e^2/|z|$, as demonstrated by (6.10). Therefore, by analogy with the three-dimensional Coulomb problem, we shall write down the longitudinal energy

$$E^z_{NMn^*}=-\frac{\mathrm{Ry}}{n^{*2}}\;,\tag{6.11}$$

where the effective "principal quantum number" n^* is obtained from the boundary conditions.

Now we introduce a variable $x=(mc^2/\hbar)^{1/2}[(a+z)/n^*]$ and retain only the first term in the potential of (6.10). We can reduce (6.8) to the form

$$\frac{d^2\chi}{dx^2}-\left[\frac{1}{4}-\frac{n^*}{x}\right]\chi=0\tag{6.12}$$

the solution of which is in the form of Whittaker functions $W_{n,1/2}(x)$. Bearing in mind also that the potential $V(z)$ does not change as a result of the substitution $z\to-z$, we find that the solutions of (6.12) should be either even or odd with respect to z. The requirement of continuity of the functions and of their derivatives at $z=0$ yields the following condition for the odd states

$$W_{n^*;1/2}\left(2\frac{a}{a_0 n^*}\right)=0\;,\tag{6.13}$$

whereas in the case of the even states, we obtain

$$\frac{d}{dz} W_{n^*,1/2}\left(2\frac{a+z}{a_0 n^*}\right)\bigg|_{z=0} = 0 . \tag{6.14}$$

The conditions (6.13, 14) give the values of the numbers n^* governing the number of nodes of wave functions and the sequence of the energy levels. In the limit $\gamma \to \infty$ all of the energy levels of the longitudinal motion are hydrogenic, i.e., we have $n^* = 1, 2, \ldots$, with the exception of the ground state the energy of which decreases logarithmically with the increasing γ (see [6.2], Problem 3 in § 112; also (6.70, 71)).

There is a unique relationship between the states in weak and strong magnetic fields. This relationship is found by calculating the number of nodes of a wave function in both limits of weak and strong fields. In fact, an increase in the magnetic field deforms the spherical symmetry of the hydrogen atom to the cylindrical symmetry. Bearing in mind that a free atom is characterized by $n_\varrho = n - l - 1$ nodes of the radial wave function, corresponding to n_ϱ nodal spheres, and that there are $l - M$ nodes of the angular function corresponding to cones with the z axis. Recall also that in a strong magnetic field we have correspondingly $N - (|M| + M)/2$ nodal cylinders ($\varrho = $ const.) and $2n^*$ (for even states) or $2n^* - 1$ (for odd states) nodal planes intersecting the $0z$ axis. Thus, following *Praddaude* [6.6] we find:

$$l - M = 2n^* - \text{even} ,$$

$$= 2n^* - 1 - \text{odd} ,$$

$$n - l - 1 = N - \frac{|M| + M}{2} , \quad m = M . \tag{6.15}$$

For example, the lower even state with $N = 0$, $M = 0$, and $n^* = 0$ corresponds to $n = 1$, $l = 0$, and $M = 0$ of the hydrogen atom.

The dependence of the energy E on the parameter γ is described by the formula

$$\frac{\bar{E}(\gamma)}{\text{Ry}} = \frac{1}{\text{Ry}}(E_N + E^z_{NMn^*} - \gamma M) = \gamma\left(N - M + \frac{1}{2}\right) - \frac{1}{2n^{*2}} . \tag{6.16}$$

The behavior of the first few levels was calculated by *Galindo* and *Pascual* [6.5] using approximate formulas based on interpolation between the limits $\gamma \to 0$ and $\gamma \to \infty$. It should be pointed out that the problem of correspondence of the terms and the possibility of their exact crossing has not yet been solved. A magnetic field does not separate the variables (in contrast to an electric field) and the only quantities (apart from the energy) which are conserved are the projection of momentum and parity. In this case it would seem that the Wigner-Neumann theorem on noncrossing of terms should apply [6.2]. However, in the case of an atom in a magnetic field there is an additional approximate integral of motion (Sect. 6.3) which can give rise to an exact (or negligibly repelling) crossing of terms. Until the problem is finally solved, we shall follow the identification of the terms given above. This topic is discussed in [6.7].

6.2 Adiabatic Theory for Highly Excited Atomic States in a Strong Magnetic Field

The results of Sect. 6.1 enable an interesting generalization to large quantum numbers n corresponding to the motion of a Coulomb well or to the rapid motion in a magnetic field ($N \gg 1$). This was done by *Zhilich* and *Monozon* [6.9] and we shall follow their treatment. The approach adopted in [6.9] is based on the slowness (adiabaticity) of the motion of an electron along magnetic field (z axis) compared with its motion in a transverse plane. Comparing the classical frequency of motion in a Coulomb field $\omega_n = me^4/\hbar^3 n^3$ with the Larmor frequency $\omega_L = eB/mc$, we obtain the condition

$$\frac{\omega_n}{\omega_L} = \frac{m^2 e^3 c}{\hbar^3 n^3 B} \ll 1 \ . \tag{6.17}$$

The condition (6.17) of the slowness of motion of the z coordinate ensures the retention of a parametric dependence on z in the wave functions describing the transverse motion, i.e., we can assume that

$$\chi_{Nn}(\varrho, z) = R_N(\varrho, z) W_{Nn}(z) \ . \tag{6.18}$$

Then, (6.4) yields the following equations for R_N and W_{Nn}:

$$\frac{1}{\varrho} \frac{d}{d\varrho} \left(\varrho \frac{dR_N}{d\varrho} \right) + \left[-\frac{M^2}{\varrho^2} + \frac{4}{\sqrt{\varrho^2 + z^2}} - 4\gamma^2 \varrho^2 + q_N^2(z) \right] R_N = 0 \ , \tag{6.19}$$

$$\frac{d^2 W_{Nn}}{dz^2} + [K^2 - q_N^2(z)] W_{Nn} = 0 \ , \quad K^2 = 2E - \gamma M \ , \tag{6.20}$$

where the eigenvalues $q_{2N}(z)$ can be found in the $n \gg 1$ case from the Bohr-Sommerfeld quantization conditions

$$\int_{\varrho_1(z)}^{\varrho_2(z)} \sqrt{q_N^2(z) + \frac{4}{\sqrt{\varrho^2 + z^2}} - \frac{M^2}{\varrho^2} - 4\gamma^2 \varrho^2} \, d\varrho = \pi \left(N + \frac{1}{2} \right) \ . \tag{6.21}$$

($\varrho_1(z) < \varrho_2(z)$) are the classical turning points representing the roots of the integrand above).

Simple results are obtained from (6.21) in the two limiting cases of $|z| \gg \varrho_2(0)$ and $|z| \ll \varrho_1(0)$. If $z \gg \varrho_2(0)$ then (6.21) yields an expansion for the eigenvalues

$$q_N^2(z) = -\frac{2}{|z|} + \left(\gamma + \frac{2}{\gamma}\frac{1}{|z|^3}\right)(2N + |m| + 1) \ , \tag{6.22}$$

which determine the form of the effective potential in (6.20). We can replace this potential by a more general expression of the type

$$\frac{2}{\sqrt{z^2 + b_N^2}} \ , \quad b_N^2 = \frac{2}{\gamma}(2N + |m| + 1) \tag{6.23}$$

which is identical with (6.22) apart from terms of the order of $|z|^{-3}$ if $z \gg b_N$. After this substitution (6.20) becomes

$$\frac{d^2 W_{Nn}(z)}{dz^2} + \left[\frac{2}{\sqrt{z^2 + b_N^2}} - P_{Nn}^2\right] W_{Nn}(z) = 0 \ , \tag{6.24}$$

$$P_{Nn}^2 = -K^2 + \gamma(2N + |m| + 1) \ . \tag{6.25}$$

If $B \to \infty$, (6.24) clearly reduces to (6.12) with a one-dimensional Coulomb potential which has the solutions given by (6.13, 14). The parameter b_N determines the size of that region along z in which the potential is close to the Coulomb form. Clearly, the size of the Coulomb well further decreases as the number N is increased.

Inclusion of corrections of the next order with respect to the adiabaticity parameter (6.17) makes it possible to find the quantum defects $\delta n_{u,g}$ for levels due to deviations of the field from the pure Coulomb form

$$(P_{Nn}^2) = \frac{1}{(n + \delta n_{u,g})^2} \ , \quad n = 0, 1, 2 \dots . \tag{6.26}$$

The quantities $\delta n_{u,g}$ are found for even and odd states on the basis of the solution of (6.24) (which reduces after the substitution $x^2 = 4p_{Nn}^2(z^2 + b_N^2)$ to an equation in terms of Whittaker functions), subject to the quantization conditions of (6.17). The result is [6.9]

$$\delta n_u = 2b_N \ , \tag{6.27}$$

$$\delta n_g = -1/\ln(2b_N/n) \ . \tag{6.28}$$

Therefore, each doublet level of the hydrogenic atom in a magnetic field splits into two levels corresponding to the quantum defects described by (6.27, 28). In the limit $B \to \infty$ the two conditions merge to form a doubly degenerate hydrogenic level of (6.16).

If $z \ll \varrho_1(0)$, we can substitute in the quantization condition of (6.21) an expansion directly in powers of $(z/\varrho)^2$. It is easiest to obtain the solution for the case when $M = 0$ [corresponding to $\varrho_1(0) = 0$] by assuming that the parameter q_N^2 (proportional to N) is large:

$$q_N^2 = \frac{2}{\varrho_2(0)} - \frac{z^2}{\varrho_2^3(0)} = \gamma(2N+1) ,$$

$$\varrho_2(0) = 2q_N/\gamma .$$
(6.29)

Solving (6.29) by the method of successive approximations, we obtain

$$q_N^2(z) = \gamma(2N+1) - \frac{\gamma}{2(2N+1)^{1/2}} + \frac{\gamma^{3/2}}{8(2N+1)^{3/2}} z^2 .$$
(6.30)

We can see that the effective potential along the z axis coincides in this case with the oscillator potential. The equation for motion along the z axis becomes

$$\frac{d^2 W_{Nn}}{dz^2} - \Omega_N^2 z^2 W_{Nn} + P_{Nn}^2 W_{Nn} = 0 ,$$
(6.31)

where the characteristic frequency of electron motion along the z axis is

$$\Omega_N = \frac{\gamma^{3/4}}{2\sqrt{2}(2N+1)^{3/4}} \text{ (a.u.) ,}$$
(6.32)

$$P_{Nn}^2 = K^2 - \gamma(2N+1) + \frac{\gamma^{1/2}}{(2N+1)^{1/2}} .$$
(6.33)

The spectrum of energy levels E is identical in this case with the oscillator spectrum [6.9, 10]

$$E = \frac{\hbar e B}{2mc}(2N+1) - e^2 \left(\frac{eB}{4\hbar c(2N+1)}\right)^{1/2}$$

$$+ \hbar \left(\frac{e^7 B^3}{m^2 \hbar^3 c^3 2^6 (2N+1)^3}\right)^{1/4} \left(n + \frac{1}{2}\right) .$$
(6.34)

The spectrum of (6.34) is obtained if the frequency is small compared with the Larmor frequency, i.e., when

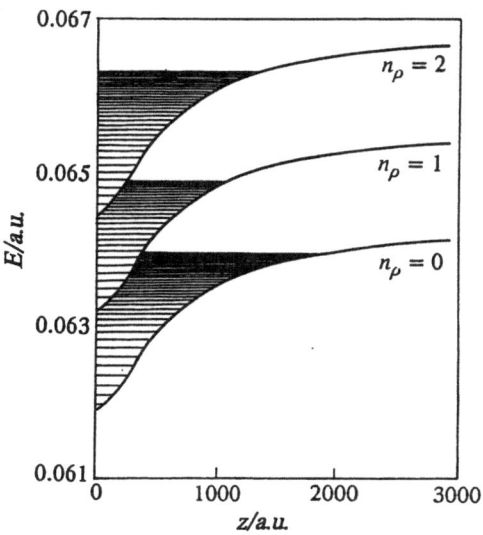

Fig. 6.1. Energy levels of an excited atom in a strong magnetic field corresponding to a sufficiently high Landau level

$$\Omega_N/\omega_L \sim (m^2 e^3 c/32\hbar^3 BN^3)^{1/4} \ll 1 \ . \tag{6.35}$$

Therefore, the spectrum of bound electrons at a sufficiently high Landau level $N \gg 1$ varied continuously from the oscillator type in the case of low-lying levels to highly excited hydrogenic levels ($n \gg 1$), which become more dense at the limit of the series (Fig. 6.1).

6.3 "Latent" Symmetry of an Atom in a Magnetic Field

The Hamiltonian H of a hydrogen atom in a magnetic field B directed along the z axis is

$$\hat{H} = \frac{\hat{p}^2}{2} - \frac{1}{r} + \omega^2 \varrho^2/2 + i\omega \hat{l}_z \ , \tag{6.36}$$

where $\hat{l}_z$ is the operator representing the projection of the orbital angular momentum l along the direction of the magnetic field B (0z axis); $\omega = B/c$ (a.u.). The equations of motion of an atomic electron in a magnetic field do not allow the separation of variables in any coordinate system (in contrast to the electric field case) and, consequently, these equations do not contain additional integrals of motion of the type represented by constants of the separation of variables. Therefore, when the Zeeman structure of one of the levels overlaps the structure of another level at the crossing points, we cannot expect exact coincidence of the energies (as found in an electric field). Nevertheless,

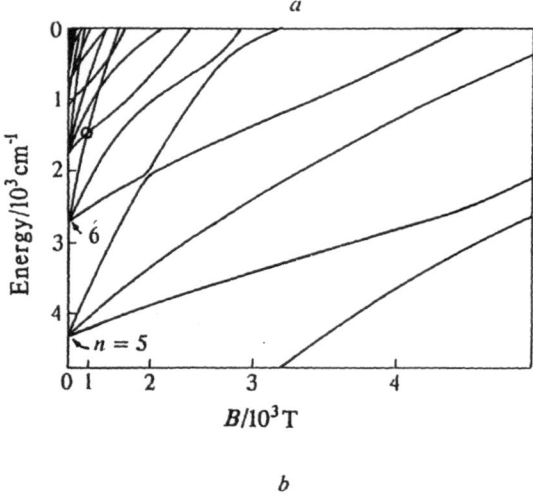

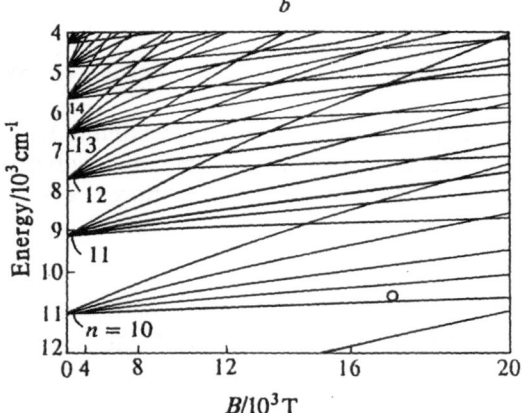

Fig. 6.2 a, b. Crossing of Zeeman energy sublevels E (cm^{-1}) of an atom with the increase in magnetic field B in the case of small (a) and large (b) values of the principal quantum numbers [6.11]

numerical calculations of the Zeeman structure carried out by *Zimmerman* et al. [6.11] revealed an approximate symmetry of the hydrogen atom in the applied magnetic field. It was manifested by a strong (exponential) fall of the splitting ΔE_n at the points of crossing of the Zeeman sublevels as a function of the principal quantum number n. Figure 6.2 (taken from [6.11]) shows the pattern of crossing of the Zeeman sublevels at low and high values of the quantum number n. Clearly, "anticrossing" at low values of n changes to a pattern of almost complete crossing on increase in n. Figure 6.3 shows how the value of ΔE_n varies on increase in n in the case of crossing of the outer components (continuous line) and of the outer with middle (dashed line) Zeeman components. The dependence on the level number is clearly exponential.

Among the many proposed explanations [6.11–14] of the approximate symmetry, we shall consider the results of *Solov'ev* [6.13] who attributed the observed change in ΔE_n to the presence of an additional integral of motion

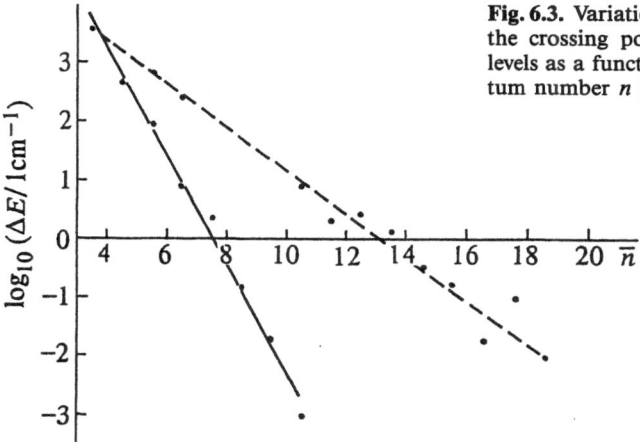

Fig. 6.3. Variation of the splitting ΔE_n at the crossing points of the Zeeman sub-levels as a function of the principal quantum number n [6.11].

$\hat{\Lambda}$ for a hydrogen atom in a magnetic field (*Herrick* [6.14]). The integral can be obtained, following [6.13] using classical equations of motion for the orbital momentum and the Runge-Lenz vector $A = [p \times l] - (r/r)$ [6.12] in a magnetic field ($\varrho = x + y$)

$$\frac{dl}{dt} = -\omega^2 r \times \varrho \ , \quad \frac{dA}{dt} = -\omega^2 (p \times [r \times \varrho] + \varrho \times [r \times p]) \ . \tag{6.37}$$

Averaging (6.37) over the period of motion along an unperturbed trajectory (Kepler ellipse), we obtain a system of equations describing the change in the trajectory under the influence of a magnetic field. These equations can be used to demonstrate the existence of an integral of motion

$$\hat{\Lambda} = 4A^2 - 5A_z^2 \tag{6.38}$$

which is conserved together with the energy E and the projection of the orbital angular momentum l_z. Conservation of $\hat{\Lambda}$ applies right up to terms of the order of ω^4. Bearing in mind that A^2 varies in the range $0 < A^2 < 1$, we can find the range of variation of Λ: $-1 \leqslant \Lambda \leqslant 4$.

The integral of motion Λ gives rise to additional conditions on quantization of angular variables. Let us assume that Θ is the angle between the vectors B and A; then,

$$\Lambda = A^2 (4 - 5 \cos^2 \Theta) \ . \tag{6.39}$$

If $\Lambda = 0$, the vector A is on a double conical surface described by the condition $\cot \Theta_0 = 2$. The conversion of Λ means that all of the trajectories of motion can be divided into two classes; trajectories within a double cone (if $0 \leqslant \Theta < \Theta_0, \pi - \Theta_0 \leqslant \Theta \leqslant \pi$) or outside of it (if $\Theta_0 \leqslant \Theta \leqslant \pi - \Theta_0$). We can write down the quantization conditions if we introduce a generalized momentum, which is canonically conjugate to the coordinate Θ. This is clearly the component of the orbital angular momentum $l_\perp(\Theta)$, perpendicular to the plane of the vectors $\underline{B}$ and A. Expressing $l_\perp$ in terms of the integrals of motion $m, E(n = \sqrt{-2E})$, and A, we obtain [6.13]:

$$l_\perp(\Theta) = [n^2 - U_{\mathrm{eff}}(\Theta)]^{1/2} , \quad U_{\mathrm{eff}}(\Theta) = \frac{n^2 \Lambda}{1 - 5\sin^2\Theta} - \frac{m^2}{\sin^2\Theta} . \tag{6.40}$$

The effective "angular" potential $U_{\mathrm{eff}}(\Theta)$ is plotted in Fig. 6.4 for the cases $\Lambda < 0\,(a)$ and $\Lambda > 0\,(b)$. We can see that the presence of an additional integral of motion results in sharply divided regions of classical motion, defined by the roots of the effective potential $\Theta_1\ldots,\Theta_6$. If $\Lambda < 0$ the Bohr-Sommerfeld quantization conditions can be written down separately for the upper and lower parts of a double cone:

$$I_1(\Lambda) = \int_{\Theta_1}^{\Theta_2} l_\perp(\Theta)\,d\Theta = \pi\left(K + \frac{1}{2}\right) , \tag{6.41}$$

$$I_2(\Lambda) = \int_{\Theta_5}^{\Theta_6} l_\perp(\Theta)\,d\Theta = \pi\left(K + \frac{1}{2}\right) , \quad K = 0,1,2,\ldots . \tag{6.42}$$

If the potentials in this case are identical, the resultant equations are doubly degenerate. From these states, localized in the upper and lower parts of the cone, we can construct wave functions which are symmetric and antisymmetric relative to the (x, y) plane. If $\Lambda > 0$, the states are nondegenerate and the quantization condition becomes

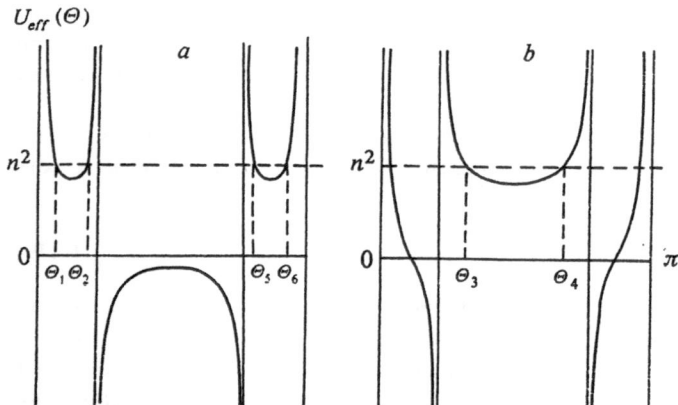

Fig. 6.4. Effective angular potential $U_{\mathrm{eff}}(\Theta)$ for different values of the integral of motion $\Lambda < 0$ **(a)** and $\Lambda > 0$ **(b)** based on [6.13]

$$I_3(\Lambda) = \int_{\Theta_3}^{\Theta_4} l_\perp(\Theta)\,d\Theta = \pi\left(K+\frac{1}{2}\right) . \tag{6.43}$$

The integrals (6.41–43) cannot be calculated analytically in general. The maximum value of $I_i(\Lambda)$ is obtained for $\Lambda = 0$ and it determines the total number of states

$$N = \frac{1}{\pi}[I_1(0)+I_2(0)+I_3(0)] = n-m , \tag{6.44}$$

which is identical with the exact quantum number of states with given values of n and m.

In the first (in respect of ω^2) order of perturbation theory the energy is expressed in terms of the period-average value of $\varrho^2 = x^2+y^2$. Calculating this average with the aid of equations for an unperturbed trajectory, we find that [6.13]:

$$E = E_0+\frac{\omega^2}{2}\overline{\varrho^2} = -\frac{1}{2n^2}+\frac{\omega^2 n^2}{4}(n^2+m^2+n^2\Lambda_K) . \tag{6.45}$$

The greatest interest lies in the outer Zeeman components which are the first to experience crossing. These sublevels correspond to the lower levels in the effective potentials in Fig. 6.4 and they can be determined using the parabolic approximation of the potential near its minimum. This gives

$$E = -\frac{1}{2n^2}+\omega^2 n^2\sqrt{5}\left(K+\frac{1}{2}\right)[(5(2K+1)^2+n^2)^{1/2}-\sqrt{5}(2K+1)] ,$$

$$K = 0,1,2,\dots . \tag{6.46}$$

In transfering to quantum mechanics, the integral of motion is replaced by the operator $\hat{\Lambda}$ which commutes with the Hamiltonian in the subspace of the wave functions with given values of n. This can be demonstrated by expressing the operator ϱ^2, which occurs in the Hamiltonian, in terms of $\hat{\Lambda}$. This relationship makes it possible to also construct wave functions which diagonalize the Hamiltonian of (6.36) in the subspace of states with a given n. Such a construction is possible because of the separation of variables in elliptic-cylindrical coordinates for the hydrogen atom in the case when the independent variables are the operators l_z and the quadratic combinations of the Runge-Lenz vector. We shall not consider the explicit form of these functions, but direct the reader to [6.13, 14], and to the works cited therein.

The approximate symmetry of the hydrogen atom associated with the presence of an additional integral of motion Λ, found in [6.13], accounts for the relationships deduced from numerical calculations (Figs. 6.2, 3). Indeed, as can be seen from Fig. 6.4, in the case of states with different values of Λ the splitting at the level-quasicrossing points is governed by the penetrability of a classical barrier and if $n \gg 1$, the splitting should be exponentially small. Calculations of the splitting E_n carried out in this way give the result [6.13]:

$$\Delta E_n \approx \exp\{-n \ln [(\sqrt{5}+2)(\sqrt{5}+1)/2]\} \approx \exp(-1.92n) . \tag{6.47}$$

Equation (6.47) indeed demonstrates an exponential decrease of the splitting with an increase in n, which is close to the calculated data in Figs. 6.2, 3.

An interesting approach to a calculation of the quadratic Zeeman effect was developed by *Braun* [6.17] and by *Kazantsev* et al. [6.18]. It is based on the following property of the matrix elements of the perturbation operator $V = x^2 + y^2$ expressed in terms of parabolic coordinates (when the axis $0z$ is parallel to $\boldsymbol{B}$):

$$
\begin{cases}
V_{n_1 n_2} \equiv \dfrac{n^2}{2} w_{n_1} = \dfrac{n^2}{2} [3n^2 - m^2 + 1 + 12(a - n_1)^2] \\[4mm]
V_{n_1 n_1 - 1} \equiv \dfrac{n^2}{2} P_{n_1} = \dfrac{n^2}{2} [n_1(n_1 + |m|)(2a - n_1 + 1)(2a - n_1 + |m| + 1)]^{1/2} \\[4mm]
V_{n_1 n_1'} = 0 \quad \text{with} \quad |n_1 - n_1'| \geq 2 , \quad 2a \equiv n_1 + n_2 = n - |m| - 1 . \tag{6.48}
\end{cases}
$$

Writing down the wave function with a given value of n as an expansion in terms of parabolic functions $\psi_{n_1 n_2 m}$ with the coefficients C_{n_1} and using the property described by the system (6.48), we obtain recurrence relationships for the coefficients C_{n_1}

$$P_{n_1} C_{n_1 - 1} + (w_{n_1} - \varepsilon) C_{n_1} + P_{n_1 + 1} C_{n_1 + 1} = 0 , \tag{6.49}$$

where the eigenvalues ε are related to the energy E of an atom by

$$E = -\frac{1}{2n^2} + m \omega_{\mathrm{L}} + \frac{1}{4} \omega_{\mathrm{L}}^2 n^2 \varepsilon + 0(\omega_{\mathrm{L}}^4) . \tag{6.50}$$

Following [6.17], we shall obtain the quasiclassical solution of (6.49) at high quantum numbers $n_1 \gg 1$. We shall do this by representing C_k in the form of the product:

$$C_k = \prod_{s=k_0}^{k} \mu_s = \exp\left(\mathrm{i} \sum_{s=k_0}^{k} \Pi_s\right) \approx \exp\left(\mathrm{i} \int_{k_0}^{k} \Pi_s ds\right) , \tag{6.51}$$

where the functions Π_s play the role of the classical momentum in the space of parabolic quantum numbers. Substituting (6.51) into (6.49) and using the condition $n_1 \gg 1$, we reduce the recurrence relationships to a quadratic equation of the type

$$P_{n_1} + (w_{n_1} - \varepsilon) \mu_{n_1} + P_{n_1+1} \mu_{n_1}^2 = 0 \ , \tag{6.52}$$

the discriminant D_{n_1} of which has the approximate form

$$D_{n_1} \simeq (U_{n_1}^+ - \varepsilon)(U_{n_1}^- - \varepsilon) \ . \tag{6.53}$$

The function $U_{n_1}^\pm$ plays a role similar to the potential energy in the Schrödinger equation and, after allowance for (6.48), these functions are described by

$$U_{n_1}^\pm = 3n^2 - m^2 + 1 - 12(a-n_1)^2 \pm 8 \{[(a+\tfrac{1}{2})^2 - (a-n_1)^2]$$
$$\times [(a+|m|+\tfrac{1}{2})^2 - (a-n_1)^2]\}^{1/2} \ . \tag{6.54}$$

If the energy ε lies within the interval $U_{n_1}^- < \varepsilon < U_{n_1}^+$ then $D_{n_1} < 0$ and the classical momenta Π_s in (6.51) become

$$\Pi_k \simeq \arccos \frac{\varepsilon - w_k}{2P_k + 1/2} \equiv \arccos B_k \ . \tag{6.55}$$

The expressions obtained for the "momenta" Π_k can be used also to find the energy by applying the Bohr-Sommerfeld quantization rules:

$$\int_{k_i}^{k_i'} \left(s + \frac{1}{2} \right) \frac{dB_s}{ds} \frac{ds}{(1-B_s^2)^{1/2}} = \pi \left(N + \frac{1}{2} \right) \ , \tag{6.56}$$

where $N = 0, 1, 2, \ldots$ is an integer.

The nature of the spectrum depends on the potential curves governed by the projection of the momentum m. For example, if $m = 0$, these functions become

$$U_{n_1}^+ \approx 5n^2 - 20(n_1-a)^2 \ , \quad U_{n_1}^- \approx n^2 - 4(n_1-a)^2 \ , \tag{6.57}$$

i.e., they represent two inverted parabolas with the center at the point $n_1 = a$. Classical motion occurs in the range limited by the upper (U^+) and lower (U^-) parabolas. At values of the energy ε less than the maximum of the lower parabola ($U_{n_1}^- < \varepsilon < \max U_{n_1}^- = n^2$) the motion occurs in two symmetric potential wells separated by a maximum $U_{n_1}^-$. In view of this symmetry the Zeeman sublevels are doubly degenerate. The difference between the energies of these levels $\varepsilon_g - \varepsilon_u$ is determined by the barrier penetrability and is described, as in the ordinary coordinate space, by a phase integral between the turning points in the subbarrier region. The doublet splitting of the levels disappears for a sufficiently large value of $|m| > n/\sqrt{5}$, corresponding to the region of motion without a maxima. Calculations [6.17] of the energy ε_N and of the splitting in the case when $m = 0$ give the following results. The energy of the lower (doublet) levels deduced from (6.56) is ($N = 0, 1, 2, \ldots$)

$$\varepsilon_N = 1 + 4n\sqrt{5}\,(N + \tfrac{1}{2}) - 12(N + \tfrac{1}{2})^2 + \ldots \quad . \tag{6.58}$$

The corresponding splitting is given by

$$\varepsilon_g - \varepsilon_u \approx (-1)^n \left(\frac{2-\sqrt{5}}{2}\right)^2 (20n^2)^{N+1} \frac{2}{\pi} (e/2N+1)^{2N+1}$$

$$\times \exp\left[-\frac{3}{\sqrt{5}} \frac{2N+1}{n} - \frac{9}{4\sqrt{5}} \frac{(2N+1)^2}{n} \right] . \tag{6.59}$$

Equations (6.58, 59) are strictly valid if $N \ll n$, but a comparison with numerical calculations [6.19] shows that they are highly accurate even for $N \sim n$.

6.4 Oscillator Strengths of Atomic Transitions in Strong Magnetic Fields

A calculation of the oscillator strengths in a weak magnetic field was carried out by *Clark* and *Taylor* [6.12] by perturbation theory methods. Evolution of the oscillator strengths $f_{nn'}$ of the Zeeman components with the increase in the magnetic field B is such that when the Zeeman structures of different levels overlap, there is no significant change in $f_{nn'}$: the components "penetrate" each other freely. As pointed out in Sect. 3.3, this is one of the proofs of the existence of an additional symmetry of an atom in a magnetic field.

The oscillator strengths in ultrahigh magnetic fields have a strong anisotropy due to the existence of a preferred direction $B_{\parallel}$ oz (related to the direction of revolution of an electron). A calculation of the oscillator strengths in fields in the range $B \gg B_0 \sim 10^5$ T based on a general adiabatic theory in Sects. 6.1, 2 was carried out by *Hasegawa* and *Howard* [6.4]. Using [6.4] as a guide, we shall consider the oscillator strengths for the absorption and emission of circularly polarized light in the xy plane

$$f_{ij}^{\pm} = \frac{2m}{\hbar^2} (E_j - E_i) |\langle \psi_j, r_{\pm} \psi_i \rangle|^2 , \quad r_{\pm} = x \pm iy , \tag{6.60}$$

where ψ_i and ψ_j are wave functions of the type described by (6.18) and E_i are the energies of the levels.

It is convenient to introduce generalized momenta $\pi_\pm$ and the coordinates X, Y of the center of a cyclotron orbit of an electron

$$\pi_a = \left(p + \frac{e}{2c} B \times r\right)_a , \quad \pi_\pm = \frac{1}{2}(\pi_x \pm i\pi_y) , \tag{6.61}$$

$$X = x - \frac{\lambda^2}{\hbar}\pi_y , \quad Y = y + \frac{\lambda^2}{\hbar}\pi_x , \tag{6.62}$$

where $\lambda = c\hbar/eB$ is the radius of the cyclotron orbit. The variables π_a, X, and Y obey the commutation relations

$$[\pi_x \pi_y] = \hbar^2/i\lambda^2, \quad [X, Y] = i\lambda^2, \quad [\pi_a X] = [\pi_a Y] = 0 . \tag{6.63}$$

The wave function of the ground state (usually employed in variational calculations) has the structure

$$\psi_0 \sim \exp\left(-\frac{x^2 + y^2}{4a_\perp^2} - \frac{z^2}{4a_\parallel^2}\right) , \tag{6.64}$$

where $a_\perp$ is the transverse size of the orbit approaching the value λ in the limit $B \to \infty$ and $a_\parallel$ is the longitudinal size, which is of the order of the Bohr radius a_0. Using the operators π_a and the completeness of the system of functions ψ_i, we readily derive the following rule for the oscillator strengths:

$$\sum_j f_{ij}^\pm = \frac{1}{i\hbar}\langle\psi_i, (r_\mp \pi_\pm - \pi_\mp r_\pm)\psi_i\rangle . \tag{6.65}$$

In the case of the ground state ($i = 0$), using the function (6.64), we obtain

$$\sum_j f_{0\to j}^\pm = \frac{1}{2} \pm \frac{a_\perp^2}{2\lambda^2} . \tag{6.66}$$

We can see that in the limit $B \to \infty$ the sum of the oscillator strengths for the left-hand circular polarization (LCP) tends to unity, whereas for the right-hand circular polarization (RCP) it tends to zero.

The matrix elements of the wave functions Φ_{NM} of an electron in a magnetic field are found using the standard properties of the operators $\pi_\pm$, X, and Y which are used to describe the relevant coordinate

$$\pi_\pm \Phi_{NM} = \frac{\hbar}{\lambda\sqrt{2}}\sqrt{\frac{N+1}{N}} e^{\pm i\Theta}\Phi_{N\pm1, M\pm1} , \quad R_\pm \Phi_{NM} \propto \Phi_{N, M\pm1} . \tag{6.67}$$

The remaining factors of f_{ij} are governed by the overlap integrals of the wave functions F_{NMn} of one-dimensional quasi-Coulomb motion along the z axis. For example, in the case of transitions from the ground state 0 ($N = M = n = 0$) to the first excited states ($N = 1$, $M = 1$, n), we obtain [6.4]

$$f_{0\to 11n}^{+} = \langle F_{11n}, F_{000}\rangle^2 - \frac{m\lambda^2}{\hbar^2}(\varepsilon_{11n} - \varepsilon_{000})\langle F_{11n}, F_{000}\rangle \ , \quad \text{LCP} \ , \tag{6.68}$$

$$f_{0\to 0-1n}^{-} = \frac{m\lambda^2}{\hbar^2}(\varepsilon_{11n} - \varepsilon_{000})\langle F_{11n}, F_{000}\rangle \ , \quad \text{RCP} \ , \tag{6.69}$$

where ε_{NMn} are the energies of longitudinal motion and $\langle F_{11n}, F_{000}\rangle$ are the overlap integrals of the "longitudinal" wave functions.

Calculations of the functions F_{NMn} (and of the associated overlap integrals) are based on a general system for matching the solutions in the range of high and low values of the z coordinate. At high values of z the functions $F_{NMn}(z)$ are identical, according to the results obtained in Sect. 6.1, with the functions in a one-dimensional Coulomb well. At low values of z they can be found by a perturbation method and a characteristic logarithmic singularity is then encountered in the integration of the Coulomb potential. Matching the two solutions, we find the energy levels

$$\varepsilon = -\frac{\hbar^2}{2m^2}\frac{1}{a^2 n^2} \tag{6.70}$$

and the corresponding form of the wave functions.

The energy of the ground state ($n\to 0$) decreases logarithmically with the increase in B

$$\frac{1}{n} \simeq \log\frac{a^2}{4\lambda^2} + \alpha_{NM} + \dots \ . \tag{6.71}$$

The wave functions of the ground state $n = 0$ are concentrated near the origin of the coordinate system so that their overlap integrals are large:

$$\langle F_{i,n=0}, F_{i,n=0}\rangle \simeq 1 - 4/\log\frac{a^2}{4\lambda^2} + \dots \ . \tag{6.72}$$

Substitution of (6.72) into (6.68) yields the following expression for the oscillator strength involving the ground state

$$f_{0\to 110}^{+} = 1 - 4/\log\frac{a^2}{4\lambda^2} + \dots \ , \tag{6.73}$$

which is in agreement with the sum rule of (6.66).

The oscillator strengths for the transitions to states with $n > 1$ can be found similarly. In the case of high values $n > 1$ the oscillator strengths $f_{0 \to 11n}$ are proportional to a normalization factor n^{-3}. Similarly, we can use the oscillator strength per unit energy interval dn/dE. The wave functions characterized by $n > 0$ have logarithmically small overlap integrals with the function F_{000} concentrated in a region $z \simeq \ln^{-1}(a^2/4\lambda^2)$.

Therefore, the corresponding oscillator strengths are small:

$$\frac{dn}{d\varepsilon} f_{0 \to 11n}^{+} = 4I^{-3/2} \left[\log \left(\frac{a^2}{2\lambda^2} \right) - \text{const.} \right]^{-1} , \tag{6.74}$$

where

$$I = -\varepsilon_0 = \frac{\hbar^2}{2ma^2} \left(\ln \frac{a^2}{2\lambda^2} - 0.577 \right) . \tag{6.75}$$

The oscillator strengths for the RCP transitions include an additional small power-law factor. The general transition scheme, based on [6.4], is shown in Fig. 6.5. The strongest transitions are of type A and these are followed by logarithmically suppressed transitions of type B, and finally by transitions of type C suppressed in a power-law manner.

Detailed calculations of the oscillator strengths for an atom of hydrogen in a magnetic field, including the intermediate range $B \sim B_0$, were carried out by *Forster* et al. [6.19]. Figure 6.6, based on [6.19], shows the behavior of the oscillator strengths $f_{\tau\tau'}$ predicted for various transitions between the lower states of hydrogen. We can see that the transitions allowed in the absence of

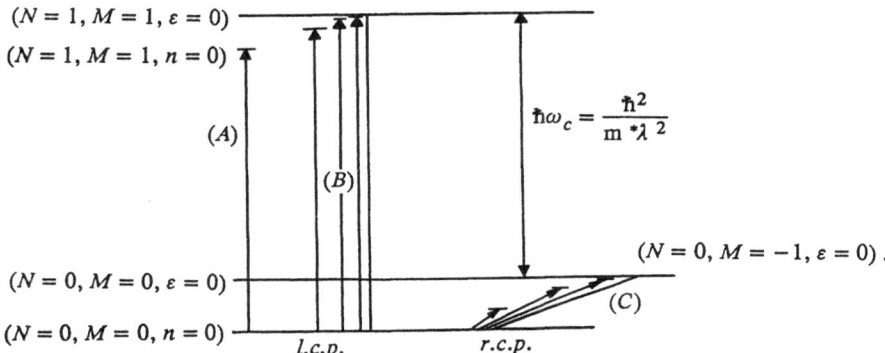

Fig. 6.5. Schematic diagram showing various types (A, B, C) of transitions in an atom subjected to an ultrahigh magnetic field [6.4]. The quantum numbers N, M and n of the states are given for right-handed and left-handed polarizations (r.c.p. and l.c.p.)

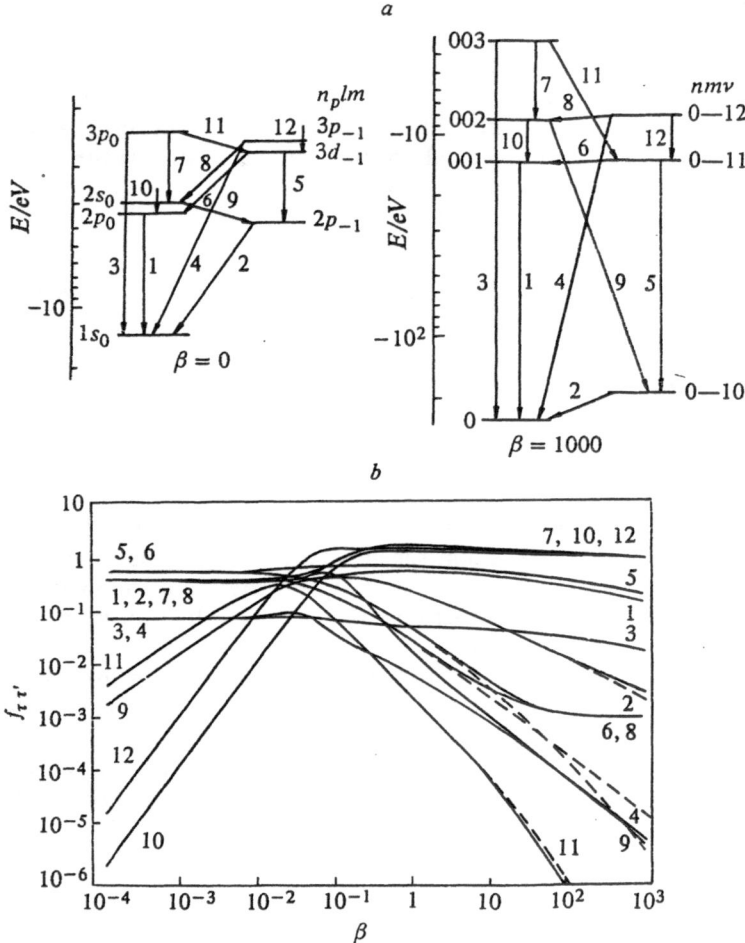

Fig. 6.6a,b. Changes in the oscillator strengths $f_{\tau\tau'}$ of transitions between the lowest states of the hydrogen atom with increasing magnetic field [6.19] (parameter $\beta = B/B_0$). The dashed curves correspond to the approximation of an infinite proton mass (**b**). The scheme of transitions corresponding to $\beta = 0$ and $\beta = 10^3$ is shown in (**a**)

a magnetic field (of type $1-8$) produce little change in the value of $f_{\tau\tau'}$ whereas in the case of other transitions ($9-12$) the oscillator strengths change by several orders of magnitude.

6.5 Classical Trajectories of an Atomic Electron in a Magnetic Field. Stochastization Effects

6.5.1 Calculation of Classical Trajectories

An increase in the intensity of a magnetic field B alters the energy spectrum of an atom from a pure Coulomb (Rydberg) type to a Landau oscillator spectrum with an adjoining one-dimensional quasi-Coulomb spectrum. It is quite difficult to follow in detail such a transition within the quantum theory framework, as found in Sects. 6.1 – 3. However, this can be done on the basis of classical mechanics valid in the case of sufficiently highly excited atomic states. In this case the trajectory of an atomic electron should evolve with the increase in B from a Kepler ellipse to Larmor circles. We have shown in Sect. 5.5 that the classical description of the motion of an electron in an electric field F close to the critical value F_c gives good results for probability of ionization. We can therefore expect that in the case of a magnetic field a classical description can serve as a satisfactory basis for a future quantum theory.

The classical motion case was investigated in detail by Delos et al. [6.20] who numerically solved the classical equations of motion of an electron in Coulomb and magnetic fields. We shall follow the treatment given in [6.20]. The equations for classical trajectories in cylindrical coordinates $\hat{\varrho}$ and $\hat{z}$ (when the $0z$ axis is parallel to B) can be obtained using a Hamiltonian H containing the Coulomb potential $-e^2(\hat{\varrho}^2+\hat{z}^2)^{-1/2}$ the centrifugal potential $L_z^2/2m\hat{\varrho}^2$ and a "diamagnetic" term $e^2B^2\hat{\varrho}^2/8mc^2$. The Hamiltonian equations of motion for canonically conjugate coordinates $\hat{p}$ and $\hat{z}$ and momenta $\hat{p}_\varrho$ and $\hat{p}_z$ can be reduced to a dimensionless form containing just one parameter

$$L = L_z(e^2B^2/mc^2)^{1/6}m^{-1/2}e^{-4/3} , \tag{6.76}$$

which is a combination of the parameters of the Coulomb (e^2) and magnetic (proportional to B^2) interactions. This form of equations is obtained after the substitution of variables [6.20]

$$\varrho = \hat{\varrho}/\alpha , \quad z = \hat{z}/\alpha , \quad p_\varrho = \hat{p}_\varrho/\beta , \quad p_z = \hat{p}_z/\beta , \quad t = \hat{t}/\gamma , \tag{6.77}$$

where

$$\alpha = (mc^2/B^2)^{1/3} , \quad \beta = m^{1/2}e^{2/3}(e^2B^2/mc^2)^{1/6} , \quad \gamma = mc/eB . \tag{6.78}$$

In terms of new variables the Hamiltonian H contains a single parameter which is the effective z component of the angular momentum L. The corresponding equations of motion are

$$\frac{d\varrho}{dt} = p_\varrho \;,\quad \frac{dz}{dt} = p_z \;,$$

$$\frac{dp_\varrho}{dt} = -\frac{\varrho}{(\varrho^2+z^2)^{3/2}} - \varrho/4 + L^2/\varrho^3 \;,$$

$$\frac{dp_z}{dt} = -z/(\varrho^2+z^2)^{3/2} \;.$$

(6.79)

The trajectories of an electron in terms of the variables p, z, p_ϱ, and p_z are still complex. However, we can obtain a full picture of these trajectories by considering a section obtained by cutting with the $z = 0$ plane (Poincaré section [6.21]). In fact, it is clear from the system of equations (6.79) that d^2z/dt^2 and z always have opposite signs. Therefore, this system must be intersected by the $z = 0$ plane throughout the whole duration of motion $-\infty < t < +\infty$.

Numerical calculations of the trajectories reported in [6.20] were carried out as follows: it was assumed that $z = 0$ and for a given H and L, selection was made of twenty random values of the variables p and p_ϱ; p_z was deduced from the Hamiltonian and this was followed by solution of the equations of motion given by the system (6.79). We shall first consider the general aspects of motion. An electron moves in an effective potential

$$V(\varrho,z) = -(\varrho^2+z^2)^{-1/2} + L^2/2\varrho^2 + \varrho^2/8$$

(6.80)

which has two characteristic values: a minimum at $z_0 = 0$ at the point ϱ_0:

$$\frac{\varrho_0^4}{4} + \varrho_0 - L^2 = 0 \;,\quad V(\varrho_0) = E_{\min}(L) \;,$$

(6.81)

as well as an energy E_s for the detachment of an electron from a nucleus in the limit $z \rightarrow \infty$:

$$\varrho_s = (2L)^{1/2} \;;\quad V(\varrho_s,\infty) = L/2 = E_s(L) \;.$$

(6.82)

It is convenient to describe the motion of an electron of energy E by introducing a dimensionless energy

$$f = [E - E_{\min}(L)]/[E_s - E_{\min}(L)]$$

(6.83)

which vanishes at $E = E_{\min}$ and becomes unity at $E = E_s$.

Figure 6.7 schematically shows the regions characterized by the different types of motion in the (f, L) plane. Elliptical motion at low values of L (corresponding to weak B fields) changes to helical motion at high values of L (strong fields B). Phase trajectories (paths) in the $z = 0$ plane are plotted in

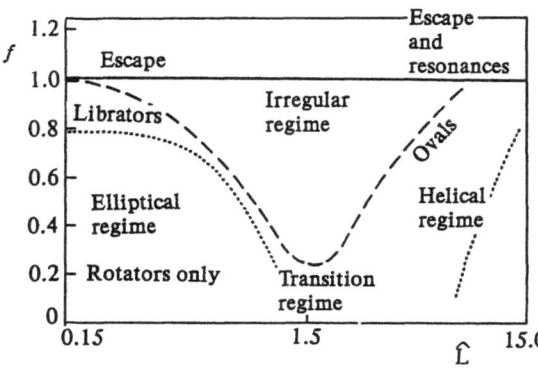

Fig. 6.7. Schematic representations of the regions of electron motion in the plane of the dimensionless energies $f = [(E - E_{min}(L)][E_s(L) - E_{min}(L))]^{-1}$ and of the angular momentum L [6.20]. The ranges of existence of various types of motion are indicated

Fig. 6.8 for several values of f and L. Following [6.20], we shall describe the motion of an electron in each of the regions in the (f, L) plane.

a) Elliptical trajectories correspond to the usual motion along Kepler ellipses. Among these we can identify ellipses elongated along the positive or negative axes (known as "librators"). The motion along "librators" occurs [6.20] if $E \gtrsim -1/10L^2$, i.e., in a narrow region of the (f, L) plane in Fig. 6.7. These trajectories apparently play an important role in the transition to unstable motion (see below). The bulk of the trajectories (known as "rotator") correspond to motion along ellipses close to the (x, y) plane.

b) Helical trajectories occur in a strong magnetic field and correspond, as in the quantum theory in Sect. 6.2, to a sharp division of periods of motion along (parallel to the $0z$ axis) and across the applied magnetic field. As in Sect. 6.2, we can adopt adiabatic separation of the motion in the Hamilton-Jacobi equation retaining a parametric dependence of the potential $V(\varrho, z)$ on z.

The nature of the energy spectrum obtained using the semiclassical quantization conditions is of the kind shown in Fig. 6.1.

c) Irregular motion occurs when the interactions of an electron with the Coulomb and magnetic fields are comparable, and it is manifested by the fact that a trajectory continuously fills the phase (p_ϱ, ϱ) space (Fig. 6.8). The mechanism of appearance of stochastic motion in this case is not yet fully understood, but it is obviously related to the resonances of two existing types of periodic motion: along a Kepler ellipse and along a Larmor circle.

d) Transition motion is observed also for comparable intensities of the interactions, but at lower energies f and it represents stable motion. If $L \sim 1.5$, this motion is however different from ellipses and circles. At low energies f the motion occurs near a minimum of the effective potential and can be investigated by a quadratic expansion of this potential $[U(\varrho) \propto (\varrho - \varrho_0)^2]$. Consequently, the potential becomes oscillatory in respect of the variables ϱ and z; the trajectory is then close to that of a two dimensional oscillator.

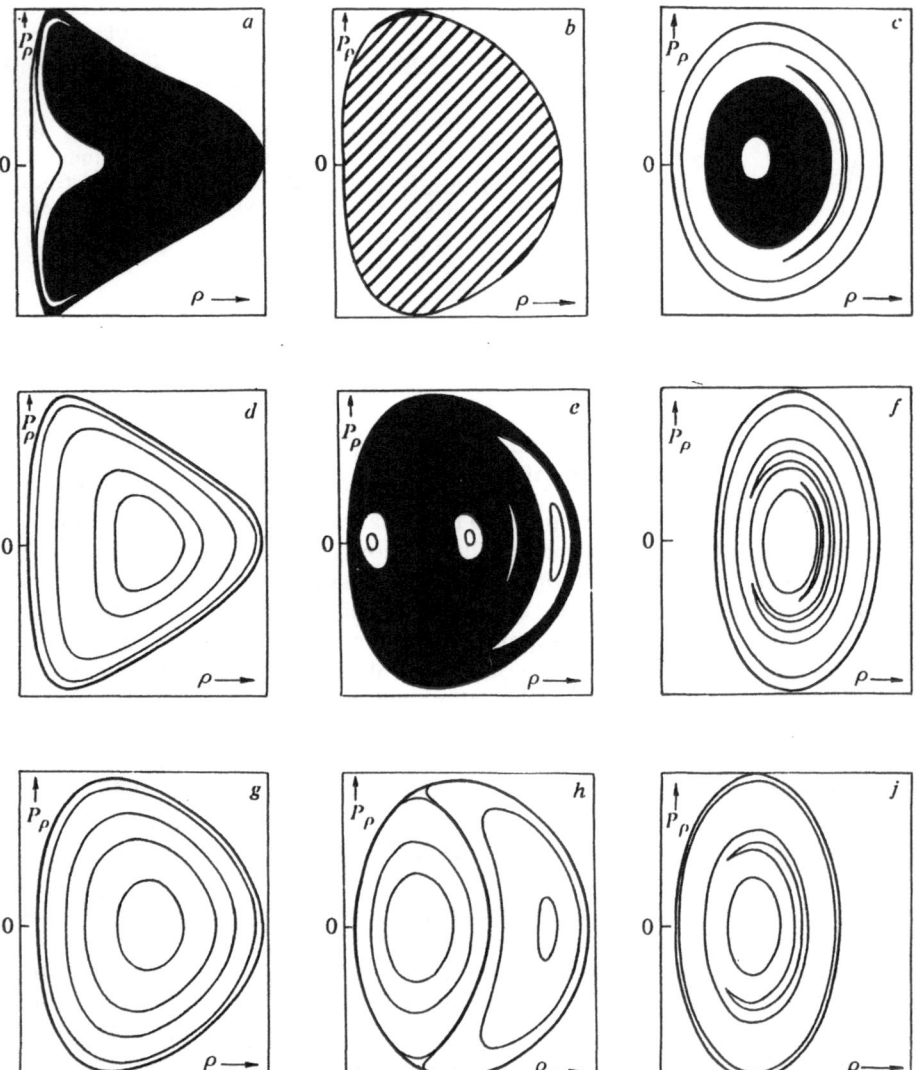

Fig. 6.8. Phase trajectories (P_ϱ, ϱ) of an electron in the $z = 0$ plane obtained for different values of f and L [6.20]. The columns from left to right correspond to $L = 0.50$, 1.51 and 5.03, respectively, whereas the rows counting upwards from the bottom correspond to $f = 0.1$, 0.4 and 0.8, respectively. The black regions represent stochastic trajectories

We shall conclude our account of classical calculations by estimating the regions of appearance of transition effects in the case of a hydrogen atom in a magnetic field. If we express the parameter L in terms of the initial parameters B (in teslas) and L_z (in units of $\hbar$), we obtain $L_z B^{1/3} = 61.7 L$. In a field of $B \sim 10$ T for an intermediate value of $L \sim 1.5$, we find that $L_z \sim 40 \hbar$. These values of L_z can however decrease to a few units in the case of hydrogenic excitons in a medium with a permittivity ~ 10 and an effective electron mass $0.1 \, m_e$.

6.5.2 Stochastization of Electron Motion in Coulomb and Magnetic Fields

Following *Robnik* [6.15] we shall consider in greater detail the appearance of unstable motion of an electron in Coulomb and magnetic fields. Such motion is due to a resonance interaction of modes representing the motion in these fields and the range of its manifestation becomes narrower when one of the interactions becomes stronger. Therefore, there is a definite range of parameters (representing projections of the orbital angular momentum L, the energy E, and the field B/B_0) in which electron trajectories uniformly (but randomly) cover the range of allowed motion in phase space (Figs. 6.6, 9).

The transition to random motion had been investigated by both *Robnik* [6.21] and *Delos* et al. [6.20], and in both cases this was accomplished numerically using classical mechanics. Figure 6.9 shows the behavior of the points of intersection of electron trajectories by the $z = 0$ plane (Poincaré section; [6.21] for various values of the energy E) and the parameters $L = 1$ and $\gamma = B/B_0 = 1$. The motion on a trajectory represents mapping of points on the phase plane (p_ϱ, ϱ) representing periodic intersection of the trajectory by the $z = 0$ plane. It is of interest to consider fixed points and invariant curves which are not affected by successive mapping. The minimum and maximum values of the energy in Fig. 6.9 are $E_{min} = -0.394 \ldots$ and $E_{max} = 0.5$. We can see that when the energy is low (Fig. 6.9 a) the phase space consists of invariant curves corresponding to periodic motion of an electron along the trajectories. The existence of such curves is proof of the existence of an additional (third) integral of motion $I_3(p, q)$ which defines an invariant surface $I_3(p, q) = $ const. and the points of intersection of this surface with the plane $(\varrho, p_\varrho, z = 0)$ form invariant curves. At the centers of these curves there is a fixed imaging point corresponding to totally periodic motion.

An increase in the energy E (Fig. 6.9 c) results in a bifurcation that gives rise to a second fixed point surrounded by a family of closed curves. When the energy is $E = -0.04$ (Fig. 6.9 d), the structure of the trajectories changes drastically: curves with multiple intersections appear and in the corners of such intersections there is an accumulation of points (with a nonzero measure), so that these curves are no longer the usual lines and on further increase in E (Fig. 6.9 e) these curves broaden into a uniformly filled layer. A further increase in E produces a more or less uniform broadening of the layer (alternating with regions of regular motion) and subsequently gives rise to uniform filling of the whole region of allowed motion in phase space. The value $E = E_c = -0.04$ at which there is an abrupt change in the nature of the trajectories is called the critical energy. Therefore, if $E < E_c$, the motion occurs mainly along the invariant curves corresponding to a different value of the conserved invariant $I_3(p, q)$. If $E > E_c$, this invariant is lost and the motion in-

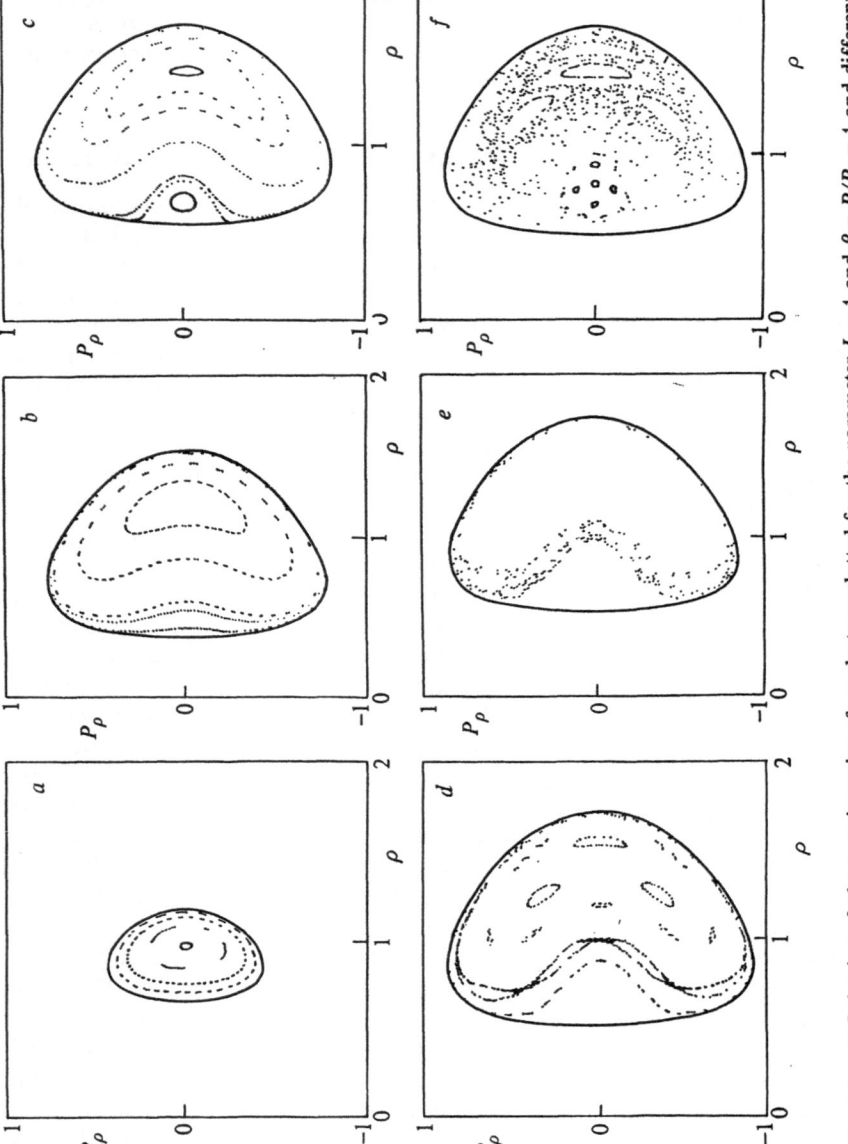

Fig. 6.9a–f. Behavior of phase trajectories of an electron plotted for the parameter $L = 1$ and $\beta = B/B_0 = 1$ and different energies E (a.u.). (**a**) -0.3; (**b**) -0.1; (**c**) -0.05; (**d**) -0.04; (**e**) -0.01; (**f**) -0 [6.15]

side the allowed region is uncorrelated (chaotic). A detailed mechanism of stochastization (of the type representing overlap of the resonances [6.21]) has not yet been finally identified and, moreover, the explicit form of the integral $I_3(p, q)$ is not known. Therefore, it is not clear whether this integral corresponds to the approximate classical integral found in [6.13, 14] (Sect. 6.3).

Nevertheless, numerical calculations [6.22] make it possible to determine the unstable region. We note that the Hamiltonian H and, consequently, the parameters $E_{\min}$ and E_s depend only on the combination γL^3 (after change of all variables to the dimensionless form with the aid of L), so that calculations carried out for $L = 1$ can be used to obtain results for any value of L. In [6.20] the ratio f of (6.83) is calculated near the critical fields $E = E_c$. The minimum value of $f_{\min}$ is 0.22 for $\gamma L^3 \simeq 2.7$. Above the curve $f(E_c, \gamma L^3)$ we have irregular stochastic motion of an electron, whereas below it the motion is regular and quasiperiodic (Fig. 6.9). It follows from the scaling parameter γL^3 that the magnetic fields in which stochasticity appears decrease rapidly (proportionally to L^{-3}) on increase in L. A reduction or an increase in the parameter γL^3 results in a predominance of either the Coulomb or the magnetic interaction, thereby narrowing the range of stochastic behavior. The relationship between this classical description of the motion of an electron and numerical quantum calculations is still unclear in many respects. This applies to both the region of regular classical motion and (even more so) the region of stochastic motion. In any case, here we are dealing with the case of quantization of motion with unseparable variables, which is of fundamental and practical importance.

6.5.3 Numerical Calculations of Spectra of an Atom in a Magnetic Field

Large numbers of numerical calculations [6.22–27] of the spectra of the hydrogen atom in a strong magnetic field have been executed either by perturbation theory methods or on the basis of asymptotic expansions in terms of the B field. Some results of the calculation and interpolation formulas for lower excited states based on these calculations are given in Sect. 6.2. Sufficiently convenient universal data for arbitrary atomic levels in any B field are not yet available, although the results of calculations carried out for a number of states in various ranges of variation B are in good mutual agreement [6.27].

It is appropriate to mention here a simple circumstance [6.27] associated with the behavior of atomic terms in the region of transition from ultrahigh magnetic B fields to low fields. A direct comparison of calculations carried out for low fields in a spherical basis subject to a diamagnetic perturbation with calculations for high magnetic fields yields results very similar to more rigorous numerical calculations. A comparison of the results obtained by direct joining of the results corresponding to the two limiting cases of low and high B fields with those obtained by more rigorous calculations is made in Fig. 6.10 [6.27]. We can clearly see that the curves agree throughout the full range of B. This demonstrates that modification of the atomic basis of the states from Kepler orbits to cyclotron revolution occurs in a narrow range of

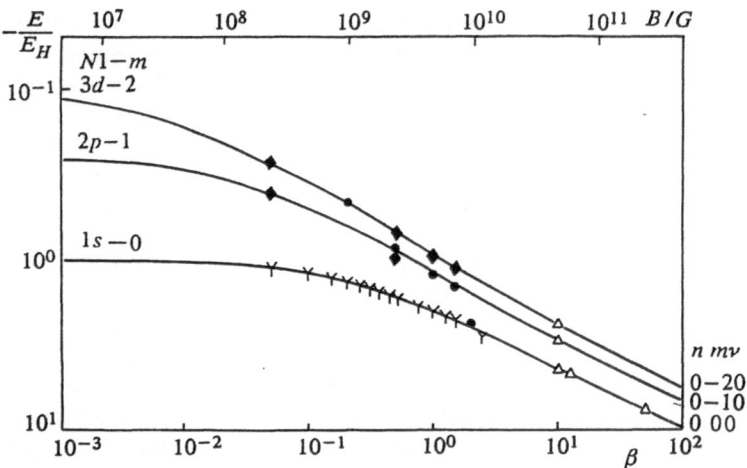

Fig. 6.10. Comparison of the behavior of atomic terms found by direct combination of the results for $\beta \ll 1$ and $\beta \gg 1$ with the results of more accurate numerical calculations [6.26]: ♦ [6.6]; Y [6.8]; △ [6.7]. The dots represent places where the terms were combined

B. This is also supported by the results of a classical calculation of electron trajectories (Sect. 6.5) demonstrating the feasibility of simultaneous co-existence of Kepler and cyclotron orbits. Clear ideas on the transformation of wave functions of the energy states can be provided by the results of numerical calculations of *Rosner* et al. [6.27].

6.6 Hydrogen Atom in Crossed Electric and Magnetic Fields

6.6.1 First-Order Theory

Simultaneous action of electric and magnetic fields on an atom is encountered in many practical applications. In particular, such simultaneous action occurs when an atom is situated in a magnetized plasma where an electric field F is created by the surrounding charged particles [6.28, 29]. Crossed $F-B$ fields also appear when atoms move across a magnetic field because in the system of coordinates linked to an atom there is an electric (Lorentz) field $F_L = [v \times B]/c$. This gives rise to a dependence of the energy levels of the atom on its velocity, which is of considerable interest for an atom in a plasma and for excitons in a solid. The dependence of the energy levels of an exciton in a strong magnetic field on its momentum across the magnetic field was first investigated by *Gor'kov* and *Dzyaloshinskii* [6.30].

Frequently the appearance of an effective magnetic field is due to a change to a rotating system of coordinates. Such a change is convenient in many physical problems, including magnetic resonance [6.31], an atom in a rotating electric field [6.32, 33], or in a field of circularly polarized light [6.34], and an atom in the field of a moving charge [6.35, 36]. We must distinguish the problems where there is a true interaction with a magnetic field and an effective interaction due to the rotation of the coordinate system. In the former case we always have a diamagnetic perturbation, whereas in the latter case there are only quadratic corrections to the Zeeman effect. However, this distinction is unimportant in the first order of perturbation theory.

The problem of the behavior of an atom in F and B fields was considered long ago using classical mechanics [6.37]. We shall consider the period-average characteristics of the motion of an electron in F and B fields. We shall do this by using an additional integral of motion in a Coulomb field (Runge-Lenz vector) related to the period-average value of the coordinate $\langle r \rangle$:

$$A = -\frac{2}{3}\frac{e^2}{a}\langle r \rangle \ , \quad (a = e^2/2|E|) \ . \tag{6.84}$$

In the case of period-average values of the orbital angular momentum M of an atom and of the vector A in a static electric field F, we obtain the following equations [6.38]

$$\dot{M} = \frac{3}{2}\frac{a}{e^2}[FA] \ , \quad \dot{A} = \frac{3}{2m}[FM] \ . \tag{6.85}$$

In a uniform magnetic field a classical particle rotates (revolves) at an angular velocity $\Omega_B = -eB/2mc$, which corresponds to the equations

$$\dot{M} = [\Omega M] \ , \quad \dot{A} = [\Omega A] \ . \tag{6.86}$$

If a particle is subjected simultaneously to both F and B fields, the corresponding equations of motion for M and A are obtained by adding (6.85, 86). Introducing new vectors representing the angular momentum

$$J_{1,2} = \frac{1}{2}\left(M \pm \sqrt{\frac{m}{a}}\,A\right) \tag{6.87}$$

and the frequency

$$\omega_{1,2} = \Omega \pm \frac{3}{2}F\sqrt{\frac{a}{m}} \ , \tag{6.88}$$

we rewrite the equations of motion in the form [6.37]

$$\dot{J}_1 = [\omega_1 J_1] \ , \quad \dot{J}_2 = [\omega_2 J_2] \ . \tag{6.89}$$

It is clear from (6.89) that the new angular momenta J_1 and J_2 precess at frequencies ω_1 and ω_2 independently of one another. The correction V_1 to the energy of a particle in fields F and B expressed in terms of the variables J and ω is

$$V_1 = J_1 \cdot \omega_1 + J_2 \cdot \omega_2 . \tag{6.90}$$

Therefore, the change in the energy is determined by the projections of the vectors $J_{1,2}$ along the directions $\omega_{1,2}$. The form adopted in (6.90) provides a simple opportunity for the generalization of the results to the quantum case. This can be done by independent quantization of the projections of the momenta J_1 and J_2 identified by quantum numbers n' and n'':

$$V_1 = n'\hbar|\omega_1| + n''\hbar|\omega_2| . \tag{6.91}$$

The numbers n' and $n'n''$ assume, in accordance with the definitions of J_1 and J_2, half-integral values: $-(n-1)/2 \ll n'$, $n'' < (n-1)/2$.

Consistent quantum-mechanical generalization of the classical results was provided by Demkov et al. [6.39]. The "correct" wave functions $\psi_{nn'n''}$ corresponding to the diagonalized Hamiltonian of (6.90) can be obtained from parabolic wave functions $\psi_{ni_1 i_2}$ corresponding to specific projections i_1 and i_2 of the vectors J_1 and J_2 along the electric field. This can be accomplished by rotation through angles β_1 and β_2 between the vectors $\omega_{1,2}$ and the directions of the field F [6.39]:

$$\psi_{nn'n''} = \sum_{i_1,i_2} D_{n'i_1}^{(n-1)/2}(0,\beta_1,0) D_{n''i_2}^{(n-1)/2}(0,\beta_2,0) \psi_{ni_1 i_2} . \tag{6.92}$$

Here, $D_{k,l}^{(j)}(0,\beta,0)$ are the Wigner rotation matrices [6.39] describing rotation by angles $\beta_{1,2}$ given by the relationships (in the $F \perp B$ case)

$$\tan\beta_2 = \Omega / \frac{3}{2\hbar} nea_0 F , \quad \beta_1 + \beta_2 = \pi . \tag{6.93}$$

If $B = 0$ the angles $\beta_1 = \pi$ and $\beta_2 = 0$ and the functions $\psi_{mn'n''}$ are identical with the usual parabolic functions $\psi_{ni_1 i_2}$ corresponding to the Stark effect. If $F = 0$, then the angles $\beta_{1,2} = \pi/2$ and the functions $\psi_{nn'n''}$ transform into parabolic functions oriented along the magnetic field B. The relationship between them and spherical functions is considered in Sect. 6.1.1.

It is of special interest to consider the case of mutually perpendicular fields F and B, when the change in the energy is

$$V_1 = \hbar(n'+n'')|\omega_{1,2}| = \hbar(n'+n'')\sqrt{\left(\frac{3}{2}n\frac{a_0}{\hbar e}\right)^2 F^2 + \Omega_B^2} . \tag{6.94}$$

Here, we are dealing with an additional degeneracy of the levels due to the fact that the energy V_1 depends only on the sum of the quantum numbers $n'+n''$ and not on each of the numbers n' and n'' separately.

6.6.2 Second-Order Corrections

Calculations in the second order of perturbation theory on a hydrogen atom subjected to $F-B$ fields are much more difficult. They were considered by *Solov'ev* [6.40]. One has to allow here for the second-order corrections due to the perturbation V_1 of (6.90) and for the first order of the diamagnetic perturbation $V_2 = [B \times r]^2/8c^2$. The magnetic interaction included in V_1 makes no contribution because the resultant matrix elements of this interaction, which are off-diagonal with respect to n, vanish. Consequently, the effective operator $\hat{A}$ allowing for the second-order corrections is [6.40]

$$\hat{A} = \hat{V}_2 - F^2 z G_n z \equiv \hat{V}_2 + \hat{W} F^2 , \tag{6.95}$$

where G_n is the Coulomb Green's function including summation over all intermediate states of the investigated atom. The operator V_2 associated with the diamagnetic interaction is expressed above (Sect. 6.3, also [6.40]) in terms of the angular momentum operator L and the Runge-Lenz vector A:

$$\hat{V}_2 = \frac{n^2 B^2}{16c^2}(n^2+3+L_B^2+4A^2-5A_B^2) , \tag{6.96}$$

where L_B and A_B are the projections of these operators along the magnetic field direction. A similar expression in the space of states with a given value of n can be obtained also in the case of W (when the $0z$ axis is parallel to the field F):

$$W = n^4(5n^2+31+24L^2-21L_z^2+9A_z^2)/16 . \tag{6.97}$$

Expressing next the operators L and A in terms of new angular momentum operators J_1 and J_2 and using the wave functions $\psi_{nn'n''}$ of (6.92), which correspond to specific projections of these operators, we obtain the second-order correction to the energy [6.40]:

$$E^{(2)} = -\frac{n^4 F^2}{16}[17n^2+19-12(n'^2+n'n'' \cos \gamma + n''^2)]$$

$$+ \frac{n^2 B^2}{48c^2}[7n^2+5+4n'n'' \sin \gamma_1 \sin \gamma_2 + (n^2-1)(\cos^2 \gamma_1 + \cos^2 \gamma_2)$$

$$-12(n'^2 \cos^2 \gamma_1 - n'n'' \cos \gamma_1 \cos \gamma_2 + n''^2 \cos^2 \gamma_2)] , \tag{6.98}$$

where γ_1 and γ_2 are the angles between the vector B and the vectors ω_1 and ω_2; $\gamma = \gamma_1 + \gamma_2$.

The results are valid only for the condition that the degeneracy of the levels is lifted in the first order of perturbation theory. This condition breaks down in the case of mutually perpendicular F and B fields when the frequencies ω_1 and ω_2 are equal: $\omega_1 = \omega_2 = \omega$, and when the first-order correction depends only on the sum of the quantum numbers $n' + n''$. Therefore, (6.98) gives the correct result if the difference between ω_1 and ω_2 exceeds the correction $E^{(2)}$. The case $F \perp B$ cannot be considered analytically in the second order of perturbation theory. The general calculation procedure in this case involves numerical diagonalization of bilinear combinations of the angular momenta J_1 and J_2 in the subspace of quantum numbers n' and n'', and it is given in [6.40].

The behavior of the ground state of the hydrogen atom subjected simultaneously to electric and magnetic fields was considered by *Turbiner* [6.41] by the methods of perturbation theory. The expansion for the energy can be represented conveniently in the form

$$E = E_{sz} + E^{(\|, \perp)} , \tag{6.99}$$

where E_{sz} is the sum of the energies of the fields F and B taken separately:

$$E_{sz} = -1 - \frac{9}{2} F^2 + \frac{B^2}{2} - \frac{3555}{32} F^4 - \frac{53}{96} B^4 + \dots , \tag{6.100}$$

and $E^{\|, \perp}$ contains the hitherto unknown cross terms for mutually parallel ($E^\|$) and perpendicular ($E^\perp$) directions of the F and B fields

$$E^\| = \frac{159}{16} F^2 B^2 - \frac{1\,742\,009}{26\,880} F^2 B^4 + \dots , \tag{6.101}$$

$$E^\perp = \frac{93}{4} F^2 B^2 - \frac{22\,770\,991}{107\,520} F^2 B^4 + \dots . \tag{6.102}$$

The results represented by (6.99–102) allow us to qualitatively understand the characteristics of the behavior of an atom in F and B fields. Indeed, if we assume that the electric field F is constant, we can find the magnetic susceptibility of the investigated atom [6.41]:

$$\left. \begin{array}{l} \chi^\| = -1 - \dfrac{159}{8} F^2 + \dfrac{53}{24} B^2 + \dfrac{1\,742\,009}{6720} F^2 B^2 + \dots , \\[3mm] \chi^\perp = -1 - \dfrac{731}{24} F^2 + \dfrac{53}{24} B^2 + \dfrac{15\,308\,863}{30\,240} F^2 B^2 + \dots . \end{array} \right\} \tag{6.103}$$

We can see that the sign of the term with the electric field (proportional to F^2) is opposite to the sign of the usual "diamagnetic" term (proportional to B^2). Therefore, the presence of the electric field increases the magnetic susceptibility of the atom. It should be pointed out that this effect is stronger in the case when $F \perp B$. On the other hand, if we fix the value of B, we can find the influence of this field on the polarizability of the atom in the field F [6.41]:

$$\begin{aligned}
\alpha^{\parallel} &= 9 + \frac{3555}{8} F^2 - \frac{159}{8} B^2 + \frac{1\,742\,009}{13\,440} B^4 + \dots , \\[2ex]
\alpha^{\perp} &= 9 + \frac{3555}{8} F^2 - \frac{731}{24} B^2 + \frac{15\,308\,863}{60\,480} B^4 + \dots .
\end{aligned} \qquad (6.104)$$

It is seen that the magnetic field decreases the polarizability of the atom.

6.6.3 Atom in Electric and Strong Magnetic Fields

The limits of a strong magnetic field ($B \gg B_0$) and a weak magnetic field ($F \ll B$) were considered in [6.42] in the specific case, mentioned above, of energy levels of an exciton moving across a magnetic field. A detailed description of the structure of such spectra is outside the scope of the present review. We shall consider only an interesting feature of the spectrum of an hydrogen atom in crossed F and B fields, which was investigated by *Burkova* et al. [6.42].

The characteristics of the motion of a free charge in F and B fields are related, as is known, to the drift of the charge at a velocity

$$v_{\mathrm{D}} = c \frac{F \times B}{B^2} , \qquad (6.105)$$

which is the same for an ion (of mass m_i) and for an electron (of mass m_e). The drift of an electron may give rise to new bound states in an atom localized at a certain distance y_0 equal to the drift displacement during an effective cyclotron oscillation period [6.42]

$$y_0 = v_{\mathrm{D}} \left/ \frac{eB}{(m_i + m_e)c} \right. = Mc^2 F/eB^2 , \quad (M \equiv m_i + m_e) . \qquad (6.106)$$

The spectrum of an atomic electron can be obtained conveniently in this case by transforming the wave function to a system of coordinates linked to the drift motion [6.42]

$$\psi = \Phi \exp \{ i \gamma Mc [BF] r / 2\hbar B^2 \} , \qquad (6.107)$$

where $\gamma = (m_i - m_e)/M$. If the z axis is directed along the field B and the y axis along the field F, and if the origin of the coordinate system is shifted along the y axis by an amount y_0, the Schrödinger equation becomes

$$\left\{ -\frac{\hbar^2}{2\mu}\Delta_\varrho - \frac{\hbar^2}{2\mu}\frac{\partial^2}{\partial z^2} + \frac{ie\hbar y}{2mc}[\varrho\nabla_\varrho]B + \frac{e^2}{8\mu c^2}B^2\varrho^2 \right.$$

$$\left. -e[x^2+(y+y_0)^2+z^2]^{-1/2} - Mc^2F^2/2B^2 \right\}\Phi = E\Phi , \qquad (6.108)$$

where $\mu = m_e m_i/M$ is the reduced mass.

We shall consider the effective potential energy U along the y axis:

$$U = \frac{e^2 B^2}{8\mu c^2}y^2 - \frac{e^2}{|y+y_0|} - \frac{Mc^2F^2}{2B^2} . \qquad (6.109)$$

We can see that the potential U can have two wells: one Coulomb well at $y \approx y_0$ and the second at $y = 0$. Such a structure of the potential is realized for a sufficiently large value of the following parameter

$$\left(\frac{M}{\mu}\right)^{1/2}\left(\frac{\hbar c}{e^2}\right)\left(\frac{F}{B}\right)\left(\frac{B}{B_0}\right)^2 > 1 \qquad (6.110)$$

when the bottom of the well at $y = 0$ is higher than the Coulomb binding energy $me^4/\hbar^2$.

The energy spectrum can be calculated subject to the conditions (6.110) if, as in Sect. 6.1, we separate the variables of the longitudinal (along the z axis) and transverse motion, and if we reduce the Schrödinger equation of (6.108) to the one-dimensional form with the effective potential $U(x,y,z)$ obtained by averaging the initial potential over the transverse coordinates p:

$$U = -e^2\int\frac{\Phi(\varrho)d\varrho}{[x^2+(y+y_0)^2+z^2]^{1/2}} , \qquad (6.111)$$

where $\Phi(\varrho)$ is the wave function of the transverse motion with a characteristic scale length a_B (Sects. 6.1, 2).

We shall consider the bound states near $y = 0$ and use the condition $y_0 \gg a_B$, which allows us to set $x = y = 0$ in (6.111) and this gives

$$U = -\frac{e^2}{(z^2+y_0^2)^{1/2}} \approx -\frac{e^2}{y_0} + \frac{e^2}{y_0^2}z^2 . \qquad (6.112)$$

Consequently, the Schrödinger equation for the motion along the z axis becomes

$$\frac{d^2f}{dz^2} + \frac{2\mu}{\hbar^2}[E-E_0-U(z)]f = 0 , \qquad (6.113)$$

where

$$E_0 = e\hbar B/2\mu c - Mc^2 F^2/2B^2 \tag{6.114}$$

is the energy corresponding to the limit of the continuous spectrum. In the case of the lower energy levels in a well the effective values of z_{eff} are small compared with y_0, allowing us to use the expansion of (6.112) which obviously leads to an oscillator potential. The energy spectrum which is then determined is of the form

$$\frac{E_n - E_0}{2\,\mathrm{Ry}} = -\lambda^{-1} + \lambda^{-3/2}\left(n + \frac{1}{2}\right), \quad \lambda = \frac{M}{\mu}\frac{\hbar c}{e^2}\frac{F}{B}\frac{B}{B_0} . \tag{6.115}$$

In the case of the hydrogen atom subjected to a field $B \sim B_0$ and also to fields F compatible with (6.110) the binding energy is of the order of 0.55 eV when the separation between the levels is 0.1 eV.

6.7 Conclusions

The above review demonstrates the recent rapid growth of interest in the Stark and Zeeman effects. On the one hand, this is due to the numerous possible practical applications of these effects and, on the other, it is due to problems of a fundamental nature associated with the dynamics of systems with nonseparable variables. In applications of the Stark and Zeeman effects it is usually essential to know not just one parameter, but all of the characteristics of an atom in fields F and B, such as the splitting of levels, line intensities, probabilities of radiative and autoionization decay, etc. The range of F and B field intensities, and of the quantum number n of atoms is very wide. Very frequently the values of F and B are determined by the parameters of the ambient medium, such as the temperature T and density N of a plasma, which can also range within very wide limits. Therefore, it is very important to have analytic results for the parameters of an atom in F and B fields, particularly in a clear form that would be suitable for the use in practical applications.

Many of the problems discussed above have not been finally solved. This applies particularly to an atom in a magnetic field, when the inability to separate variables greatly complicates the situation. Investigations of the behavior of an electron in this case by the methods of classical mechanics and the discovery of stochastic regions of motion leave open the problems of the nature of quantum motion and its correspondence to classical motion. One would hope that this review will draw attention to these problems.

7. Atom in a Nonresonant Oscillating Electric Field

7.1 The Types of Oscillating Fields in Plasmas. Quasi-energetic Level Structure

The interaction of atoms with periodically oscillating fields $E(t) = E_0 \cos \Omega t$ is the subject of very extensive investigations in laser physics [7.1, 2]. We shall dwell on some particular aspects of the interaction of the atom with such fields which is of interest for plasma applications. Oscillating fields in plasmas are produced by both laser fields used for plasma creation or diagnostics (laser produced or high frequency plasmas) and the fields of plasma oscillations themselves, for example Langmuir oscillations. This section covers the dynamics of atomic interaction with a strong electromagnetic (E.M.) field with no regard for kinetic effects, which results in different relaxation mechanisms. We are dealing here with properties and spectra of the spontaneous emission of the system "atom plus external E.M. field".

The influence of the harmonic E.M. field on the atom results in a much more complicated variation of the atomic structure as compared with the case of constant electric or magnetic fields. Strictly speaking it would not be right to consider that there are any stationary energetic levels because the perturbation depends on time. However, due to the harmonically oscillating nature of the perturbation, one can introduce the idea of quasi-energy [7.3] as an analog to the idea of quasimomentum for the space periodicity.

The quasi-energy ε for the perturbation of the $E_0 \cos \Omega t$ type is introduced according to relation

$$\psi_\varepsilon \left(t + \frac{2\pi}{\Omega} \right) = e^{-i2\pi\varepsilon/\Omega} \, \psi_\varepsilon(t) \; . \tag{7.1}$$

The wave function $\psi_\varepsilon(t)$ corresponds to a definite quasi-energy ε and is no longer periodic. Spontaneous transitions from these quasi-energetic levels to lower lying atom levels are possible. This generally leads to a complicated spectrum of spontaneous emission with plenty of maxima and minima. The deciphering of such spectra gives information on frequencies and intensities of oscillations in plasmas.

The exact solution of the problem of the atom's behavior in an oscillating field is hardly impossible. Therefore, we shall dwell on a number of practically important approximations. First of all this concerns the two-level approximation for the transitions between only two isolated atomic levels. These levels are supposed to be nondegenerate as for orbital momentum projections, although it is clear that for real systems this approximation is invalid. Nevertheless, considering degeneration this leads, as a rule, to a more accurate definition of the numerical coefficients and does not introduce any new parameter into the problem. Perturbation $E_0 \cos \Omega t$ may be in resonance to the transition if the frequency Ω is close to ω_0 of the two level system and $|\Omega - \omega_0| \ll \omega_0$. Usually in plasmas the frequencies Ω of harmonic oscillations are much smaller than the atomic transition frequency ω_0, so $\omega_0/\Omega \gg 1$. In this case the perturbation is called adiabatic. It is of great importance whether in this case the condition of multiphoton resonance $\omega_0 = n\Omega$ is observed, where n is an integer number. For example, if other atomic levels lie between two considered transitions, the probability increases sharply when the multiphoton resonance takes place for any of these levels.

Modern investigators usually deal with sufficiently strong oscillating fields E_0 which reveal effects that are nonlinear in light intensity $I_0 \sim E_0^2$. The essence of the term "nonlinear" can be briefly expressed as the absence of the proportionality of the probabilities of the radiative processes to the nth power of intensity I_0^n. The later dependence is realized only in the weak field case $(I_0 \sim E_0^2 \Rightarrow 0)$, when multiquantum processes are not likely. On the contrary, for strong fields multiquantum processes are not at all negligible.

A special part is played by degenerate hydrogen-like levels corresponding to frequency $\omega_0 = 0$. Here it is necessary to distinguish the ranges of frequencies Ω small as compared with the distance from neighbouring levels $(\omega_{nn'} \sim \mathrm{Ry}\, n^{-3})$ on the one hand, and those which can be compared with them on the other hand. In the first case it is enough to take into account only the interaction of degenerate states belonging to a given n. In the second case it is necessary to consider all atomic levels. We shall examine the first case of small frequencies $\Omega \ll \omega_{nn'}$.

7.2 The Blokhintsev Spectrum

Let us consider an excited level n of an hydrogen atom immersed in a linear polarized electric field $E(t) = E_0 \cos \Omega t$, the frequency Ω of which is small compared with the distance $\omega_{nn'}$ from the other levels. The problem was first considered by *Blokhintsev* [7.4]. What we are interested in is the spectrum of spontaneous transitions of such an atom to some lower state, for example, the ground state. Putting the $0z$ axis along the field E_0 and neglecting transitions to other levels one can see that the energy levels possess oscillating displacements

$$\kappa(t) = CE(t) = CE_0 \cos \Omega t \tag{7.2}$$

due to the linear Stark effect, C being the Stark constant characterizing a displacement of a Stark component. *Blokhintsev* [7.4] considered the spectrum of one Stark component connected by a dipolar allowed transition with the ground atomic state. The radiation spectrum $I_n(\Delta\omega)$ observed at the frequency ω ($\Delta\omega \equiv \omega - \omega_0$) is a discrete one owing to the oscillating nature of the perturbation, and it consists of harmonics with intensities

$$I_n(\omega) = \left| \int\limits_{-2\pi/\Omega}^{2\pi/\Omega} dt \exp\left[i(n\Omega)t - iCE_0 \int\limits_0^t \cos\Omega t \right] \right|^2 , \tag{7.3}$$

where n is the harmonic's number, $\Delta\omega \equiv n\Omega$.

It is easy to evaluate (7.3) analytically and to obtain the following result for the dimensionless spectrum $S(\Delta\omega/\Omega)$ defined by the relation $S(\Delta\omega/\Omega)d(\Delta\omega/\Omega) = I(\omega)d\omega$

$$S\left(\frac{\Delta\omega}{\Omega}\right) = \sum_{n=-\infty}^{\infty} J_n^2\left(\frac{CE_0}{\Omega}\right) \delta\left(\frac{\Delta\omega}{\Omega} - n\right) . \tag{7.4}$$

Thus, the *Blokhintsev* spectrum (7.4) consists of a number of satellites spaced at a frequency Ω from each other, and their intensities are defined by Bessel functions $J_n(CE_0/\Omega)$. The argument of these functions contains the so called modulation depth (magnitude of the Stark displacement CE_0) and frequency.

Let us consider what this spectrum is like within the large modulation index (that is, low perturbation frequency $\Omega \Rightarrow 0$). Here the limits of integration in (7.3) may be extended to infinity. Then for the small frequency Ω the main contribution in (7.3) will be points t_k of the stationary phase where

$$\Delta\omega = \kappa(t_k) . \tag{7.5}$$

Making an expansion in the exponent near the point and summing up all the points t_k, one finds [7.1]

$$I(\omega) \sim \sum_k \frac{1}{|\dot{\kappa}(t_k)|} , \tag{7.6}$$

where $\dot{\kappa}(t_k)$ is the time derivative of the instanious frequency shift $\kappa(t)$. The magnitude $|\dot{\kappa}(t_k)|^{-1}$ is obviously the relative time of frequency shift staying near its value of $\kappa = \Delta\omega$ (7.5). The relative time is essentially the probability for the shift to have a definite value near $\Delta\omega$. Using the analytical form of $\kappa(t)$ from (7.2) with the help of (7.6) one obtains

$$I(\omega) = \frac{1}{\pi} [(CE_0)^2 - (\Delta\omega)^2]^{-1/2} . \tag{7.7}$$

The spectrum in (7.7) coincides with the shift of the distribution function $W(\kappa)$ and correspondingly, can be obtained by the averaging of the static spectrum in a constant field $\delta(\Delta\omega - CE)$ according to the distribution function $W(E) = \pi^{-1}(E_0^2 - E^2)^{-1/2}$. From this point of view the result (7.7) may be called static (not containing the frequency Ω). However, the form of the function $W(E)$ itself is entirely determined by the oscillating nature of the perturbation. The function (7.7) increases when $\Delta\omega \Rightarrow CE_0$, or considering $\Delta\omega = n\Omega$, when $n \Rightarrow CE_0/\Omega$. It is for these values of n that the *Blokhintsev* satellite in the spectrum (7.4) acquires the maximum intensity:

$$\max J_n\left(\frac{CE_0}{\Omega}\right) = J_{CE_0/\Omega}\left(\frac{CE_0}{\Omega}\right) = \frac{6^{1/3}\,\Gamma(1/3)\sin\pi/3}{3\,\pi\,n^{1/3}}\ . \tag{7.8}$$

The ratio of this satellite intensity to the intensity of the zero order satellite $J_0(CE_0/\Omega)$ increases monotonically:

$$[J_{CE_0/\Omega}(CE_0/\Omega)/J_0(CE_0/\Omega)] \sim (CE_0/\Omega)^{1/3}\ . \tag{7.9}$$

Thus, when $CE_0/\Omega \to \infty$, the shape of the curve of the satellites approaches the static distribution (Fig. 7.1). In the weak field case, satellite intensities go down rapidly with the increase of their number. Therefore the line intensity is determined by a number of first satellites:

$$I(\omega) \approx \left[1 - \frac{1}{2}\left(\frac{CE_0}{\Omega}\right)^2\right]\delta(\Delta\omega) + \left(\frac{CE_0}{2\Omega}\right)^2 [\delta(\Delta\omega+\Omega) + \delta(\Delta\omega-\Omega)]\ . \tag{7.10}$$

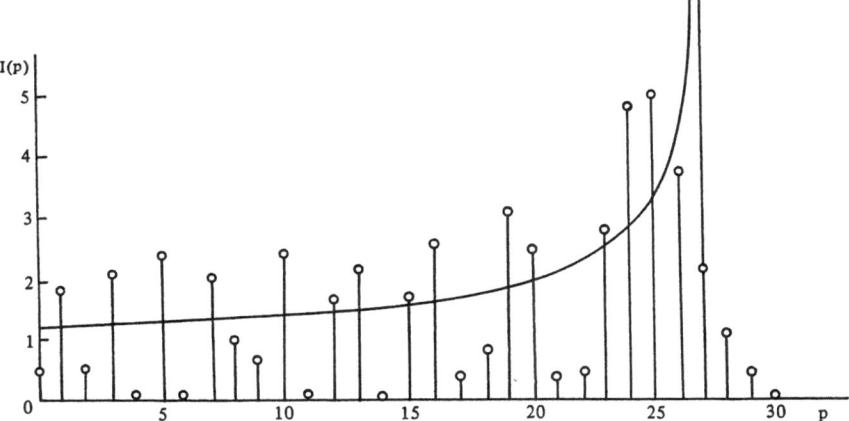

Fig. 7.1. The spectrum $I(p)$ of a Stark component with the given quantum numbers n, n_1, n_2 in an oscillating electric field $E_0\cos\Omega t$ (Blokhintsev spectrum) for the index of modulation $\frac{3}{2}n(n_1 - n_2)\hbar E_0/me\Omega = 27$ [7.6]

The *Blokhintsev* spectrum (7.4) responds to the perturbation (7.2) with a rigorously fixed phase of oscillation. However, in plasmas the atom may experience the influence of the superposition of fields with close frequencies but accidental phases. In this case the perturbation can be written in the form

$$E(t) = E_0(t) \cos [\omega t + \varphi_0(t)] \;, \qquad (7.11)$$

where $\varphi_0(t)$ and $E_0(t)$ are accidental variables experiencing instant jumps in accidental time moments. If one considers these accidental processes to be Gaussian then it is easy to obtain the spectrum which is analogous to (7.4):

$$I(\Delta\omega) = \sum_{n=-\infty}^{+\infty} I_n(C^2\bar{E}^2/\Omega^2)e^{-C^2\bar{E}^2/\Omega^2}\delta(\Delta\omega - n\Omega) \;, \qquad (7.12)$$

where the $I_n(Z)$ are modified Bessel functions, and $\bar{E}^2$ is the average square of the amplitude fluctuation.

The spectrum (7.12) obtained by *Lifshitz* [7.5] differs from the *Blokhintsev* spectrum (7.4) for large modulation indices $C\bar{E}^2/\Omega \to \infty$ by having the Gaussian form of the shift distribution as distinct from (7.4). For small $\bar{E} \to 0$, both spectra are close. The realization of two types of spectra given by (7.4, 12) depends on the relation of the typical time τ_c between phase jumps and the time of a given spectrum section formation $(n\Omega)^{-1}$. However, the question concerns the perturbation statistics and deviates from the limits of the problems considered here, but is discussed in [7.6], for example.

7.3 Hydrogen Atom in a Rotating Electric Field

The *Blokhintsev* spectrum considered above responds to a magnitude variation electric field with a fixed direction only. It is of interest to consider the atomic spectrum in the field changing in its direction only with the fixed magnitude. The situation is realized for the atom in a rotating electric field [7.7, 8], for example, in the field of a circularly polarized light wave. Let the rotation frequency Ω be small compared to the distance between atomic states. The Hamiltonian of the hydrogen atom in such a field form

$$\hat{H}(t) = \hat{H}_0 + d F e^{i\Omega t} \;, \qquad (7.13)$$

where H_0 is an unperturbed Hamiltonian.

We shall solve this problem in the coordinate system rotating over the $0x$ axis together with the electric field. It is obtained by the wave function transformation:

$$\psi(t) = e^{iL_x\Omega t}\psi'(t) \;, \qquad (7.14)$$

where ψ and ψ' are wave functions in motionless and rotating systems, and L_x is the x component of the atomic orbital momentum. The Schrödinger equation for the wave function ψ' in the rotating coordinate system becomes stationary:

$$(H_0 + d_z F + L_x \Omega) \psi' = E \psi' \ . \tag{7.15}$$

Here due to the coordinate system rotation the additional interaction $L_x \Omega$ has appeared. The solution of the problem (7.15) is completely equivalent to the determination of atomic energy levels in crossed electric and magnetic fields (Chap. 4). The role of the former is obviously played by the rotation velocity Ω.

The function ψ' is determined as in Chap. 4 by the rotation of parabolic wave functions with the $0z$ axis parallel to F at the angles α_1, α_2 determined by the ratio of the electrical $d_z F$ and magnetic $L_x \Omega$ interactions. The radiation intensity $I(\omega)$ is defined by the matrix elements of the atomic dipolar momentum $d_{if}(t)$. For their determination it is sufficient to use the connection between the components of the vector $\exp(-iL_x \Omega t)d$ $\exp(+iL_x \Omega t)$ rotated at the angle Ωt between the rotating and fixed coordinate systems. As a result the atomic spectrum in the rotating field takes the form:

$$I(\omega) = \sum_{n'_{i,f} n''_{i,f}} \sum_{l=0,\pm 1} | \sum_{k_1^i {}^f k_2^i {}^f} D_{n'_i k_1^i}^{(n_i-1)/2}(0,\alpha_1^i,0)$$

$$\times D_{n'_i k_2^i}^{(n_i-1)/2}(0,\alpha_2^i,0) D_{n'_f k_1^f}^{(n_f-1)/2}(0,\alpha_1^f,0) D_{n'_f k_2^f}^{(n_f-1)/2}(0,\alpha_2^f,0)$$

$$\times \langle n_i k_1^i k_2^i | d_l | n_f k_1^f k_2^f \rangle |^2 \delta(\Delta\omega - \Delta\omega_{if} + l\Omega) \ . \tag{7.16}$$

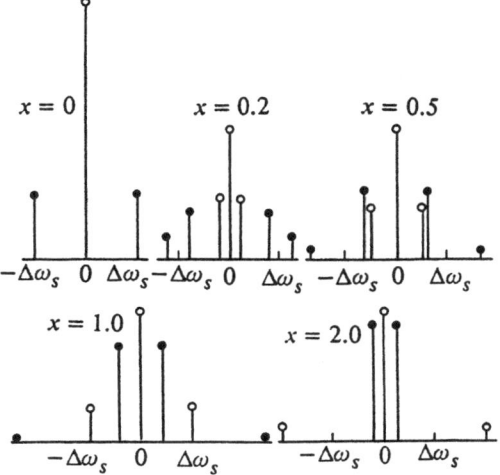

Fig. 7.2. The L_a-line spectrum in a rotating electric field for different ratios $x = \Omega/cF$ of the angular velocity Ω to the Stark splitting CF [7.7] ($\Delta\omega_s$ is the position of the Stark components in the constant electric field F)

Here $D_{i_1' i_2'}^{(n-1)/2}(0, \alpha_1, 0)$ are the Wigner rotation matrices [7.8], and the quantum numbers n', n'' run values from $-(n-1)/2$ to $+(n-1)/2$. The spectrum acquires an especially simple shape for the L_α line (transition $2 \to 1$) when matrices $D_{km}^{(1/2)}$ are two dimensional. It is shown on Fig. 7.2, taken from *Ishimura's* work [7.7]. It is seen that unlike the *Blokhintsev* spectrum (7.4), the quantity of spectral components are finite. However, their number exceeds the atomic state numbers n^2 for a given level n. The discrepancy is obviously due to the appearance of the combined frequencies $\pm \Omega$ connected with the rotation of the atomic dipole $d(t)$.

7.4 Multiphoton Transitions in a Two-Level System

Let us consider the influence of an oscillating perturbation $F_0 \sin \omega t$ on a two-level system for which the distance ω_0 between levels 1 and 2 is large as compared with perturbation frequency: $\omega_0/\omega \gg 1$. Substituting into the Schrödinger equation the system's wave function $\psi(t)$ described by the superposition $a_1(t) \cdot |1\rangle + a_2(t) \cdot |2\rangle$ of the wave functions of levels $|1\rangle$ and $|2\rangle$, one obtains equations for the amplitudes:

$$
\left.
\begin{aligned}
i\dot{a}_1 &= -\frac{\omega_0}{2} a_1 + d_{12} F_0 a_2 \sin \omega t \;, \\
i\dot{a}_2 &= +\frac{\omega_0}{2} a_2 + d_{12} F_0 a_1 \sin \omega t \;.
\end{aligned}
\right\}
\tag{7.17}
$$

The system (7.17) is invariant as for the translation in time "t" by the period $2\pi/\omega$ as well as by half of the period π/ω with simultaneous changing of the sign in front of the nondiagonal element.

Since the frequency ω is small, one can find the eigenvalues of energy $E_{1,2}(t)$ of the system as well as the eigenfunctions $\varphi_{1,2}(\tau)$ in every moment of time. Indeed, substituting $a_{1,2}$ into (7.17) in the form $\exp[+i\int^t E(\tau) d\tau]$ one finds:

$$
E_{1,2} = \pm \frac{\omega_0}{2} (1 + \beta^2 \sin^2 \omega t)^{1/2} \equiv \pm \Omega(t) \;, \quad \beta \equiv 2 d_{12} F_0 / \omega_0 \;.
\tag{7.18}
$$

The instantaneous (adiabatic) wave functions $\varphi_{1,2}(\tau)$ corresponding to a definite energy $E_{1,2}(\tau)$ are equal to

$$
\left.
\begin{aligned}
\varphi_1(t) &= |1\rangle \cos\frac{\chi}{2} - |2\rangle \sin\frac{\chi}{2} \;, \\
\varphi_2(t) &= |1\rangle \sin\frac{\chi}{2} + |2\rangle \cos\frac{\chi}{2} \;,
\end{aligned}
\right\} \quad \tan \chi = \beta \sin \omega t \;.
\tag{7.19}
$$

The total wave function $\psi(t)$ of the system may be expressed in terms of a superposition of adiabatic wave functions $\varphi_{1,2}$:

$$\Psi(t) = b_1(t)\varphi_1(t)\exp\left[i\int^t \Omega(\tau)d\tau\right] + b_2(t)\varphi_2(t)\exp\left[-i\int^t \Omega(\tau)d\tau\right] .$$

Coefficients $b_{1,2}(t)$ are slowly time varying functions and an proper adiabatic approximation corresponding to constant values of b_1 and b_2.

Let us find the simplest characteristic of the system: the transition probability in unit time $W_{1,2}$ from state $|1\rangle$ into state $|2\rangle$, absorbing n-photons. Suppose that our system meets the n-photon resonance condition: $\omega_0 = n\omega$. The probability $W_{1,2}$ is easily found with the help of perturbation theory assuming that the parameter of interaction with a field β is as small as one likes [7.9]

$$W_{1,2}^0 = 2\pi\omega^2\left(\frac{n\beta}{4}\right)^{2n}\frac{1}{[(n-1)!!]^4}\delta(\omega_0 - n\omega) . \tag{7.20}$$

It is seen that the greater the number of photons n that are necessary for the excitation of the system, the lower is the probability.

Now let us take the probability $W_{1,2}$ outside of the perturbation theory limits [7.9]. The adiabatic case $\omega_0/\omega \gg 1$ under consideration above is obviously the total analog to the quasiclassical approximation, with the only difference being that time t appears in this case instead of a space variable. In correspondence with general quasiclassical results, the transition probability $w_{1,2}$ between two adiabatic energy levels $E_1(t)$ and $E_2(t)$ is determined by the points t_k of their crossing in the plane of the complex values of t. From (7.18) one has:

$$E_1(t_k) = E_2(t_k); \; t_k = \frac{k\pi}{\omega} + \frac{i}{\omega}\,\text{arsh}\,(1/\beta), \; k = 0, \pm 1, \pm 2, \ldots . \tag{7.21}$$

So the contribution to $w_{1,2}$ of a point is equal to

$$w_{1,2} = \exp\left[-2\,\text{Im}\int^{t_0}\Omega(\tau)d\tau\right] , \tag{7.22}$$

where the lower limits may be chosen in an arbitrary point on the real axis. The integral in (7.22) is expressed in terms of a total elliptic integral of the third kind $D(k) = [F(k) - E(k)]/k^2$, where $F(k)$ and $E(k)$ are the total elliptic integrals of the first and second kind. As a result, one has [7.9]:

$$w_{1,2} = \exp(-\delta) , \quad \delta \equiv \frac{2n}{\sqrt{1+\beta^2}}D\left(\frac{1}{\sqrt{1+\beta^2}}\right) . \tag{7.23}$$

It is seen that the magnitude of $w_{1,2}$ is exponentially small according to the multiphoton parameter $n = \omega/\omega_0 \gg 1$.

To obtain the transition probability in a unit time $W_{1,2}$, it is necessary to consider the contribution of all points t_k from (7.21). The total transition amplitude $A_{1,2}$ accumulates separate amplitudes a_k arising from each point t_k. When passing from one point t_k to another the additional phase factor $\exp(iS)$ appears each time in the amplitude, where

$$S = n \int_0^\pi \sqrt{1+\beta^2 \sin^2 \varphi}\, d\varphi = n\sqrt{1+\beta^2}\, E\left(\frac{\beta}{\sqrt{1+\beta^2}}\right) \tag{7.24}$$

is the accumulation of the action between neighboring points t_k.

At the time interval T the contribution to the probability will be given by $N = \omega T/\pi$ points t_k. Summing up the amplitudes from these N points and squaring the modulus $A_{1,2}$ one finds

$$|A_{1,2}|^2 = w_{1,2}\frac{\sin^2 [N(s-\pi)/2]}{\sin^2 [(s-\pi)/2]} \ . \tag{7.25}$$

Directing $N \to \infty$, one finds for the transition probability in a time unit $W_{1,2} = |A_{1,2}|^2/T$ [7.9]:

$$W_{1,2} = \frac{2\omega^2}{\pi} w_{1,2}\delta\left[\frac{\omega_0}{\pi}\int_0^\pi \sqrt{1+\beta^2 \sin^2 \varphi}\, d\varphi - K\omega\right] , \tag{7.26}$$

where $K = 2m+1$ is an odd integer number designating the harmonics at which the absorption is allowed according to dipole selection rules. It is seen from (7.26) that the shift of energy levels caused by the external field F_0 (Stark effect) enters into the energy conservation law.

For small field strengths $K\beta^2 \ll 1$, the perturbation theory transition probability is obtained from (7.26)

$$W_{1,2}^0 = \frac{2\omega^2}{\pi}\left(\frac{e\beta}{4}\right)^{2K} \delta(\varepsilon - K\omega) \ . \tag{7.27}$$

Equation (7.27) coincides with (7.20) for $K \gg 1$. To ensure this, you use the Stirling formula for factorials in (7.20). It is of interest that the adiabatic formula of the WKB approximation produces good accuracy even for $K = 1, 3, 5$, where its ratio to the precise solution is equal to 0.75, 0.89 and 0.95, accordingly. The dependence of the probability on the field strength parameter β is conveniently described by the value

$$\mathscr{L} = \frac{1}{2K}\ln\left(\frac{\pi W_{1,2}}{2\omega^2}\right) \ . \tag{7.28}$$

The dependence of $\mathscr{L}(\beta)$ on β is shown in Fig. 7.3, taken from [7.9].

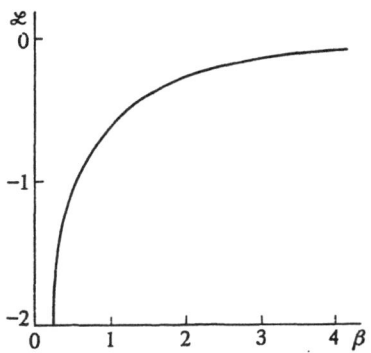

Fig. 7.3. The universal dependence of the transition probability logarithm $\mathscr{L}$ on parameter β [7.9]

7.5 The Quasi-energy Spectrum of a Two-Level System. Intensities of Satellites

The concept of quasi-energy is convenient for the interpretation of the spectra of atomic systems experiencing the influence of strong oscillating perturbations. As a rule, these perturbations substantially intermix neighboring atomic levels, for example, levels belonging to the same quantum number n. What is usually observed is the spontaneous radiation spectra from the system of such "mixed" sublevels to some lower lying, usually unperturbed state. The classic objects for the observation of such spectra in plasmas are the transitions between the levels $n = 4$ and $n = 2$ of helium. The close levels 4^1D, 4^1F of helium are mixed by a low frequency field of plasma (Langmuir) oscillations. The initially forbidden transition $4^1F - 2^1P$ becomes allowed due to this mixing. Therefore, near the helium forbidden line $4^1F - 2^1P$ one observes plasma satellites shifted from the frequency of the forbidden transition at a value of $\pm n\omega_{pe}$ where ω_{pe} is the frequency of plasma oscillations. For the first time the possibility of plasma satellite observation near the forbidden helium lines was substantiated by *Baranger* and *Mozer* [7.10]. However, their calculation was based on perturbation theory according to which the satellite intensities sharply decrease with the increase of their number n (as it was for *Blokhintsev* satellites in Sect. 7.2). Strong E.M. fields generated both by intense plasma oscillations and external sources (by lasers and high frequency generators) are of great interest for modern plasma physics investigations. Under these conditions the satellite spectra become drastically complicated, their intensity becoming the complex function of the field strength F_0 and frequency ω of the external E.M. field.

At first we shall find the spectrum of the quasi energy ε for a two-level system experiencing the influence of a low frequency (in general, nonresonant) perturbation $\omega_0/\omega \gg 1$. For this purpose, following [7.11] let us use the quasi-energy definition (7.18) and substitute instead of $\psi_\varepsilon(t)$ its expansion (7.19) according to the adiabatic wave functions. To determine the quasi-energy it is necessary to know the matrix $\hat{F}$, which connects the coefficients $b(t)$ during the translation t at π/ω. Thus, if matrix $\hat{F}$ is a bound matrix

$$\begin{bmatrix} b_1(t+\pi/\omega) \\ b_2(t+\pi/\omega) \end{bmatrix} = \hat{F} \begin{bmatrix} b_1(t) \\ b_2(t) \end{bmatrix} , \tag{7.29}$$

then the value of ε is determined from the solubility condition for the system of equations

$$\left.\begin{aligned} [F_{11}b_1(t)+F_{12}b_2(t)]\, e^{iS/2} &= \pm e^{-i\pi\varepsilon/\omega} b_1(t) , \\ [F_{21}b_1(t)+F_{22}b_2(t)]\, e^{-iS/2} &= \pm e^{-i\pi\varepsilon/\omega} b_2(t) . \end{aligned}\right\} \tag{7.30}$$

This is the consequence of the quasi-energy definition (7.1) and the expansion (7.19) on the adiabatic basis.

The matrix elements F_{ik} are obviously connected to the probabilities of the transitions between adiabatic states. The probability $w_{1,2}$ for the resonance case ($\omega_0/\omega = n$) were calculated in Chap. 7.4 (7.26). Analogous calculations with regard to state phases give [7.11]:

$$\left.\begin{aligned} F_{11} &= F_{22}^* = (1-e^{-\delta})^{1/2} \exp[-i\varphi(\delta)] , \\ F_{12} &= -F_{21}^* = \exp(-\delta-iS) . \end{aligned}\right\} \tag{7.31}$$

Formulas (7.23, 24) after substituting ω/ω_0 for n and phase $\varphi(\delta)$ become equal to

$$\varphi(\delta) = \frac{\delta}{\pi} - \frac{\delta}{\pi} \ln \frac{\delta}{\pi} + \arg \Gamma\left(i\frac{\delta}{\pi}\right) + \frac{\pi}{4} . \tag{7.32}$$

According to (7.30), the squares of the modulus $|F_{12}|^2$ and $|F_{11}|^2 = |F_{22}|^2$ determined for $\omega_0/\omega = n$ the probabilities of undergoing a transition into another state, or remaining in each of the states.

Assuming the condition for solving the system (7.30), equating its determinant to zero and using the explicit form (7.31) of the matrix F_{ik}, one finds [7.11]:

$$\sin \pi \frac{\varepsilon}{\omega} = \pm(1-e^{-2\delta})^{1/2} \sin\left[\frac{S}{2} - \varphi(\delta)\right] . \tag{7.33}$$

For fields that are not too strong when the magnitude $\delta \gg 1$ and the function φ takes the form: $\varphi(\delta) \approx -\pi/12\delta \to 0$, the magnitude ε coinciding with the function $S(\beta)$ far from the resonances, where $S \neq (2n+1)\pi$. The result cor-

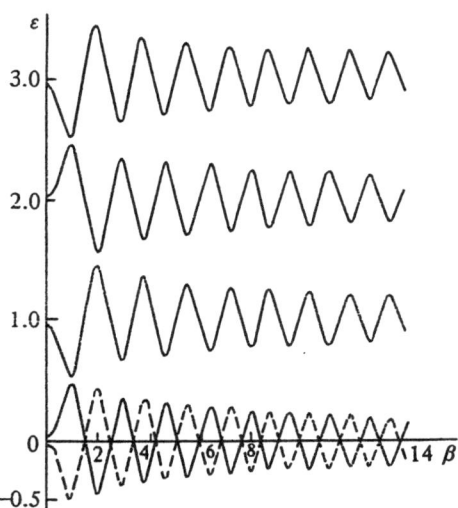

Fig. 7.4. The dependence of the two-level system quasi-energy ε on the reduced field strength β for $\omega_0 = 4.098$. Wide lines are the quasi-energy branches reduced to the first Brillouin zone [7.11]

responds to the adiabatic limit where the coefficients $b_{1,2}$ in (7.30) are constant. Near the resonances, factors $\exp(-\delta)$ result in a splitting of the quasi-energetic levels and phase $\varphi(\delta)$, resulting in the displacement of the resonance location with regard to its adiabatic limit. For strong fields when $\delta \ll 1$, the function $\varphi(\delta)$ tends to $-\pi/4$, and the quasi energy as a function of field strength oscillates (parameter β) and tends to zero:

$$\varepsilon \approx \pm \left(\frac{\omega_0}{2\pi d_{12}F_0}\right)^{1/2} \sin\left(\frac{2d_{12}F_0}{\omega} + \frac{\pi}{4}\right) . \tag{7.34}$$

The dependence of the quasi energy ε/ω on the field β is shown in Fig. 7.4 taken from [7.11]. For $\beta \ll 1$ the dependence is conditioned by the quadratic Stark effect. With the increase of β, multiquantum resonances are observed and their shapes and locations are determined by the parameter δ. The transition to the asymptotic formula (7.23) takes place for a further increase of β.

Let us write down the wave function of the system with the help of the expansion in terms of the satellites of initial states $|1\rangle$ and $|2\rangle$:

$$\Psi_\varepsilon(\tau) = e^{-i\varepsilon\omega t} \sum_{n=-\infty}^{\infty} (P_n^\varepsilon |1\rangle + Q_n^\varepsilon |2\rangle) e^{-in\omega t} . \tag{7.35}$$

The coefficients P_n^ε, Q_n^ε defining the satellite intensities are projections of the function $\psi_\varepsilon(\tau)$ on these states. With regard to the symmetry properties of the system (7.30), the coefficients take the form:

$$P_n^\varepsilon = \frac{1}{2\pi}(1\pm\cos n\pi)\int_0^\pi \langle 1| \Psi_\varepsilon(\tau)\rangle \exp{(i\varepsilon\tau+in\tau)}d\tau \ , \tag{7.36}$$

$$Q_n^\varepsilon = \frac{1}{2\pi}(1\pm\cos n\pi)\int_0^\pi \langle 2| \Psi_\varepsilon(\tau)\rangle \exp{(i\varepsilon\tau+in\tau)}d\tau \ , \tag{7.37}$$

where the signs $\pm$ correspond to different signs of quasi energies in (7.18). It is seen that there are either odd or even satellites in the spectrum. The function (7.35) describes the pair of neighboring states mixed by a strong, low-frequency oscillating perturbation. The observed spectrum of spontaneous transitions to a lower state with a frequency ω and energy E is easily found by perturbation theory:

$$I(\omega_f)\propto \sum_n |Q_n^\varepsilon|^2\delta\left(\varepsilon+n+\frac{E_0-\hbar\omega_f}{\hbar\omega}\right) . \tag{7.38}$$

To be more definite we assume here that the dipole transitions to the lower state $|0\rangle$ are allowed from the state $|2\rangle$ only, which corresponds to the afore-mentioned formulation of the forbidden line satellites problem. As seen from (7.35) the spectrum of satellites consists of a series of equally spaced lines corresponding to different quasi energies. It is convenient to present the satellite intensities in terms of the adiabatic expansion (7.35) for the wave function $\psi^\varepsilon(\tau)$ in the form

$$P_n^\varepsilon = \frac{1}{2}(1\pm\cos n\pi)[A(p)+B(p)] \ , \tag{7.39}$$

$$Q_n^\varepsilon = \frac{1}{2}(1\pm\cos n\pi)[-C(p)+D(p)] \ , \tag{7.40}$$

where $p = \varepsilon+n$, and

$$A(p) = \pi^{-1}\int_0^\pi d\tau b_1(\tau)\cos{[\chi(\tau)/2]}\exp\left[i\int_0^\tau \Omega(\tau)d\tau+ip\tau\right] , \tag{7.41}$$

$$B(p) = \pi^{-1}\int_0^\pi d\tau b_2(\tau)\sin{[\chi(\tau)/2]}\exp\left[-i\int_0^\tau \Omega(\tau)d\tau+ip\tau\right] , \tag{7.42}$$

$$C(p) = \pi^{-1}\int_0^\pi d\tau b_1(\tau)\sin{[\chi(\tau)/2]}\exp\left[i\int_0^\tau \Omega(\tau)d\tau+ip\tau\right] , \tag{7.43}$$

$$D(p) = \pi^{-1} \int_0^\pi d\tau \, b_2(\tau) \cos \left[\chi(\tau)/2\right] \exp\left[-i \int_0^\tau \Omega(\tau) d\tau + ip\tau \right] . \qquad (7.44)$$

Formulas (7.39−44) make it possible to give analytic estimations for the satellite intensities using the large value of the adiabaticity parameter $\omega_0/\omega \gg 1$. Indeed, the main contribution from the integrals (7.36,37) is effected by points of constant phase t_k, where

$$\frac{\omega_0}{\omega} \frac{1}{2} (1 + \beta^2 \sin^2 \omega t_k)^{1/2} = p . \qquad (7.45)$$

The dependence of the instantaneous adiabatic system levels on the parameter n is shown in Fig. 7.5 from [7.11]. The main contribution to the radiation of the satellite with a given value of n is given by the points where the satellite location crosses the instantaneous level location (points τ_1 and τ_2 in Fig. 7.5), as in Sect. 7.2. The intensity magnitude is determined by the relative time during which the system remains near these points. Since the system spends most of the time near the maximum $\left(\dfrac{\hbar \omega_0}{2}(1+\beta^2)^{1/2}\right)$ and minimum $(\hbar \omega_0/2)$ energy values, the greatest intensity corresponds to those satellites whose frequencies coincide with the extrema.

A simple estimation of the satellite intensity with the help of the relative time of location $|d\Omega/dt|^{-1}$ of the adiabatic energy at the given value $\Omega(t_k) = p$ gives [compare with (7.7)]:

$$I_p \propto |d\Omega/dt|^{-1}_{t_k} \propto \frac{p}{\omega_0/\omega} \left(\frac{4p^2}{(\omega_0/\omega)^2} - 1\right)^{-1/2} \left(1 + \beta - \frac{4p^2}{(\omega_0/\omega)^2}\right)^{-1/2} , \qquad (7.46)$$

where $p = n\omega_0 + \varepsilon$.

As has been pointed out, (7.46) shows that the most intensive satellites must be for $p \sim \omega_0/2$ and for $p \sim \omega_0/2(1+\beta^2)^{1/2}$. When $\beta \Rightarrow 0$ only the second

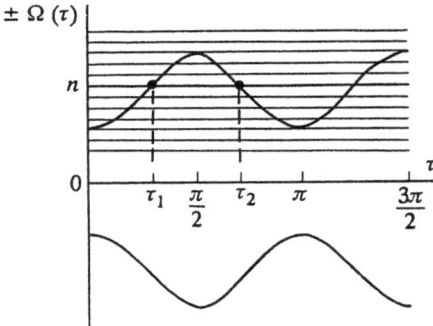

Fig. 7.5. The temporal dependence of the adiabatic levels of the system. Horizontal lines mark the position of satellites [7.11]

maximum is taken into account and (7.46) makes the transition to the static distribution (7.7) for *Blokhintsev* satellites.

Expression (7.46) describes only an approximate shape of satellites with no regard for the details of their distribution according to the intensities inside of this shape. A more detailed investigation of the satellite intensity dependence on their numbers was performed in [7.11] on the basis of a detailed analysis of the special points t_k in a complex t plane.

The detailed satellite structure depends on how its number $p = \varepsilon + n\omega$ correlates with the maximum and minimum values of the adiabatic energy. The general formula for the satellite distribution according to their intensities has the form[1]

$$|D(p)|^2 \approx |b_2|^2 \frac{4}{|p^2 - \omega_0^2/4|^{1/2}} \frac{(p + \omega_0/2)}{|p^2 - \omega_0^2(1 + \beta^2)/4|^{1/2}} Q \mathrm{Ai}^2(\pm Q^2) ,$$
(7.47)

$$|B(p)|^2 \approx |b_2|^2 \frac{4}{|p^2 - \omega_0^2/4|^{1/2}} \frac{|p - \omega_0/2|}{|p^2 - \omega_0^2(1 + \beta^2)/4|^{1/2}} \cdot \frac{1}{Q} \mathrm{Ai}'^2(\pm Q^2) ,$$
(7.48)

where Ai and Ai' are Airy functions and their derivatives [7.12].

The arguments Q in the Airy functions in (7.48) assume different values depending on the domain of the parameters under consideration. Their magnitudes are determined by the phases running between the different points t_k and have the following forms [7.11]:

$$\frac{2}{3} Q_1^3 = p \operatorname{arch} \varrho - \frac{\omega_0}{2} \int_0^{\operatorname{arch} \varrho} dz (1 + \beta^2 \cosh^2 z)^{1/2} , \quad p > \frac{\omega_0}{2} (1 + \beta^2)^{1/2} , \quad (7.49)$$

$$\left. \begin{aligned} \frac{2}{3} Q_2^3 &= \frac{\omega_0}{2} \int_0^{\arccos \varrho} dz (1 + \beta^2 \cos^2 z)^{1/2} - p \arccos \varrho \\ \frac{2}{3} Q_3^3 &= -\frac{\omega_0}{2} \int_0^{\arcsin \varrho} dz (1 + \beta^2 \sin^2 z)^{1/2} + p \arcsin \varrho \end{aligned} \right\} \frac{\omega_0}{2} < p < \frac{\omega_0}{2} (1 + \beta^2)^{1/2} ,$$
(7.50)

$$\frac{2}{3} Q_4^2 = \frac{\omega_0}{2} \int_0^{\operatorname{arsh} \varrho} dz (1 - \beta^2 \sinh^2 z) - p \operatorname{arsh} \varrho , \quad 0 < p < \frac{\omega_0}{2} , \quad (7.51)$$

e

$$\varrho = \left| \frac{p^2 - \omega_0^2/4}{\omega_0^2 \beta^2/4} \right|^{1/2} .$$
(7.52)

---

[1] Coefficients $A(p)$ and $C(p)$ are exponentially small for $p > 0$.

The values Q_1 and Q_4 are used in the Airy functions arguments (7.48) with positive signs, whereas Q_2 and Q_3 have negative signs. The selection of one out of two values in the range $\omega_0/2 < p < \omega_0(1+\beta^2)^{1/2}/2$ depends on the realization of the following conditions:

$$\left| \frac{\omega_0}{2} \int\limits_0^{\arcsin \varrho} dz \, (1+\beta^2 \sin^2 z)^{1/2} - p \arcsin \varrho \right| \gg 1 - \text{phase } Q_2 \,, \qquad (7.53)$$

$$\left| \frac{\omega_0}{2} \int\limits_{\arcsin \varrho}^{\pi/2} dz (1+\beta^2 \sin^2 z)^{1/2} - p \left(\frac{\pi}{2} - \arcsin \varrho \right) \right| \gg 1 - \text{phase } Q_3 \,. \qquad (7.54)$$

The functions Q_i may be expressed at will in terms of total elliptic integrals [7.12].

Thus, the general expressions for satellite intensities (7.47, 48) contain factors analogous to a static shape (7.46) multiplied by the Airy functions with complicated arguments Q taking into account the intensity variations under the shape. Although the arguments Q_i are complicated enough, one must remember these formulas give the intensity distribution for arbitrary external field strengths F_0 and they are limited by the adiabaticity condition $\omega_0/\omega \gg 1$ only.

For $\beta \to 0$ the results of perturbation theory are obtained from (7.38). For example, for $p > \omega_0(1+\beta^2)^{1/2}/2$ one obtains:

$$|D(p)|^2 \approx \frac{1}{\pi} \frac{1}{2p-\omega_0} \left[\frac{\beta^2}{16} \frac{p+\omega_0/2}{p-\omega_0/2} \right]^{p-\omega_0/2} \left(\frac{\omega_0}{p+\omega_0/2} \right)^p \,, \qquad (7.55)$$

that is, we have a power decrease of the satellite intensity while its number increases.

For $\beta \to \infty$ (strong fields), the *Blokhintsev* results are obtained from (7.47, 48). In fact, for $p \sim \omega_0 \beta/2$, it is easy to get from (7.49–52):

$$|Q_n^\varepsilon|^2 \approx 2^{4/3} (\omega_0 \beta)^{-2/3} \, \text{Ai}^2 \left[2^{2/3} \frac{(p-\omega_0 \beta/2)}{\sqrt{\omega_0 \beta}} \right] \,, \qquad (7.56)$$

which coincides with the asymptote of the *Blokhintsev* spectrum $|Q_n^\varepsilon|^2 \propto I_n^2$ $(\omega_0 q/2)$ for $n \sim \omega_0 q/2 \gg 1$. It is only natural because in a strong field F_0 it is possible to consider a two-level system as degenerated and therefore exhibiting a linear Stark effect, as in the *Blokhintsev* case.

7.6 Highly Excited Atom in a Low Frequency, Nonresonant Electric Field. Quasiclassical Solution

The consideration in Sects. 7.2–5 was generally limited to cases of a strong E.M. field influencing the levels with the same value of n. For highly excited atomic levels $n \gg 1$ the external perturbation frequency ω may be of a comparable value or even dominate over the distance between levels approximately as Ry n^{-3}. Experimental investigations [7.13] of the transitions between hydrogen atomic levels $n_i = 10$ and $n_f = 50$ in the presence of a strong microwave field together with the main resonance at the frequency $\omega_i - \omega_f$ have shown a large number of satellite resonances. These satellites correspond to the transition from "i" to "f" with the simultaneous emission or absorption of "k" low frequency photons. We consider transitions between the system of discrete quasi-energy levels belonging to the upper or lower atomic levels. To calculate the intensities of such transitions we shall utilize the theory developed by *Berson* [7.14].

Highly excited atomic levels with $n \gg 1$ are hydrogen-like and the electron wave function ψ of the electron moves simultaneously in a Coulomb field r^{-1} and in field $F_0 \sin \omega t$ of linearly polarized, low frequency radiation with the frequency value ω. Let us expand the ψ function according to the spherical functions $Y_{lm}(\Theta, \varphi)$ as well as in a Fourier series according to radiation field harmonics:

$$\Psi = \sum_{N=-\infty}^{+\infty} \sum_{l=|M|}^{\infty} r^{-1} f_{Nl}(r) Y_{lM}(\Theta, \varphi) \exp\left[-it(E+N\omega)\right] . \tag{7.57}$$

The radial functions $f_{Nl}(r)$ satisfy the system equation:

$$[d^2/dr^2 + K_{Nl}^2(r)]f_{Nl}(r) = -F_0 r \{[(l^2-M^2)/(2l-1)(2l+1)]^{1/2}$$

$$\times [f_{N-1,l-1}(r) + f_{N+1,l-1}(r)] + [((l+1)^2-M^2)/(2l+1)(2l+3)]^{1/2}$$

$$\times [f_{N-1,l+1}(r) + f_{N+1,l+1}(r)]\} , \tag{7.58}$$

where

$$K_{Nl}^2(r) = 2[E+N\omega+1/r-l(l+1)/2r^2] . \tag{7.59}$$

Here E is the electron's energy, M is the magnetic quantum number defining the orbital momentum projection on the field direction F.

We shall search for a solution of the system (7.58) in the form of quasiclassical wave functions:

$$f_{Nl}(r) = -iK_{Nl}^{-1/2}/2\{a_{Nl}^+(r) \exp[i(S_{Nl}(r) + \pi/4)]$$

$$- a_{Nl}^-(r) \exp[-i(S_{Nl}(r) + \pi/4)]\} , \tag{7.60}$$

where

$$S_{Nl}(r) = \int_{r_1}^{r} dr' K_{Nl}(r') ,$$

and r_1 is the turning point.

Let us suppose following [7.14] that the oscillating field weakly influences the electron motion along the classical trajectory in a Coulomb field. This implies the following: first, the orbital momentum l may be taken equal at some average value $L \gg 1$ in all of the channels. Second, item $N\omega$ in (7.59) must be small compared with at least one of the remaining members. In this action S_{Nl} takes an approximate form:

$$S_{Nl} \approx S_0 + N\omega\tau - m\varphi(\tau) , \quad m = l - L \ll L , \tag{7.61}$$

where $S_0 \gg 1$ is the part of the action identical for all the channels conditioned by the classical motion along the trajectory, $\varphi(\tau)$ is the classical angle depending on time τ of the motion along the trajectory; later on cases $\tau > 0$ or $\tau < 0$ will be marked by $+$ or $-$ signs, accordingly.

Substituting (7.60) into (7.58), retaining members containing the small difference $S - S'$, and using the afore mentioned simplifications, one finds [7.14]:

$$i\, da_{Nm}(\tau)/d\tau = -\tfrac{1}{4} F_0 r(\tau)(1 - M^2/L^2)^{1/2} \sum \exp\,[i(N'-N)\omega\tau$$
$$+ i(m-m')\varphi(\tau)]\, a_{N'm'}(\tau) . \tag{7.62}$$

The elliptical trajectory parameters entering into (7.62) are connected with the time variable τ by the usual parametric relations:

$$\left.\begin{array}{l} r = v_0^2(1 - \varepsilon \cos u) , \quad \tau = v_0^3(u - \varepsilon \sin u) , \\[2mm] \varphi = 2\arctan\,[(1+\varepsilon)^{1/2}/(1-\varepsilon)^{1/2} \tan u/2] , \quad \varepsilon = (1 - L^2/v_0^2)^{1/2} \\[2mm] v_0 = (-2E)^{-1/2} , \quad -\pi \le u \le \pi . \end{array}\right\} \tag{7.63}$$

It is necessary to supplement (7.62) with boundary conditions

$$\left.\begin{array}{l} a_{Nm}(-\pi)\exp\,(-2i\pi v_N) = a_{Nm}(\pi) , \\[2mm] v_N = (-2E - 2N\omega)^{-1/2} , \end{array}\right\} \tag{7.64}$$

which provide an exponential damping of the radial functions in the forbidden domains of the motion.

Equation (7.62) describes transitions in the system of equidistant levels, the matrix elements of which are dependent on quantum number differences $N-N'$ and $m-m'$ only. Such a system is precisely solved by the generation function method [7.12]. Its general solution has the form:

$$\left.\begin{array}{l} a_{Ns}(u) = R \displaystyle\sum_{N's'=-\infty}^{\infty} b_{N's'} J_{N-N'+s-s'}(h_-(u)) J_{s'-s}(h_+(u)) \\[4mm] \qquad\qquad \times \exp\,[i\gamma_-(u)(N-N'+s-s') + i\gamma_+(u)(s'-s)] , \\[4mm] h_\pm(u) = [C_\pm^2(u) + d_\pm^2(u)]^{1/2} , \quad \tan\gamma_\pm(u) = C_\pm(u)/d_\pm(u) , \end{array}\right\} \tag{7.65}$$

$$C_{\pm}(u) = [F_0 v_0^3 (1 - M^2/L^2)^{1/2}/2]$$

$$\times \int_{-\pi}^{u} du' r(u')(1 - \varepsilon \cos u') \cos [\omega \tau(u') \pm \varphi(u')] ,$$

$$d_{\pm}(u) = [F_0 v_0^3 (1 - M^2/L^2)^{1/2}/2] \qquad (7.66)$$

$$\times \int_{-\pi}^{u} du' r(u')(1 - \varepsilon \cos u') \sin [\omega \tau(u') \pm \varphi(u')] ,$$

$$m = N + 2s .$$

Here R is a normalized factor, arbitrary constants b_{Ns} must-be determined for the boundary condition (7.64). In point $\tau = -\pi$ the arguments of the Bessel function in (7.65) turn to zero so that the only remaining members are those with nonzero indices which leads to:

$$a_{Ns}(-\pi) = R b_{Ns} . \qquad (7.67)$$

Substituting (7.67) into the boundary condition (7.64) one obtains a system of homogeneous differential equations for the coefficients b_{Ns}

$$b_{Ns} \exp [-2i\pi(v_0 + N\omega v_0^3)] =$$

$$\sum_{N's' = -\infty}^{\infty} i^{N-N'} J_{N-N'+s-s'}[C_-(\pi)] J_{s'-s}[C_+(\pi)] b_{N's'} , \qquad (7.68)$$

where phase v_N in (7.62) is replaced by the first members of the series expansion according to $N\omega$.

Let us find a solution to the system (7.68) in the form:

$$b_{Ns} = J_{N+s}(g_-) J_s(g_-) \exp [i(N\varphi_1 + S\varphi_2)] . \qquad (7.69)$$

Substituting (7.69) into (7.68) one finds that the system is satisfied for

$$v_0 = n, \ \varphi_1 = \pi(1 + \omega n^3), \ \varphi_2 = \pi, \ g_{\pm} = \frac{C_{\pm}(\pi)}{2 \sin (\pi \omega n^3)} , \qquad (7.70)$$

where n is a principal quantum number of the level.

Substituting coefficients b_{Ns} from (7.69) into (7.68) and using the addition theorem for the Bessel functions one finally finds [7.14]

$$a_{Ns}(u) = R J_{N+s}[w_-(u)] J_s[w_+(u)] \exp \{iN[\gamma_-(u) + \Psi_-(u)]$$

$$+ is[\gamma_-(u) - \gamma_+(u) + \Psi_-(u) - \Psi_+(u) + \pi]\} , \qquad (7.71)$$

$$w_\pm(u) = \{h_\pm^2(u) + g_\pm^2 - 2h_\pm(u)g_\pm \cos[\gamma_\pm(u) - \pi\omega n^3]\}^{1/2} ,$$

$$g_\pm \sin[\gamma_\pm(u) - \pi\omega n^3] = w_\pm \sin\Psi_\pm .$$

$$(7.72)$$

Thus, the formulas (7.71, 72) define the quasiclassical atomic wave function with large values of quantum numbers n, L, M in the radiation field. The energy levels in the limits of the approximation used are not shifted or splitted in the presence of the field. The only thing that appears is a system of quasi-energy levels spaced at $N\omega$. Frequency ω may be of the order of the distance between levels but must not become in resonance. The interaction with an E.M. field results in the mixing of the states with different orbital momenta, but on the average the states of orbital momenta experience little change.

With the help of the wave function (7.71, 72) one can find the spectrum of spontaneous emission for the transition between high (n_f) and lower (n_i) atomic levels in the field of low frequency emission with frequency ω. For reasons of simplicity, let us restrict ourselves to the case of the identical polarization of the spontaneous and external fields. The transition probability W_{if} is defined by the matrix element

$$W_{if} = -i \int_0^t d\tau \langle \Psi_{n_f L_f}^{M_f} | z e^{i\Omega t} | \Psi_{n_i L_i}^{M_i} \rangle ,$$

$$(7.73)$$

where Ω is the frequency of the spontaneous emission observed, and z is a component of the atomic radius vector.

Using the wave functions (7.71, 72) in (7.73), performing angular integrations and dividing by the total transition intensity we find the shape $I(\Omega)$ of the spontaneous emission between the highly excited levels of the atom immersed in the external low frequency E.M. field:

$$I(\Omega) \propto \sum_k \delta\left(\frac{1}{2n_i^2} - \frac{1}{2n_f^2} - \Omega + k\omega\right) W_k ,$$

$$(7.74)$$

$$W_k = (1 - M_i^2/L_i^2) |R_k^+ + R_k^-|^2 ,$$

$$(7.75)$$

where $R_k^\pm$ are the overlap integrals of the radial wave functions

$$R_k^\pm = \sum_{N_i, l_i} \int dr\, r f_{N_i+k, l_i\pm 1}(r) f_{N_i l_i}(r) .$$

$$(7.76)$$

This result indicates that the spectral structure consists of a system of satellites, the intensities of which are determined by the overlap integrals (7.76). To estimate the integrals $R_k^\pm$ let us use the fact that the observed transition frequency Ω is usually much higher than the distance between the atomic levels: $\Omega n_i^3 \gg 1$. Under this condition the main contribution to the transition probability is effected by strongly curved close parabolic trajectories for which the condition $L \ll n(\varepsilon \approx 1)$ is valid. Using these simplifications and making the transition to the integral according to the coupling parameter of the Coulomb trajectory, one finds

$$R_k^{\pm} = \tfrac{1}{4} \sum_{N_i l_i} \int d\tau' r(\tau') a_{N_i l_i}(v) a_{N_i + k, l_i \pm 1}(v) \exp\left[-i\Omega\tau' \pm i\varphi(\tau')\right] , \qquad (7.77)$$

$$r = L_i^2(1 + v^2)/2, \quad \tau' = L_i^3(v + v^3/3)/2, \quad \varphi = 2\arctan v . \qquad (7.78)$$

It is easy to see that the main contribution to the integrals (7.77) is given by the domain of small magnitudes of variable v so that functions $a_{N_i s_i}(v)$ may be replaced by their values at $v = 0$. With regard to (7.66) this gives

$$
\left.
\begin{aligned}
a_{N_i s_i}(0) &= (2/\pi n_i^3)^{1/2} J_{N_i + s_i}(w_i \cos\beta/2) J_{s_i}(w_i \cos\beta/2) \\
&\quad \times \exp\left[i(2N_i \gamma_i + s_i\pi)\right] , \\
w_i &= C_i \cot \pi\omega n_i^3 - d_i ,
\end{aligned}
\right\} \qquad (7.79)
$$

$$
\left.
\begin{aligned}
C_i &= -F_0 n_i^5 \int_0^{\pi} du\, (1 - \cos u)^2 \cos(\omega n_i^3)(u - \sin u) , \\
d_i &= F_0 n_i^5 \int_0^{\pi} du (1 - \cos u)^2 \sin(\omega n_i^3)(u - \sin u) , \\
\tan \gamma_i &= C_i/d_i, \quad \sin\beta = M_i/L_i .
\end{aligned}
\right\} \qquad (7.80)
$$

After carrying out the functions $a_{Nl}(v = 0)$ the integral (7.77) obviously results in to the Airy functions and their derivatives. Performing the summation in (7.77) with the help of the addition theorem for Bessel functions, one finds

$$
\left.
\begin{aligned}
R_k^+ &= P_+ J_{k+\kappa}(x \cos\beta/2) J_\kappa(x \cos\beta/2) \exp\left[ik(\Psi + 2\gamma_f) + i\kappa\pi\right] , \\
x &= [w_i^2 + w_f^2 - 2 w_i w_f \cos 2(\gamma_i + \gamma_f)]^{1/2}, \quad w_i \sin 2(\gamma_i + \gamma_f) = x \sin\Psi, \\
P_\pm &= (n_i n_f)^{-3/2}(4/\Omega^5)^{1/3}[-Ai'(y) \pm y^{1/2} Ai(y)] , \\
\kappa &= (L_i - L_f - k + 1)/2, \quad y = (\Omega L_i^3/2)^{2/3} .
\end{aligned}
\right\} \qquad (7.81)
$$

The radial integral R_k^- is obtained from (7.80) by the substitution of $\kappa - 1$ instead of κ and P_- instead of P_+. The magnitude κ takes number values only.

Further on it is necessary to substitute overlap integrals $R_n^{\pm}$ into (7.74) and to perform the averaging according to M_i and summing according to L_i. The first of them may be replaced by the integration by β from $-\pi/2$ to $+\pi/2$, and the second integral by integration according to the parameter y. The averaging mentioned is performed in an analytic form after which it follows

$$W_k(x) \propto \frac{1}{n_i^5 n_f^3} F_k(x), \quad F_k(x) = 2\int_0^1 dz \left(\frac{1}{3} + z^2\right) J_k^2(xz) . \tag{7.82}$$

Thus, the satellite intensity distribution $W_k(x)$ is expressed in terms of one universal function of parameter x (7.81). The limiting form of the function $F_k(x)$ for $x \gg k$ is

$$F_k(x) \approx [2\ln 2x - 2\psi(|k| + 1/2) + 3 - (-1)^k 4\cos(2x)/x + 0(x^{-2})]/3\pi x , \tag{7.83}$$

where $\psi(z)$ is the ψ function [7.12].

For small values of x (weak fields F_0) the satellite intensities are proportional to $x^{2|k|}$, which is only natural. With the increase of the field F_0 (parameter x) their intensity increases sharply and reaches, the maximum value for some $x_{max}(k)$ and then falls. As the value of k increases the maximum of the corresponding $F_k(x)$ becomes farther from zero. So, for $k = 1, 2 \ldots 7$, the values of $x_{max}(k)$ are equal to: 2.38, 3.9, 5.1, 6.3, 7.5, 8.7, 9.9 accordingly. After the maximum, $F_k(x)$ also displays weak oscillations.

The dependence of the universal parameter x on the quantum numbers n_i, n_f and frequency ω is quite complicated from (7.81). For $n_f^2 \gg n_i^2$ we have approximately:

$$x = C_f \cot(\pi\omega n_f^3) - d_f , \tag{7.84}$$

where the parameters C_f and d_f are defined above (7.80). The dependence of the relative satellite intensity with the numbers $k = -1$ and $n = 44$ on the power P of a microwave source with a frequency $\omega = 7.829$ MHz is shown in Fig. 7.6.

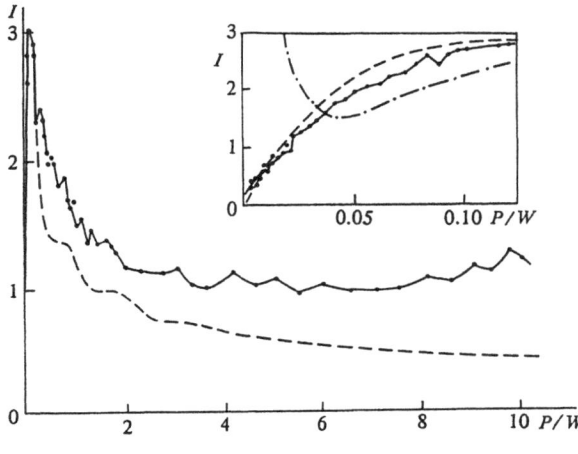

Fig. 7.6. Relative satellite intensity I for $k = -1$ and $n = 44$ vs the power P of the microwave source with the frequency $\omega = 7.829$ MHz. The domain of small power is shown in the insert. The dotted curve is the result of experiment [7.13]; the dashed curve is the function (7.82), the dash-dotted curve is the asymptotic (7.83) of the function (7.82) for large values of microwave source power [7.14]

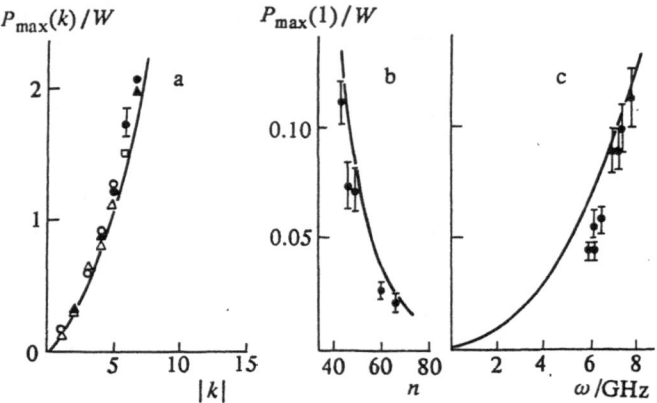

Fig. 7.7 a–c. The microwave source power $P_{max}(k)$ corresponding to the maximum of the excitation of the satellite with the number k as a function of parameters $|k|$ (**a**), n (**b**) and ω (**c**): ($n = 43$) and ($n = 44$ with $k<0$; $\Delta n = 44$) and ($n = 45$) with $k>0$ [7.14]

It is seen that the calculated curve (dotted curve) is in strict correspondence with the experimental curve (solid curve with dots).

In the domain of low frequencies $\omega n_f^3 \gg 1$, one has

$$x = 3F_0(n_f^2 - n_i^2)(1 + 35\,\omega^2 n_f^6/72)/2\omega \ . \tag{7.85}$$

Here one can establish a correspondence with the *Blokhintsev* theory (Chap. 7.2). Indeed, the parameter x in this case is equal in accordance with (7.85) to the ratio of the difference of the energies of the extreme Stark components to the perturbation frequency. Function $F_k(x)$ from (7.82) may be considered in the case as the intensity of a separate *Blokhintsev* satellite (determined by a squared Bessel function), summing according to all values of the differences of parabolic quantum numbers $n_1 - n_2$, that is averaging according to the parameter $z = |n_1 - n_2|/n$ within the limits $0 \le z \le 1$.

In conclusion, the dependence of the power $P_{max}(k)$ which corresponds to the maximum of the excitation probability of the kth satellite on the parameters $|k|$, n, ω is shown in Fig. 7.7, from [7.13]. The power is determined with the help of the maximum value $x_{max}(k)$ of the function $F_k(x)$ and equals:

$$P_{max}(k) = 1.2 \times 10^3\,\omega^2 x_{max}^2(k)/(n_f^2 - n_i^2)^2 \ , \tag{7.86}$$

where P is expressed in W and ω in GHz. In general, the calculated dependence is in good correlation with the experiments. Of course the model described does not cover all of the details of the dependencies cited. However it can represent their general features.

8. Atom in a Resonant Oscillating Electric Field. Simultaneous Influence of Constant and Oscillating Fields

8.1 Features of Resonance Conditions in Plasmas

As a rule the atoms in a plasma are under the simultaneous influence of both constant and oscillating electrical fields of different physical natures. Constant electrical fields F produced by plasma particles usually have a statistical distribution near some typical (average) value F_0. As far as oscillating electrical fields are concerned, as a rule they are due to collective plasma oscillations and have typical frequencies ω_p connected with the oscillations. As stated above, all of these frequencies are small when compared with the frequencies of transitions between different levels. Inside the atomic level with a given n these frequencies of electronic oscillations may be (as was investigated in detail in Chap. 7) rather close to the transition frequency between l-sublevels. For degenerate hydrogen-like atomic states the problem of whether the atom perceives the oscillation with frequency ω_p as static, or on the contrary as a varying oscillation, is solved by comparison of the frequency of the Stark splitting $\Delta\omega_s \sim n^2 F_0/\hbar$ in the field F_0 with frequency ω_p of the oscillation. In essence, this parameter coincides with the parameter figuring in a *Blokhintsev* spectrum (see Chap. 7.2).

The magnitude of the parameter for plasma oscillations is compared with unity (or less) for the electronic component only, and the ion oscillations are perceived by the atom as static. We should especially note the periodic oscillations connected with the influence of nonresonant, low frequency laser radiation on a hydrogen atom in plasmas. Here there may be different situations since the laser field strength $\mathscr{E}_0$ for a given frequency ω may charge within rather large limits.

Statistical properties of electronic plasma oscillations are quite diverse. We shall consider two types of oscillations: 1) oscillations of a broad spectral composition with comparatively small amplitude $\mathscr{E}_0$, and 2) oscillations with a large amplitude but with a comparatively narrow spectral composition. In the first case, the influence of oscillations on the atom is analogous to the influence of fast electron collisions which can be taken into account within the framework of perturbation theory. A parameter of the perturbation theory is the ratio of the atomic interaction $d\mathscr{E}_0$ with the field $\mathscr{E}_0$ to the oscillation spectrum width. In the second case, the influence of large amplitude fields $\mathscr{E}_0$ results in strong induced oscillations of atomic state amplitudes, the evolution of which are analogous to their evolution in the field of powerful laser radiation. Therefore, both latter cases may be considered on one and the same basis.

The considerable magnitude of the amplitudes of $\mathscr{E}_0$ the collective electronic oscillations is often accompanied by their considerable anisotropy responding to a definite direction of plasma oscillations and the polarization of their field strength vectors. In this case the oscillation spectrum may be considered as approximately one dimensional. The distribution of ion oscillation electric fields is quite often one dimensional. Therefore, one can speak about a definite mutual orientation of a constant (ion) F and varying (electron) $\mathscr{E}(t)$ electric fields. The case of a laser field influence also responds to a definite orientation of the vector $\mathscr{E}(t)$.

As is already known, the monochromatic oscillation exerts a strong influence on the atom in resonance conditions. During conditions when the atom is already under the influence of a constant electric field the question is about the resonance between the frequencies of splitting in the static field and the frequency ω (or its harmonics $n\omega$) of the influencing monochromatic field. Since the frequency shifts in a constant field F possess some statistical distribution there are always ones which fall into resonance with the monochromatic field. Below we shall basically consider hydrogen atom frequency shifts corresponding to the linear Stark effect $\Delta\omega_s = CF$ (C is a Stark constant). In this case, the resonance condition $\Delta\omega_s = CF = \omega$ is analogous to that of Doppler broadening $\Delta\omega = k \cdot v$ (v is the velocity, k is the wave vector). However, an important difference between Stark broadening and Doppler broadening is the fact that the electric field defines the wave function of the atomic state. Therefore the oscillating electric field influences results in the evolution of the atomic state in crossed constant and varying electrical fields. This problem is much more complicated than for Doppler broadening.

The influence of small amplitude and broad spectral composition periodic fields on the atom is considered in Sect. 8.2. The hydrogen atom spectrum under simultaneous influences of static S and oscillating (dynamic D) fields for their different mutual orientations is analyzed in Sects. 8.3 − 5. Section 8.6 is devoted to questions of electronic motion stochastization in an external field periodic in time.

8.2 Action of Weak Oscillating Electric Fields of Broad Spectral Composition on the Atom

Following [8.1 − 3] let us consider the influence on a hydrogen atom being in a static field produced by low frequency (LF) ionic oscillations of the high frequency (HF) stochastic electric field produced by electron oscillations

$$E_e(t) = \sum_{j=1}^{J} E_j(t) \cos \left[\Omega_j t + \varphi_j(t) \right] , \tag{8.1}$$

where phase $\varphi_j(t)$ and $E_j(t)$ are random variables submitting to some Poisson process with the average variation frequency γ_e. Parameters φ_j and E_j are constant between random jumps and submit to a definite distribution law. The frequency γ_e of the phase and amplitude jumps determines for a given oscillation frequency Ω_j the typical wave packet width; below it is assumed to be large enough. The amplitude of HF noise is defined by average value $E_0 \equiv (\langle |E_e(t)|^2 \rangle)^{1/2}$ which is considered to be small compared with the average static field strength F_0. The typical oscillation frequency $dE_0/\hbar = T^{-1}$ of the atomic state in field E_0 is considered to be small compared with frequency γ_e of the phase jumps. These conditions ensure the applicability of perturbation theory over the field amplitude E_0 for the calculation of the atomic state evolution.

Let us consider the shape $I_{\alpha\beta}(\omega, F)$ formed by the spontaneous transitions between the Stark components $\alpha \to \beta$ of the atom being in both a constant field F and the dynamic field given by (8.1)

$$I_{\alpha,\beta}(\omega, F) = \frac{1}{\pi} \mathrm{Re} \left\{ \int_0^\infty d\tau \, e^{i(\omega - \omega_0 - \omega_{\alpha\beta})\tau} \sum_{\alpha'\beta'} d_{\alpha\beta} d_{\alpha'\beta'} \{[T_{\alpha\alpha'}(\tau, 0) T_{\beta\beta'}^*(\tau, 0)]\}_{\mathrm{Av}} \right\} . \tag{8.2}$$

Here ω, ω_0 are the observed and unperturbed frequencies of the transition, $\omega_{\alpha\beta}(F) = (d_{\alpha\alpha} - d_{\beta\beta})F/\hbar$ is the Stark frequency shift of the component under consideration ($d_{\alpha\alpha'}, d_{\beta\beta'}$ are the matrix elements of the dipole momentum), $T_{\alpha\alpha'}$ is the operator of the atomic state evolution under the influence of the HF field, the brackets $\{\ldots\}_{\mathrm{Av}}$ designate the averaging according to the random variables in (8.1).

Considering the splitting $\omega_{\alpha\beta}(F)$ to be large compared with other interactions one may limit one's self to the diagonal matrix elements of the operator T for every level. Since the results for two levels simply follow from the result for one level, let us at first find the evolution operator $T_{\alpha\alpha}(\tau)$ for the upper level. Considering the actual form $dE_e(t)$ of the interaction with the HF field, one has:

$$T_{\alpha\alpha}(\tau, 0) = P \exp \left[-i/\hbar \int_0^\tau dt \, d \cdot E_e(t) \right] , \tag{8.3}$$

where P is the time ordering operator.

In correspondence with the results mentioned above, let us write down each magnitude $\varphi_j(t)$, $E_j(t)$ in the form of the superposition of random jumps. So for $\varphi_j(t)$ one obtains ($\theta(x)$ is the θ-function):

$$\varphi_j(t) = \sum_{q=0}^{Q(t)} \varphi_{jq} \theta(t - T_{jq}) \theta(T_{jq+1} - t) , \quad \varphi_{jq} = \mathrm{const.} \tag{8.4}$$

where T_{jg} is the moment of the g-th phase jump for the amplitude of the j-th wave packet, $Q(t) \simeq \gamma_e t \gg 1$ is the number of such jumps during the time interval t.

For the calculation of (8.3) let us use second order perturbation theory (the first-order corrections tend to zero after the averaging procedure)

$$[T_{\alpha\alpha}(\tau,0)-1]^{(2)} = -\frac{1}{2\hbar^2} \sum_{\alpha'} \sum_{j,j'=1}^{J} \sum_{q=0}^{\gamma_e\tau} \int_{T_{jq}}^{T_{jq+1}} dt$$

$$\times \sum_{q'=0}^{q} \int_{T_{j'q'}}^{T_{j'q'+1}} dt'(d_{\alpha\alpha'}\cdot E_{jq})(d_{\alpha'\alpha}\cdot E_{j'q'}) \exp\left[i(\omega_{\alpha\alpha'}t+\omega_{\alpha'\alpha}t')\right]$$

$$\times [\cos(\Omega_j t - \Omega_{j'}t' + \varphi_{jq} - \varphi_{j'q'}) + \cos(\Omega_j t + \Omega_{j'}t' + \varphi_{jq} + \varphi_{j'q'})] \ . \tag{8.5}$$

In the summation according to q', the second term in square brackets tends to zero and in the first term the only terms that remain are those with $j'=j$ and $q=q'$. Considering that $\omega_{\alpha'\alpha} = -\omega_{\alpha\alpha'}$ and that for an isotropic distribution of HF noise $(d_{\alpha\alpha'}\cdot E_{jq})(d_{\alpha'\alpha}\cdot E_{jq}) = |d_{\alpha\alpha'}|^2 E_{jq}^2/3$ one obtains

$$[T_{\alpha\alpha}-1]^{(2)} = -(\hbar^{-2}/12) \sum_{\alpha'} \sum_{j=1}^{J} \sum_{q=0}^{\gamma_e\tau} |d_{\alpha\alpha'}|^2 E_{jq}^2 \{i\Delta T_{jq}/(\omega_{\alpha\alpha'}-\Omega_j)$$

$$+ (1-\exp[i(\omega_{\alpha\alpha'}-\Omega_j)\Delta T_{jq}])(\omega_{\alpha\alpha'}-\Omega_j)^{-2} + i\Delta T_{jq}/(\omega_{\alpha\alpha'}+\Omega_j)$$

$$+ (1-\exp[i(\omega_{\alpha\alpha'}+\Omega_j)\Delta T_{jq}])(\omega_{\alpha\alpha'}+\Omega_j)^{-2}\} \ . \tag{8.6}$$

Let us take into account that time intervals ΔT_{jq} between phase jumps are random values distributed with the probability density $\gamma_e \exp(-\gamma_e\Delta T_{jq})$. The amplitudes E_{jq} obey some statistical distribution (Rayleigh distribution) the width of which will be considered later to be narrow enough when compared with γ_e. Then the sum of the squared amplitudes according to realizations q of a random process can be replaced by an average (for a given mode j) squared amplitude $\langle E_j^2 \rangle$. The sum according to all modes j is then replaced by the summary amplitude E_0^2 of electron oscillations. After the averaging, the result (8.6) takes the form

$$[T_{\alpha\alpha}(\tau,0)-1]^{(2)} = \left\{-(E_0^2/12\hbar^2) \sum_{\alpha'} |d_{\alpha\alpha'}|^2 \int_{\Omega_1}^{\Omega_2} d\Omega P(\Omega)\right.$$

$$\left.\times [(\gamma_e-i(\omega_{\alpha\alpha'}-\Omega))^{-1} + (\gamma_e+i(\omega_{\alpha\alpha'}+\Omega))^{-1}]\right\} \tau \ . \tag{8.7}$$

It is evident that as in the case of impact broadening theory [8.1], the increase of the evolution operator is proportional to the selected time interval τ. Therefore the coefficient in front of τ in right hand part of (8.7) can naturally be called the impact broadening operator Φ during electron oscillations. Operator Φ has been obtained using second order perturbation theory. However, it can be shown that the same result is true when one takes into account all orders of perturbation theory. So, with regard to (8.7) the following equation is true of operator $T_{aa}(\tau,0)$:

$$\frac{d}{d\tau} T_{aa}(\tau,0) = \Phi_{aa} T_{aa}(\tau,0) \tag{8.8}$$

the solution of which is the simple exponent:

$$T_{aa}(\tau,0) = \exp\left[\Phi_{aa}(F)\tau\right] , \tag{8.9}$$

where the wave impact broadening operator $\Phi = \Gamma + iD$ determines the width $\Gamma_\alpha(F) = -\mathrm{Re}\{\Phi_{aa}\}$ and the shift $D_a = -\mathrm{Im}\,\Phi_{aa}(F)$ which both depend on the field F in a resonant manner:

$$\Gamma_\alpha = (E_0^2 \gamma_\mathrm{e}/12\hbar^2)\{2d_{aa}^2/(\gamma_\mathrm{e}^2+\Omega_\mathrm{e}^2)+(|d_{aa-1}|^2+|d_{aa+1}|^2)$$

$$\times([\gamma_\mathrm{e}^2+(\omega_F^a-\Omega_\mathrm{e})^2]^{-1}+[\gamma_\mathrm{e}^2+(\omega_F^a+\Omega_\mathrm{e})^2]^{-1})\} , \tag{8.10}$$

$$D_\alpha = E_0^2((|d_{aa-1}|^2-|d_{aa+1}|^2)/12\hbar^2)\{(\omega_F^a-\Omega_\mathrm{e})/[\gamma_\mathrm{e}^2+(\omega_F^a-\Omega_\mathrm{e})^2]$$

$$+(\omega_F^a+\Omega_\mathrm{e})/[\gamma_\mathrm{e}^2+(\omega_F^a+\Omega_\mathrm{e})^2]\} . \tag{8.11}$$

Here

$$|d_{aa}|^2 = 9e^2 a_0^2 n_a^2 (n_1-n_2)_a^2/4 ,$$

$$|d_{aa-1}|^2 - |d_{aa+1}|^2 = |d_{aa}|^2/(n_1-n_2)_a ,$$

$$|d_{aa-1}|^2 + |d_{aa+1}|^2 = |d_{aa}|^2 [n^2-(n_1-n_2)-m^2-1]_a/2(n_1-n_2)_a^2 . \tag{8.12}$$

The generalization of the formulas obtained in the case of simultaneous perturbations of the upper (α) and lower (β) levels is achieved by substituting in (8.10) (instead of $|(d_a)_{aa'}|^2$) the magnitude

$$|(d_a-d_b)_{aa'\beta\beta'}|^2 = |(d_a)_{aa'}|^2 \delta_{\beta\beta'} + |(d_b)_{\beta\beta'}|^2 \delta_{aa'} - 2(d_a)_{aa}(d_b)_{\beta\beta} . \tag{8.13}$$

As a result, one obtains

$$\Gamma_{\alpha\beta} = \Gamma_\alpha + \Gamma_\beta - (d_a)_{aa}(d_b)_{\beta\beta} E_0^2 \gamma_\mathrm{e}/3\hbar^2(\gamma^2+\Omega_\mathrm{e}^2) , \quad D_{\alpha\beta} = D_\alpha - D_\beta . \tag{8.14}$$

The resonant dependence of the impact line width $\Gamma_{\alpha,\beta}$ caused by the influence of stochastic oscillations (8.1) on the static field strength F results in the appearance of some features at the static line shape of spontaneous emission. Let us analyze the mechanism of this appearance, neglecting the lower level broadening (n_β). The static shape of the Stark component considered is obtained from (8.2) by the substitution of the evolution operator (8.9) and averaging the shape for a fixed F with some field distribution function $W(F)$. As a result one has:

$$I_{\alpha\beta}(\omega) \propto |d_{\alpha\beta}|^2 \frac{1}{\pi} \int_0^\infty dF \, W(F) \frac{\Gamma_\alpha(F)}{\Gamma_\alpha^2(F) + (\omega - \omega_0 - d_{\alpha\alpha}F)^2} \, . \tag{8.15}$$

Here we have used the resonant width (8.10) and have neglected the shift D_α as compared with the Stark shift $d_{\alpha\alpha}F$.

The shape (8.15) is a superposition of Lorentz-type shapes with a fixed value of F weighed with the function $W(F)$. However, the intensities of these Lorentzian line shapes near their center are different and are determined by the magnitude of the width $\Gamma_\alpha(F)$. Near those values F^* of the electric field where the resonant impact width is a maximum, the Lorentz profile exhibits a minimum intensity in the center and a maximum intensity in the wings. If frequency shift $\Delta\omega = d_{\alpha\alpha}F^* \equiv \omega_1$ responds to the same field value F^*, then near such a shift a gap of the intensity appears. Considering the resonant dependence of the impact width Γ_α (8.10) as well as the equidistant nature of Stark splitting, one finds $F^* = \hbar\Omega_e/\frac{3}{2}n_a$. Substituting this value of F^* into the formula for Stark splitting $d_{\alpha\alpha}F^*$ one finds the frequency shift value $\omega_1 = (n_1 - n_2)_a \Omega_e$ for which the intensity minimum appears at the static line shape.

Reasoning analogous to the preceeding discussion enables us to define a general expression for the frequencies ω_1, ω_2 of the gaps expected with regard to the broadening of both atomic levels. Here it is sufficient to take into account that the total frequency shift of the Stark component is the difference of shifts of each level n_a or n_b, and are gaps responding to the resonance of frequency Ω_e with equidistant splitting interval $\frac{3}{2}n_{a,b}F$ of every level. As a result, frequencies ω_1, ω_2 of the intensity gap locations are expected take the form [8.1]

$$\omega_1 = [(n_1 - n_2)_a - n_b(n_1 - n_2)_\beta/n_a]\Omega_e \tag{8.16}$$

$$\omega_2 = [n_a(n_1 - n_2)_a/n_b - (n_1 - n_2)_\beta]\Omega_e = n_a\omega_1/n_b \, . \tag{8.17}$$

The gap locations defined by (8.16, 17) are in good agreement with experimental data. As an example, the experimentally measured hydrogen line H_δ profile [8.2] is presented in Fig. 8.1, with the theoretical data (8.16, 17) marked, determining the gap locations from different Stark components. The upper and lower indices correspond to gaps which are created due to the splitting of upper $(n_\alpha = 6)$ and lower $(n_\beta = 2)$ levels.

As it is seen from Fig. 8.1, although the well developed theory predicts the gap locations, further estimations on its basis are difficult because they are

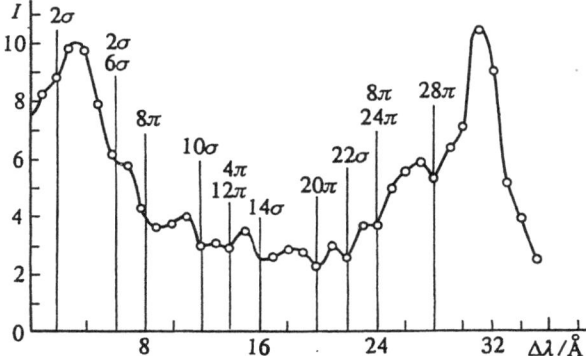

Fig. 8.1. The experimental profile [8.2] of the spectral line H_δ. Lines mark the positions of gaps due to the results of the nonadiabatic theory of HF noise-atom interaction. The Stark components related to each gap are indicated. Upper and lower indices relate to the resonances between the frequency of HF noise and Stark splitting of $n_\alpha = 6$ and $n_\beta = 2$ atomic levels

connected by the undetermined parameter γ_e, the value of which has to be extracted from additional suppositions. However, in many cases the appearance of features in the spectral line shapes is connected with strongly developed turbulence for which the magnitude E_0 of an oscillating field is large enough, so the phase φ_j in (8.1) may be considered as constant. In this case, the gap locations (8.16, 17) turn out to be invariable, however, its local width may be directly connected with the field magnitude E_0. For these purposes it is necessary to generalize the theory in the case of strong fields E_0, and Sects. 8.3–5 are devoted to these problems.

8.3 Hydrogen Atom in Static (S) and Strong Oscillating (Dynamic-D) Fields. Numerical Solutions for the Case When $S \perp D$

Let us consider the spectrum of atomic hydrogen immersed into a static electric field F and an oscillating (dynamic) field $\mathscr{E}_0 \cos \omega t$ perpendicular to it. As it is seen from the results of Chap. 7, the spectrum of a two-level system in an oscillating field is complicated. In the case under consideration the situation becomes more complicated due to the following facts. First, the system of line components split in a constant field is already not a two-level system. Second, the presence of an electric field F corresponds to a definite polarization (orientation) of the atomic state destroyed by the oscillating field $\mathscr{E}_0 \cos \omega t$.

In view of the difficulties mentioned, let us begin our consideration with the simplest case of level $n = 2$ and a mutually perpendicular orientation of the fields F and $\mathcal{E}_0 \cos \omega t$. The case of perpendicular fields is qualitatively as new one as compared with the parallel field case where the situation is reduced to the partial frequency shift of a *Blokhintsev* spectrum. In the treatment below, we adhere to the works [8.4, 5]. Let us direct the $0z$ axis of the coordinate system along the field F and the $0x$ axis along the field $\mathcal{E}_0$. The degenerate function system of level $n = 2$ represents the combination of a scalar function $|2s\rangle$ and a vector function $|2p\rangle$, the latter being a state with components $\{|2p_x\rangle, |2p_y\rangle, |2p_z\rangle\}$. For the given selection of the coordinate system, the state $|2p_y\rangle$ does not interact with other states, whereas the states $|2p_x\rangle, |2p_z\rangle$ are responsible for radiation of a definite polarization.

For brevity, we designate $|2s\rangle \equiv |1\rangle$, $|2p_z\rangle \equiv |2\rangle$ and $|2p_x\rangle \equiv |3\rangle$. Let us write down the nonzero components of the Hamiltonian

$$H_{12} = H_{21} = S, \ H_{13} = H_{31} = D \sin \omega t \ , \tag{8.18}$$

where designations are introduced for the static and dynamic component of the perturbation

$$S = F \cdot d, \ D = \mathcal{E}_0 d, \ d = e\langle 1|z|2\rangle = e\langle 1|x|3\rangle \ . \tag{8.19}$$

For the wave function

$$\Psi = \sum_j C_j(t) |j\rangle \tag{8.20}$$

the Schrödinger equation is equivalent to the system of equations:

$$i\dot{C}_j = \sum_k H_{jk} C_k \ . \tag{8.21}$$

Equation (8.21) is integrated [8.5] with the help of numerical methods for three types of initial conditions: $[C_1(0), C_2(0), C_3(0)] = [(100), (010), (001)]$. This selection of initial conditions allows determination of all matrix elements of the state evolution operator $\langle j|\hat{V}(t)|l\rangle$ with the initial conditions $V_{jl}(0) = \delta_{jl}$. The system (8.21) has periodic coefficients and its general solution takes the form

$$C_j(t) = P_j^0(t) + P_j^+(t) e^{i\varepsilon t} + P_j^- e^{-i\varepsilon t} \ , \tag{8.22}$$

where $P_j^\alpha (\alpha = +, \ 0, \ -)$ are the periodic functions of time with a period $T = 2\pi/\omega$. The magnitude ε has the sense of quasi-energy.

It is seen from (8.22) that the spectrum of the L_α line consists of a number of harmonics with frequency ω shifted from one another by the value ε. The evolution operator V_{jl} has three eigen-values $\exp(i\varepsilon\tau)$, 1, $\exp(-i\varepsilon\tau)$ and the trace which equals to trace $V_{jl}(\tau) = 1 + 2 \cos \varepsilon\tau$. The latter condition is used for numerical determination at a quasi-energy ε. It is seen that the shift of the

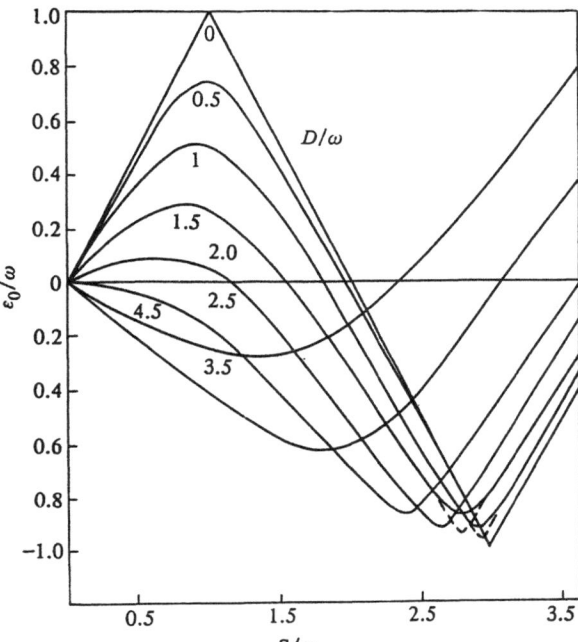

Fig. 8.2. The dependence of the quasi-energy ε_0/ω on the reduced static field S/ω for different values of the dynamic field D/ω, according to [8.5]

quasi-energy by the values $\pm n\omega$ $(n = 1, 2 \ldots)$ does not change the result, and only causes redefinition of the harmonics.

To clarify the situation, we give a quasi-energy value ε its definite branch ε_0 defined such that the branch itself, its derivative $(\partial\varepsilon_0/\partial S)$, and harmonic intensities are continuous functions of the parameter S. The results of numerical calculations [8.5] of the values of $\varepsilon_0(S/\omega, D/\omega)$ are shown in Fig. 8.2 using a reduced static field S/ω for a number of dynamic field values. When the dynamic field is absent $(D = 0)$ the selected quasi-energy bunch is a periodically broken curve of unity height and width, and is the mapping of the usual Stark splitting. For small S but an arbitrary D the following correlation is true: $(\partial\varepsilon_0/\partial S)_{S\to 0} \to J_0(D/\omega)$. The height of the first maximum for moderate values of D is well approximated by the dependence: $1 - D/2$. With increasing S, the successive extrema ε approaches closer and closer to the value ± 1. With the help of the calculated values $\varepsilon_0(S, D)$ and the representation (8.22), the periodic functions $P_j^\alpha(t)$ are defined. By calculating the squares of the Fourier coefficients one finds the equation for the L_α line intensity of spontaneous radiation:

$$I_j(\omega) = \sum_{\alpha,n} I_{j,n}^\alpha \delta(\omega - n\omega - \alpha\varepsilon_0) \ , \tag{8.23a}$$

where the coefficients $I_{j,n}^\alpha$ determine the intensity of the radiation polarized in correspondence with the value of the index j.

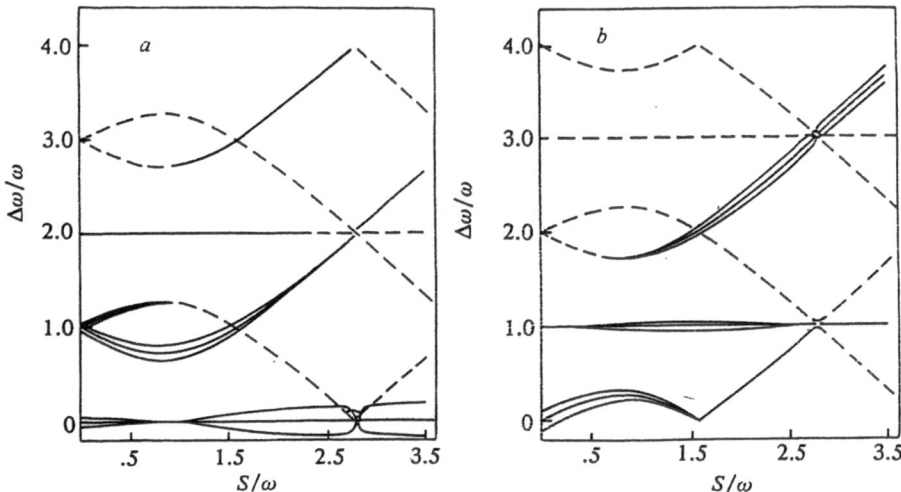

Fig. 8.3 a, b. The intensity and location of harmonics vs S/ω for $D/\omega = 1.5$ according to [8.5]. The center of each (solid) line shows the harmonic location, its intensity is proportional to the width of the line. Dashed lines are for $< 1\%$ intensity. (a) $|2p_z\rangle$; (b) $|2p_x\rangle$

The chosen field orientation harmonic intensities $I^0_{2;n}$ and $I^\pm_{3n}$ turn into zero for even values of the number n, and I^0_{3n} and $I^\pm_{2n}$ for odd values. The results of spectral calculations [8.5] are shown in Fig. 8.3 for two different polarizations. The calculated spectrum contains satellites of a different nature which only in some particular cases may be interpreted as satellites of *Blokhintsev* [8.6] or *Baranger-Mozer* [8.7] types. Thus, with $S = 0$ the $|p_x\rangle$ state spectrum is reduced to the *Blokhintsev* result. With $S \gg \omega, D$ the spectrum is close to the Stark static spectrum against the background of which weak satellites of a *Baranger-Mozer* type appear and is conditioned by a dynamic field. The locations ε_p of these satellites can be well described by perturbation theory formulae

$$\varepsilon_p = S[1 + D^2/4(\Omega - \omega)^2] \ . \tag{8.23 b}$$

In the domain of $S, D \gg \omega$ the total splitting is defined by the total static field $(S^2 + D^2/2)^{1/2}$ being the vector sum of the S and D fields.

In the general case the calculated spectrum depicts a complicated process of a satellite intensity transformation caused by the variation of the field magnitude S. The transformation is a quasiresonant process which takes place near points $S = n\omega$. Near these points in the $|p_x\rangle$ spectrum the intensity of the weakly shifted component $I^0_{3,0}$ sharply decreases and simultaneously the intensity of the strongly shifted components $I^+_{3,-1}$ and $I^-_{3,+1}$ increases sharply. Analogous "transport" of intensities takes place in the $|p_z\rangle$ spectrum, too. Here the intensity of shifted harmonics $I^+_{2,m}$ and $I^-_{2,m}$ is transported to the intensity of the central harmonic $I^0_{2,m}$.

The above spectral transformation in mutually perpendicular $S-D$ fields is already complicated enough even for the simplest case of the L_a line. It is clear that for the Balmer series the transformation of harmonic intensities will be very difficult. Therefore it is necessary to develop some analytical approximations which would allow mechanisms of resonant spectral transformation in the $S-D$ fields. Section 8.4 is devoted to the development of such a theory.

8.4 Analytical Theory of Multiquantum Resonances in $S-D$ Fields

The foundation of the analytical resonance theory developed by *Gavrilenko* and *Oks* [8.8] is based on the extraction of static components from a general perturbation in $S-D$ fields and the analysis of resonances which appear during the coincidence of frequencies caused by the extracted static perturbation with harmonics $\pm n\omega$ of the periodic field. To better understand the essence of the method, let us consider as before the case of the mutually perpendicular $S-D$ fields for the simplest spectrum of the L_α type.

We begin with the simplest case when the magnitude of the dynamic field is small enough, such that $\mathscr{E}_0 \ll F$. For the L_α-line the picture of splitting in the field $F \parallel 0z$ consists of two shifted components with an orbital momentum projection $m = 0$ as well as two unshifted (central) components with $m = \pm 1$. Let frequency ω of the periodical perturbation be close to the frequency of the Stark splitting. Let us present a periodical electric field $\mathscr{E}_0 \cos \omega t$ in the form of the sum of two fields rotating in opposite directions

$$\mathscr{E}_0 \cos \omega t = \mathscr{E}_0 \frac{e^{i\omega t} + e^{-i\omega t}}{2} . \tag{8.24}$$

We utilize the coordinate system rotating over the z axis with an angular frequency ω. In this system the interaction Hamiltonian takes the form

$$H = H_0 + d_z F - L_z \omega + \frac{d_x \mathscr{E}_0}{2} \tag{8.25}$$

where we have neglected nonresonant atoms oscillating with a frequency 2ω. Under the resonance condition both $d_z F$ and $L_z \omega$ are mutually compensated because that is the way the rotating coordinate system has been choosen. As a result the interaction of the atom with a constant (!) electric field having strength $\mathscr{E}_0/2$ and directed along the $0x$ axis remains only in the rotating coordinate system. The energy eigenvalues λ in the selected coordinate system are determined obviously by the usual formulae of the linear Stark effect theory

$$\lambda = \frac{3}{4} n (n_1 - n_2) \mathscr{E}_0 \ .$$

(8.26)

The corresponding wave functions are oriented along the $0x$ axis and are obtained from parabolic wave functions with the $0z$ axis with the help of the operator of rotation around the $0y$ axis at an angle $\pi/2$: $\varphi_x = \exp \times [-i l_y \pi/2] \varphi_z$.

Returning to the initial fixed coordinate system, let us write down the general expression for the wave function of the system near resonance $zF = \hbar \omega$:

$$\Psi(t) = e^{-i\lambda t} \exp\left[-i(H_0 + zF)t\right] \exp\left[-i l_y \frac{\pi}{2}\right] \varphi_{n n_1 n_2} \ ,$$

(8.27)

where $\varphi_{n n_1 n_2}$ is the parabolic wave function with axis $0z \parallel F$.

Thus the influence of an oscillating field $\mathscr{E}_0$ results in the additional level splitting near the resonant points by the value λ. In this simplest case the magnitude λ represents a quasi-energetic addition to the energy of the atomic levels. The action of the rotation operator $\exp (i l_y \Theta)$ on the parabolic states may be determined with the help of Wigner matrices of a finite rotation [8.10] according to the formula:

$$\exp (+i l_y \theta) \varphi_{n n_1 n_2} = \sum_{n_1' n_2'} D_{n_1 n_1'}^{((n-1)/2)} (0, \theta, 0) D_{n_2 n_2'}^{((n-1)/2)} (0, \theta, 0) \varphi_{n n_1' n_2'} \ ,$$

(8.28)

where $D_{mm'}^j$, $(\varphi_1, \Theta, \varphi_2)$ are Wigner matrices [8.10] with elements expressed in terms of trigonometric functions. A more detailed discussion and the derivation of the formula are given in Chap. 11.

An example of the L_α line spectrum calculation under resonance conditions is presented in Fig. 8.4 taken from [8.9]. The influence of the quasi-energetic splitting of the components of the magnitude $\Omega_0 = \frac{3}{2} \varepsilon_0$ is clearly seen from Fig. 8.4. One of the conditions of applicability of the above consideration requires the maximum of the quasi-energetic splitting to be small $\lambda \sim 3 n(n-1) \mathscr{E}_0/4$ as compared with the spacing $3nF/2$ between the Stark components:

$$(n-1) \mathscr{E}_0/2F \ll 1 \ .$$

(8.29)

When this condition is met it is easy to determine still more complex line spectra (e.g., L_β, H_β, [8.9]).

The single quantum resonance approximation considered is a particular case of a more general approximation of multiquantum resonances developed in [8.9]. To reveal the essence of this approximation let us consider the case $F \perp \mathscr{E}_0$ without considering the field $\mathscr{E}_0$ to be small in advance. Let us select a coordinate system with the $0z$ axis parallel to F and the $0x$ axis parallel to

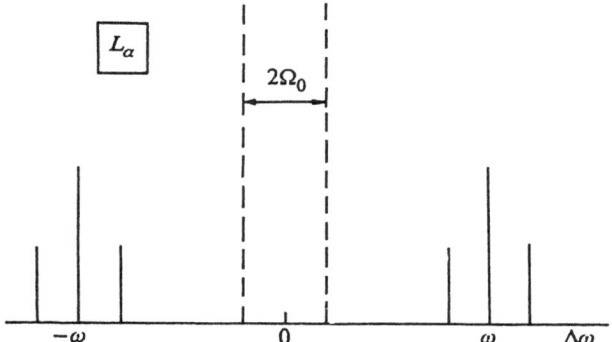

Fig. 8.4. The spectrum of hydrogen L_α line in orthogonal electric F and $\mathscr{E}_0 \cos \omega t$ fields in the case of a one-quantum resonance for levels $n = 2$ [8.8]. Solid line: polarization along vector F; dashed line: along $\mathscr{E}_0$. The splitting is equal to $\Omega = 3\mathscr{E}_0/2$

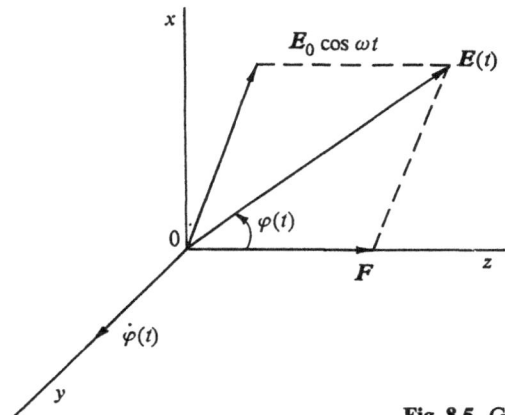

Fig. 8.5. Geometry of crossed electric fields [8.9]

$\mathscr{E}_0$. Then let us make a transition to the rotating coordinate system $x'y'z'$ of the $0z'$ axis which is at every moment of time directed along the vector of summing field $E(t)$ and makes an angle $\varphi(t)$ with the z axis, as represented in Fig. 8.5. Such a transition is produced with the help of the rotation operator $\exp(i l_y \varphi(t))$, transforming the Hamiltonian of the system to the form:

$$H = H_0 + V_{||}(t) + V_\perp(t) , \tag{8.30}$$

$$V_{||}(t) = zE(t), \quad V_\perp(t) = l_y \dot{\varphi}(t) , \tag{8.31}$$

$$E(t) = (F^2 + \mathscr{E}_0^2 \cos^2 \omega t + 2F\mathscr{E}_{0x} \cos \omega t)^{1/2} , \tag{8.32}$$

$$\dot{\varphi}(t) = -F\mathscr{E}_{0x}\omega \sin \omega t / E^2(t) . \tag{8.33}$$

Since the total field $E(t)$ and angular velocity $\dot\varphi(t)$ are functions periodic in time, they may be expanded into a Fourier series:

$$E(t) = \bar E + \sum_{q=1}^{\infty} E_{2q} \cos(2q\omega t) , \qquad (8.34)$$

$$\dot\varphi(t) = \sum_{v=1}^{\infty} b_v \sin[(2v-1)\omega t] ,$$

$$b_v = 2(-1)^{v+1}[(1+F^2\mathscr{E}_0^{-2})^{1/2} - F\mathscr{E}_0^{-1}]^{2v-1}\omega . \qquad (8.35)$$

Here $\bar E$ is the constant (time averaged) component of the total field, E_{2q} are the Fourier coefficients of the expansion permitting an evident analytical expression.

It follows from (8.11) that field $E(t)$ has a constant component $\bar E$ resulting from the value of the field average over a period. In this average field, the atomic levels experience Stark splitting. The distance ω_E between the Stark sublevels is equal to

$$\omega_E = \frac{3}{2} n\bar E = \frac{3}{\pi} n(F^2 + \mathscr{E}_0^2)^{1/2} El(k) , \qquad (8.36)$$

where $k = \mathscr{E}_0(F^2 + \mathscr{E}_0^2)^{-1/2}$, $El(k)$ is the total elliptic integral of the second type [8.11], through which the average field $\bar E$ is expressed.

In the absence of a "transversal" perturbation $V_\perp(t)$ the atomic wave function is expressed in the form of the Fourier expansion:

$$\Psi_\alpha = \exp[-i(n_1-n_2)\omega_E t] \sum_{s=-\infty}^{\infty} A_\alpha^s e^{2is\omega t} , \qquad (8.37)$$

where $A_\alpha^s(r)$ are the Fourier coefficients, $\alpha \equiv (n_1, n_2, m)$. The expansion (8.37) is a quasi-energetic atomic spectrum with a distance Q between different quasi-energy values equalling to

$$Q = \omega_E + 2n\omega \quad (n = 0, \pm1, \pm2, \dots) . \qquad (8.38)$$

If harmonics $(2v-1)\omega$ of the "transversal" perturbation $V_\perp$ are in resonance with the quasi-energetic splitting Q (8.38) then there is the situation of a multi-quantum resonance, in which many of the quasi-energy harmonics and all the harmonics of the perturbation $V_\perp$ take part [8.12]. The condition of multi-photon resonance is the equality

$$\omega_E = (2l-1)\omega , \quad l = 1, 2, 3 \dots . \tag{8.39}$$

It should be noted that the effect of resonance remains even in the case when the static field is small: $F \ll \mathscr{E}_0$, so that $\bar{E} \simeq 2\mathscr{E}_0/\pi$. In this case the resonance condition (8.39) expresses the coincidence of the dynamic field harmonics ω with splitting $\omega_E \simeq \frac{1}{2} n 2 \mathscr{E}_0/\pi$, conditioned by the same dynamic field. This circumstance allowed the authors of [8.8,9] to name the effect as "dynamic" resonance. Note that though the static field does not appear in the resonance condition itself, the account of it is vital since it is just this field that determines the difference from zero of harmonic amplitudes b_v of the transversal perturbation. It may be shown [8.9] that for $F \to 0$ the calculated spectrum coincides with the *Blokhintsev* spectrum.

Let us turn to a specific calculation of the L_α line spectrum. Designating the states belonging to level $n = 2$ as

$$(100) \equiv 1, \ (001) \equiv 2, \ (00\text{-}1) \equiv 3, \ (010) \equiv 4 , \tag{8.40}$$

and changing to the interaction representation with the Hamiltonian $V_{\parallel}(t)$ let us express the Schrödinger equation for the evolution operator $T_{\alpha\alpha'}$:

$$\left.\begin{array}{l} i\dot{T}_{\alpha\alpha''} = \sum\limits_{\alpha'} V_{\alpha\alpha'} T_{\alpha'\alpha''} , \\[2mm] V(t) = \exp\left[i\int\limits_0^t dt' V_{\parallel}(t') \right] V_{\perp}(t) \exp\left[-i\int\limits_0^t dt' V_{\parallel}(t') \right] . \end{array}\right\} \tag{8.41}$$

Using the resonance approximation (8.39), we write down the expression for the energy under the l quantum resonance condition:

$$E(t) = (2l-1)\omega/3 + \sum_{q=1}^{\infty} E_{2q}(l) \cos(2q\omega t) , \tag{8.42}$$

where the Fourier coefficients $E_{2q}(l)$ are determined by the expression:

$$\omega^{-1} E_{2q}(l) = 2(2l-1)[\pi \omega_E(F, \mathscr{E}_0)]^{-1} \int\limits_{-\pi/2}^{\pi/2} dx (F^2 + \mathscr{E}_0^2 \cos^2 x)^{1/2} \cos(2qx) . \tag{8.43}$$

Let us estimate the magnitude of the coefficients $E_{2q}(l)$. Their maximum value with $F = 0$ is equal to

$$|\omega^{-1} E_{2q}(l)|_{\max} = (2l-1)/3 \left| \int\limits_{-\pi/2}^{\pi/2} dx |\cos x| \cos(2qx) \right|$$

$$= 4(2l-1)[3\pi(4q^2-1)]^{-1} . \tag{8.44}$$

Below we shall consider a number of first resonances with $l = 1$ and $l = 2$ for which

$$\left.\begin{array}{l} 3\,|E_4(1)/\omega|_{max}/4 = 1/15\,\pi \ll 3\,|E_2(1)/\omega|_{max}/2 = 2/3\,\pi \ , \\[2mm] 3\,|E_4(2)/\omega|_{max}/4 = 1/5\,\pi \ll 3\,|E_2(1)/\omega|_{max}/2 = 2/\pi \ . \end{array}\right\} \tag{8.45}$$

Taking into account the sharp decrease of Fourier coefficients in the expansion of the energy it is sufficient to retain the value $E_2(l)$ only. Further expansion of the exponent containing the oscillating functions into the series in the Bessel functions and using the expansion (8.35) for $V_\perp$ one obtains

$$V_{\alpha\alpha'}(t) = (l_y)_{\alpha\alpha'}\exp\left[i(2l-1)(z_{\alpha\alpha}-z_{\alpha'\alpha'})\omega t/3\right]$$

$$\times \left\{\sum_{n=-\infty}^{\infty} J_n[(z_{\alpha\alpha}-z_{\alpha'\alpha'})E_2/2\omega]\exp(2in\omega t)\sum_{v=1}^{\infty} b_v \sin[(2v-1)\omega t]\right\} . \tag{8.46}$$

This expression is true of the resonance approximation if we retain in (8.46) only nonoscillating members and from the Bessel function only the members with indices $n = 0$ for $l = 1$ and $n = 0, \pm 1$ for $l = 2$ (8.45). The equations (8.41) are reduced to the form

$$\dot{T}_{1\alpha''} = \dot{T}_{4\alpha''} = a(T_{2\alpha''}+T_{3\alpha''}) \ , \quad \dot{T}_{2\alpha''} = \dot{T}_{3\alpha''} = a(T_{1\alpha''}+T_{4\alpha''}) \ ,$$

$$(\alpha'' = 1,2,3,4) \ , \tag{8.47}$$

where

$$ia(1) = -J_0 b_1/4 \approx -b_1/4 \ ,$$

$$ia(2) = (-J_0 b_2 + J_1 b_1)/4 \approx (3E_2 b_1/4\omega - b_2)/4 \ . \tag{8.48}$$

The solution of (8.47) for the initial conditions $T_{\alpha\alpha'}(0) = \delta_{\alpha\alpha'}$ takes the form:

$$T_{11} = T_{22} = T_{33} = T_{44} = (1+\cos\Omega t)/2, \quad T_{23} = T_{14} = (\cos\Omega t - 1)/2 \ ,$$

$$T_{12} = T_{13} = T_{24} = T_{34} = (-1)^{l+1}\sin\Omega t/2, \quad T_{\alpha'\alpha} = T_{\alpha\alpha'} \ , \tag{8.49}$$

where the oscillation frequency $\Omega(l) = 2\,|a(l)|$ is equal to

$$\Omega(1) = b_1/2, \quad \Omega(2) = (3E_2 b_1/4\omega - b_2)/2 \ . \tag{8.50}$$

The values $b_{1,2}$ are determined by (8.35) and for the determination of E_2 it is convenient to use the series

$$E_2 = (F^2 + \mathscr{E}_0^2)^{1/2} \left\{ k^2/4 + \sum_{s_z=2}^{\infty} [(2s-3)!!/2^s s!]^2 2s(2s-1) k^{2s}/(s+1) \right\} ,$$

$$k^2 = \mathscr{E}_0^2/(F^2 + \mathscr{E}_0^2) .$$

$$(8.51)$$

The spectrum of spontaneous emission is calculated by the formula

$$I(\Delta\omega) = \sum_{i=1}^{4} \lim_{T\to\infty} (2\pi T)^{-1} \left| \int_0^T dt \langle \Psi_i(t) | r | \Psi_{000} \rangle \exp(-it\Delta\omega) \right|^2 , \qquad (8.52)$$

where $\Delta\omega$ is the distance from the line center and $\Psi_i(t)$ is the wave function in the fixed coordinate system expressed in terms of the evolution operator $T_{\alpha\alpha'}$ by the following formula

$$\Psi_i(t) = \exp[i l_y \varphi(t)] \exp\left[-iz \int_0^t dt' E(t') \right] T(t) \Psi_i(0) . \qquad (8.53)$$

It is convenient for the calculation of the spectrum (8.52) to transfer the rotation operator action $\exp(i l_y \varphi)$ at the radius vector r using the general formulae for the vector components transformation under rotation. Using also the evolution operator form (8.53) it is easy to find for observations the transverse vectors F and $\mathscr{E}_0$ [8.9]:

$$I(\Delta\omega) \propto \sum_{p=-\infty}^{\infty} [2A_p^2 \delta(\Delta\omega - p\omega)$$

$$+ (A_p^-)^2 \delta(\Delta\omega - p\omega - \Omega) + (A_p^+)^2 \delta(\Delta\omega - p\omega + \Omega) ,$$

$$A_p^{\pm} = A_p \pm C_p , \quad C_p \equiv \sum_{s=-\infty}^{\infty} J_{p-3s}(b_1/\omega) J_s(-b_2/\omega) ,$$

$$A_p \equiv [\alpha_p - (-1)^p \alpha_{-p}]/2 ,$$

$$\alpha_p \equiv J_0(3E_2/2\omega) C_{p-2l+1} + J_1(3E_2/2\omega)(C_{p-3l+3} - C_{p-2l-1}) . \qquad (8.54)$$

An example of the L_α line spectrum calculation for the case of resonance $l=2$, $E_0 = \omega$ is shown in Fig. 8.6 taken from [8.8]. The values of the static field F and the resonance splitting for this case are equal: $F \approx 0.73\omega$, $\Omega \approx 0.23\omega$. It is evident that under resonance conditions the spectrum experiences additional splitting into components spaced by a frequency 2Ω (l).

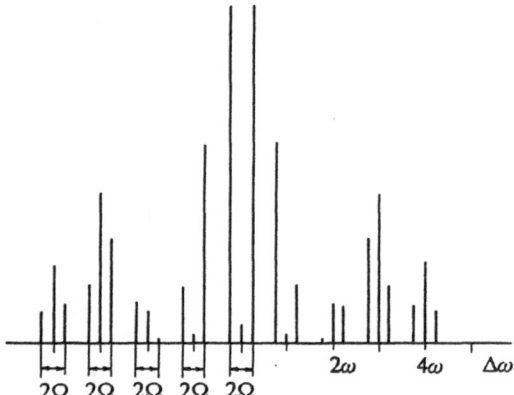

Fig. 8.6. The spectrum of L_α line in dynamic resonance for the case $l = 2$, $E_0 = \omega$ [8.8,9]

To be able to compare the analytical results with the results of numerical calculations [8.5] (Sect. 8.3) let us write down the resonance condition (8.39) in terms of the variables [8.5]:

$$D\omega^{-1} = (2l-1)\pi\{2(1-S^2D^{-2})^{1/2}E(k)\}^{-1} \ ,$$

$$l = 1,2\ldots \ , \quad [k \equiv D(S^2+D^2)^{-1/2}] \ , \tag{8.55}$$

where $S = 3F$ and $D = 3\mathscr{E}_0$.

The quasi-energies ε_0 in Sect. 8.3 and $\Omega(l)$ are connected by the relation:

$$\varepsilon_0(S,D) = (-1)^l[\Omega(F,\mathscr{E}_0)-\omega] \ . \tag{8.56}$$

To compare this with the numerical data given in Fig. 8.2 let us take a definite value of D/ω, say $D/\omega = 0.5$. For $l = 1$ from the resonance condition (8.55) one obtains the value S/D (for the case under consideration $S/D \approx 0.93$). It follows from here that $S/\omega \simeq 0.93$, which (as is seen from Fig. 8.2) precisely corresponds to the extremum of the function ε_0/ω for a given $D/\omega = 0.5$. The frequency $\Omega(1)/\omega$ is calculated with the help of (8.50) and equals $\Omega(1)/\omega \approx 0.25$, so it follows from (8.56) that $\varepsilon_0/\omega \simeq 0.75$, which precisely coincides with the data in Fig. 8.2. Thus both the location and the absolute magnitudes of the quasi-energy extrema are described perfectly well by the analytical theory of resonances.

Another extremum at the curve corresponding to $D/\omega = 0.5$ is well described by the analytical resonance theory with $l = 2$. The analogously calculated extrema locations for other values of D/ω generally coincide well enough with numerical calculations. The resonance condition with $l = 1, 3 \ldots$ corresponds to the maxima of quasi-energy ε_0, but the condition with $l = 2, 4 \ldots$ − to its minima.

The analytical model developed cannot cover all numerical results presented in Fig. 8.2 because it is bounded by certain limitations. These limitations are connected with the resonance condition and are reduced to the necessity of the determined frequency splitting $\Omega(l)$ being small as compared with the frequency of the acting perturbation. This necessity is observed under the condition

$$W(2) = 5(5\omega \, \mathscr{E}_0/36F^2)^4/9 \ll 1 \; , \tag{8.57}$$

ensuring the nonresonance effects being small.

It is seen from (8.57) that for dynamic resonance theory to be applied, a static field should not be too small. That is why the L_α line spectrum calculated under the approximation (8.57) is not transformed into a *Blokhintsev* spectrum. For the case of fields F small enough when $\omega_F \approx 3F \ll \omega$ it is also easy to develop an analytic model taking the *Blokhintsev* wave functions as a zero approximation. It is straightforward to coordinate corresponding results with numerical data [8.5] (Fig. 8.2) in the domain of small values of S (for details: [8.8, 9]).

8.5 Hydrogen Spectral Line Structure Near Resonances in $S-D$ Fields

Let us consider the structure of the hydrogen spectral lines near the resonance when the Stark level splitting CF is close to the frequency Ω of an oscillating field. It is the resonance for the very important case of an oscillating field $\mathscr{E}_0$ when the transverse amplitude is small enough, e.g., $C\mathscr{E}_{0\perp} \ll \Omega$ [8.13]. The later condition allow the perturbation theory solutions far from the resonance to be connected with the solutions in the resonance domain for the intermediate values of the parameters:

$$C\mathscr{E}_{0\perp} \ll |CF - \omega| \ll \Omega \; . \tag{8.58}$$

Thus, the evolution operator U in this case is precisely defined. The Schrödinger equation for the operator $U(t)$ takes the form:

$$\mathrm{i}\, \dot{U}^{qm}_{q'm'}(t) = Cq'[F + \mathscr{E}_{0\parallel} \cos(\Omega t + \varphi)] \; U^{qm}_{q'm'}(t)$$

$$- C\mathscr{E}_{0\perp} \cos(\Omega t + \varphi) \sum_{q'',m''} \xi^{q''m''}_{q'm'} U^{qm}_{q''m''}(t) \; . \tag{8.59}$$

Here $q = n_1 - n_2$, m are the electrical and magnetic quantum numbers, C is the Stark constant, $\mathscr{E}_{0\parallel}$, $\mathscr{E}_{0\perp}$ are the components of the dynamic field $\mathscr{E}_0$ parallel and perpendicular to F,

$$\zeta_{qm}^{q'm'} = -\frac{2}{3}\frac{\hbar}{an}x_{qm}^{q'm'} = \frac{1}{4}[n^2 - (q^2 + m^2 - q'^2 - m'^2)^2/4]^{1/2} . \tag{8.60}$$

It is considered that the field $\mathscr{E}_0(t)$ possesses a random phase φ, the $0x$ axis of the coordinate system is directed along $\mathscr{E}_{0\perp}$, the angle Θ is the angle between vectors $\mathscr{E}_0$ and F (the z axis).

One can solve (8.59) for the domain far from resonance with the help of perturbation theory. In first order perturbation theory the atomic spectrum contains Stark frequencies $CF \cdot q$ as well as frequencies $CF \cdot q \pm \Omega$ of the quasilevels spaced from the main sublevels by frequencies $\pm \Omega$. In second order perturbation theory there are sublevels with frequencies $\pm 2\Omega$, and so on. The spectrum of quasilevels in the approximation is an equidistant and unlimited one:

$$\omega_{q,j} = CFq + \Omega j , \quad j = 0, \pm 1, \pm 2 \ldots . \tag{8.61}$$

with the number j increasing as the intensity of the corresponding transition sharply decreases. Thus, in practice it is sufficient to be confined to only the first members of the perturbation theory series.

Calculations of the evolution operator under perturbation theory conditions are trivial, so we shall write down only one of its matrix elements:

$$U_{q,m\pm 2}^{qm}(t) = e^{-iCqFt} \sum_{\sigma = \pm} \frac{\cos \varepsilon_\sigma t - 1}{\theta_\sigma^2} \sum_{q',m'} \zeta_{q'm'}^{qm} \zeta_{q,m\pm 2}^{q'm'} . \tag{8.62}$$

Here we have introduced the following variables which are to be used below:

$$\varepsilon_\pm = CF + \Omega , \quad \theta_\pm = 2\varepsilon_\pm/C\mathscr{E}_{0\perp} , \quad \zeta = C\mathscr{E}_{0\parallel}/\Omega . \tag{8.63}$$

The rest of matrix elements of operator U, which may be easily obtained by the expansion over the inverse powers of $\theta_\pm$ are listed in [8.13].

Let us consider the solution of system (8.59) in the resonance domain $|CF - \Omega| \ll \Omega$. The substitution

$$U_{q'm'}^{qm}(t) = G_{q'm'}^{qm}(t) \exp[-iCq'Ft + i\varepsilon q't - i\varphi(q' - q)] \tag{8.64}$$

puts equations into the form

$$i\dot{G}_{q'm'}^{qm}(t) = \varepsilon q' G_{q'm'}^{qm}(t) - \frac{C\mathscr{E}_{0\perp}}{2} \sum_{q,m} \zeta_{q'm'}^{q''m''} G_{q''m''}^{qm}(t) , \tag{8.65}$$

$$G_{q'm'}^{qm}(0) = \delta_{qq'}\delta_{mm'} . .$$

The solution of (8.65) is found from the stationary system for the eigenvalues λ_κ and eigenvectors $H^{(\kappa\mu)}$ which have the following form:

$$(\lambda - \varepsilon q) H_{qm}^{(\kappa\mu)} + \frac{C\mathcal{E}_{0\perp}}{2} \sum_{q'm'} \zeta_{qm}^{q'm'} H_{q'm'}^{(\kappa\mu)} = 0 . \tag{8.66}$$

Here the index κ numbers eigenvalues λ_κ and μ eigenvectors corresponding to the given λ_κ.

Equating the determinant of the system (8.66) with zero one finds [8.13] for levels $n = 2, 3, 4$:

$$\lambda_\kappa = \kappa [\varepsilon^2 + (C\mathcal{E}_{0\perp}/2)^2]^{1/2} = \kappa \frac{C\mathcal{E}_{0\perp}}{2} \eta ,$$

$$\eta = \sqrt{1+\theta^2} , \quad \theta = 2\varepsilon/C\mathcal{E}_{0\perp} \equiv \theta_- . \tag{8.67}$$

Eigenvectors $H^{\kappa\mu}$ are determined by taking into account the initial conditions, symmetry relations and normalized factors [8.13]. The final result for the evolution operator takes the form

$$U_{q'm'}^{qm}(t) = \exp [i\varphi(q-q')] \sum_{\kappa\mu} H_{q'm'}^{(\kappa\mu)} H_{qm}^{(\kappa\mu)} e^{-i\omega}q'\kappa^t , \tag{8.68}$$

where frequencies

$$\omega_{q\kappa} = q\Omega + \lambda_\kappa = q\Omega + \kappa [(CF-\Omega)^2 + (C\mathcal{E}_{0\perp})^2]^{1/2} \tag{8.69}$$

determine the energy spectrum of the quasilevels near the resonance.

The energy spectrum of the quasilevels $\omega(s)$ (marked with indices $s = a, b, c \ldots$) depending on $\mathcal{E}_0$ for $CF/\Omega = 1/4$ for the level $n = 2$ is revealed in Fig. 8.7 from [8.13]. It is evident that the main Stark sublevels come nearer to the quasilevel group when F approaches the "crossing point" $\Omega/C (F \Rightarrow \Omega/C)$. Near these crossing points the repulsion between the levels occurs due to the interaction with the field component $\mathcal{E}_{0\perp}$. The minimum spacing of the quasilevels in the crossing point is equal to $C\mathcal{E}_{0\perp}/2$.

In the resonance domain $|CF-\Omega| \leq C\mathcal{E}_{0\perp}$ the amplitudes of items corresponding to quasilevels sharply increase and become comparable with the amplitudes of the principal components. When moving off of the resonance these amplitudes sharply decrease (see points in Fig. 8.7).

To calculate the intensity $I(\omega)$, let us average the evolution operator $\langle U \rangle$ over the phase of the electromagnetic field, which gives:

$$\langle U \rangle = \sum_{s=a,b,c\ldots} U_{q'm'}^{qm}(s) e^{-i\omega(s)t} , \tag{8.70}$$

where the material coefficients $U(s)$ are defined from the general formula (8.68) for the operator and its connection with solutions in the nonresonance domain.

Taking the Fourier transform of (8.70) and performing the averaging over the radiation direction and the polarization, one obtains [8.13]

$$I(\omega) = \sum_s I_s \delta[\omega - \omega(s)] \,, \tag{8.71}$$

$$I_s = \left(\pi \sum_{qm} |r_{qm}|^2 \right)^{-1} \mathrm{Re} \left\{ \sum_{qmq'm'} U^{qm}_{q'm'}(s) r_{qm} \cdot r_{q'm'} \right\} \,. \tag{8.72}$$

Thus, the spectral line splits into a number of components s in accordance with the splitting of the upper energy level $\omega(s)$, as in Fig. 8.7. The intensities of the components for the level $n = 2$ are listed in Table 8.1 taken from [8.13].

The behavior of the L_α line components, depending on the variation of the values F, is shown in Fig. 8.8 taken from [8.13]. One can see how these components are shifted by frequencies $\pm \Omega(a, a', c, c')$ and become less intense if far from the resonance, and sharply increase when approaching resonance. When angle Θ decreases, the resonant intensity variations take place in an increasingly narrow domain $|CF - \Omega| \le C\mathscr{E}_0| \sin \Theta|$, and for $\Theta \Rightarrow 0$ they disappear altogether. It is interesting that the appearance of the visible first satellites

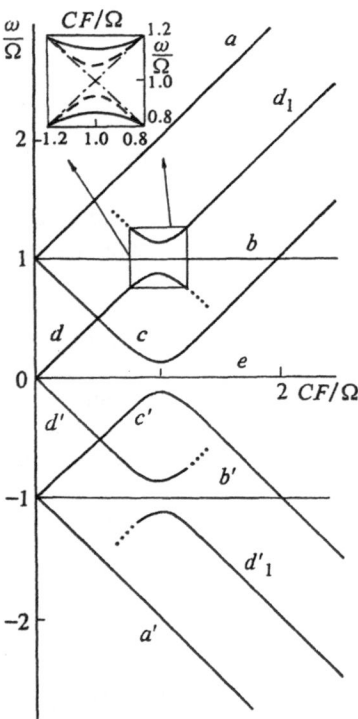

Fig. 8.7. The quasi-energy L_α line spectrum ω vs reduced constant field CF/Ω for $C\mathscr{E}_0/\Omega$, for $C\mathscr{E}_0/\Omega = 1/4$ and $\Theta = 90°$. The dotted lines designate the weak intensity of the components. In the left corner the variation of components near the crossing points is shown for different values of the angle Θ: 90° (*solid line*), 30° (*dashed line*), 0° (*dash-dotted line*) [8.13]. The intensities of different components (a, b, c, d, e) are present in Table 8.1

Table 8.1. The intensities of the Stark components of L_α line

s	F		
	$\Omega-CF\gg C\mathcal{E}_{0\perp}$	$\lvert\Omega-CF\rvert\ll\Omega$	$CF-\Omega\gg C\mathcal{E}_{0\perp}$
a,a'	$\dfrac{1}{6}\left(\dfrac{\zeta^2}{4}+\dfrac{1}{\theta_+^2}\right)$	$-$	$\dfrac{1}{6}\left(\dfrac{\zeta^2}{4}+\dfrac{1}{\theta_+^2}\right)$
b,b'	$\dfrac{1}{12}\left(\dfrac{1}{\theta_-}+\dfrac{1}{\theta_+}\right)^2$	$\dfrac{1}{12\eta^2}$	$\dfrac{1}{12}\left(\dfrac{1}{\theta_-}-\dfrac{1}{\theta_+}\right)^2$
c,c'	$\dfrac{1}{6}\left(\dfrac{\zeta^2}{4}+\dfrac{1}{\theta_-^2}\right)$	$\dfrac{1}{6\eta^2}$	$\dfrac{1}{6}\left(\dfrac{\zeta^2}{4}+\dfrac{1}{\theta_-^2}\right)$
d,d'	$\dfrac{1}{6}\left[1-\dfrac{\zeta^2}{2}-\dfrac{1}{2}\left(\dfrac{1}{\theta_-}+\dfrac{1}{\theta_+}\right)^2\right]$	$\dfrac{1}{24}\dfrac{(\eta-\theta)^2}{\eta^2}$	$-$
d_1,d_1'	$-$	$\dfrac{1}{24}\dfrac{(\eta+\theta)^2}{\eta^2}$	$\dfrac{1}{6}\left[1-\dfrac{\zeta^2}{2}-\dfrac{1}{2}\left(\dfrac{1}{\theta_-}+\dfrac{1}{\theta_+}\right)^2\right]$
e	$\dfrac{2}{3}\left(1-\dfrac{1}{2\theta_-^2}-\dfrac{1}{2\theta_+^2}\right)$	$\dfrac{2\theta^2+1}{3\eta^2}$	$\dfrac{2}{3}\left(1-\dfrac{1}{2\theta_-^2}-\dfrac{1}{2\theta_+^2}\right)$

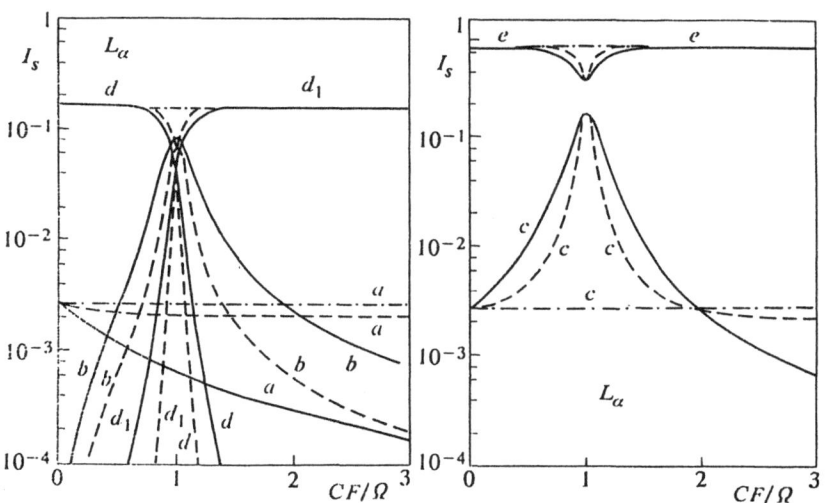

Fig. 8.8. The dependence of L_α line satellite intensities I_s on the reduced field CF/Ω for $C\varepsilon_0/\Omega$, for $C\varepsilon_0/\Omega=1/4$ and for three values of the angle Θ: 90° (*solid curves*), 30° (*dashed lines*), 0° (*dash-dotted lines*) [8.13]. The intensities of different satellites ($a-d$ — *left side*, $c-e$ — *right side*) are present in Table 8.1

of the nonshifted components of the L_α line occur near the resonance. Finally let us point out that the resonant effects may also arise near other resonant values of $F = k\Omega/Cs$, where k and s are integers.

The final line shape is obtained after averaging over the statistical distribution of values of F. As has been mentioned, the presence of resonances for certain values of F results in the appearance of typical features in the body of the spectral line. The detailed profile of these features is investigated in [8.13]. We shall not describe these effects here.

8.6 On the Stochastization of Highly Excited Electron Motion in a Periodic Field

Recently intense investigations have been conducted processes of nonlinear photoionization of atoms in a field of intense electromagnetic laser radiation [8.14]. The main feature of these processes is connected with the ionization probability experiencing a sharp threshold increase when the field amplitudes $\mathcal{E}_0$ exceed a critical value $\mathcal{E}_{0cr}$. Following [8.15] we shall briefly discuss a possible mechanism of this effect connected with the stochastization of excited electron motion under a periodic perturbation.

The Hamiltonian of the classical electron motion under the influence of a periodic field takes the form

$$H = -\frac{1}{2n^2} + \mathcal{E}_0 \cos \omega t \left(1 - \frac{M_z^2}{M^2}\right)^{1/2} \sum_{k=1}^{\infty} (x_k \sin \Psi \cos k\lambda + y_k \cos \Psi \sin k\lambda) .$$
(8.73)

Here the pairs of the canonical variables "action-angle" are introduced, namely n, λ and M, $\psi (M_z = \text{const.})$, and x_k, y_k are the Fourier components of the atomic dipole momentum.

Let us consider the electron motion near resonance

$$\omega \simeq k\Omega, \ n_\lambda \approx n_k = (k/\omega)^{1/3}, \ \Omega = n^{-3}, \ k = 1,2 \ldots , $$
(8.74)

where Ω is the classical frequency of the unperturbed Kepler motion. For $|\omega - k\Omega| \sim \Omega$ the electron motion is determined with the help of perturbation theory, and it describes small deviations from the initial values of n and M.

Near the resonance $|\omega - k\Omega| \ll \Omega$, $n \simeq n_k$ one may expand the Hamiltonian (8.73) near $n = n_k$:

$$H \simeq -\frac{3}{2}\frac{(n-n_k)^2}{n_k^4} + \frac{\mathcal{E}_0}{2}\left(1 - \frac{M_z^2}{M^2}\right)^{1/2} (x_{kr}^2 \sin^2 \Psi + y_{kr}^2 \cos^2 \Psi)^{1/2}$$
$$\times \sin (k\lambda + A) ,$$
(8.75)

where

$$x_{kr} \equiv x_k(n_k) \ , \quad y_{kr} = y_k(n_k) \ , \quad \tan A = (x_{kr}/y_{kr}) \tan \Psi \ ,$$

$$\Psi = \text{const.} \ , \quad M = \text{const.} \ .$$

The Hamiltonian (8.75) formally coincides with the Hamiltonian of a nonlinear pendulum of a finite amplitude. The motion of the system has been investigated in detail [8.16]. The type of motion depends on the relation between the frequency Ω defining the distance between resonances and frequency ν, defining the "pulsation" of particles under the influence of the force $\mathscr{E}_0$:

$$\nu = \frac{\sqrt{3}}{2} k^{1/3} \mathscr{E}_0^{1/2} \omega^{2/3} \left(1 - \frac{M_z^2}{M^2} \right)^{1/4} (x_{kr}^2 \sin^2 \Psi + y_{kr}^2 \cos^2 \Psi)^{1/4} \ . \tag{8.76}$$

For $k = 1$ the maximum frequency $\nu_{\max}$ of pulsations is equal to $\nu_{\max} = (3 \mathscr{E}_0)^{1/2}/2n$.

When the value ν approaches Ω, the overlapping of different resonances takes place and the electron motion takes the form of stochastic wandering around energy levels [8.15, 16]. The ionization threshold for systems with the Hamiltonian of the type as in (8.76) is determined by the condition [8.16]:

$$\nu \geq \Omega/6 \quad \text{or} \quad \mathscr{E}_0 \geq \mathscr{E}_{cr} = 1/27 n^4 \ . \tag{8.77}$$

Thus for fields $\mathscr{E}_0$ larger than $\mathscr{E}_{0cr}$ stochastization of electron motion over the levels takes place, which results in its ionization. Estimations of the value of $\mathscr{E}_{0cr}$ for $n = 66$ corresponding to the conditions of the experiment in [8.15] give the magnitudes $\mathscr{E}_{0cr} \simeq 10$ V/sm which correlate well with experimental data. Further discussions of the problems mentioned can be found in [8.14].

9. Decay of Atomic States

In Chap. 4, we dealt with the decay of atomic states in an electric field, and in Chap. 1 we examined auto-ionization decays. We shall now consider the general case of atomic discrete state decay as well as a number of specific decay mechanisms that are typical for plasmas. In all cases the concept of "decay" or "damping" of atomic states is connected with an interaction of the discrete states under consideration with some continuum of states in which variables are subjects for an averaging procedure. As a result decay problems are characterized by some constants of damping in time, called decay widths γ. The magnitude γ may be both the characteristic of the atomic states itself and the expression of its interaction with a medium.

It must be noted that the first adequate description of the interrelation between discrete and continuous spectra was obtained by *Leontovich* and *Mundelshtamm* in 1925 (for details, [9.1, 2]). The theory of α-decay was later developed by *Gamov* [9.3]. The modern treatment of auto-ionization decays goes back to works of *Breit* and *Wigner* [9.4] and *Fano* [5.5, 6].

9.1 Resonance of Discrete States Against the Background of a Continuous Spectrum

9.1.1 A Number of Discrete States Against the Background of One Continuum

The general calculation scheme for taking into account the interaction of discrete levels with a continuum was discussed in Chap. 1 both in the stationary (*Fano* method [9.5, 6]) and the nonstationary (*Kompaneets* method [9.7]) approaches. However, both these approaches are based on perturbation theory over the interaction which initiated the transition from an unperturbed state to ones interacting with the continuum. For a number of applied problems it is of interest to avoid such a limitation and to develop a general theory of the interaction of the discrete levels interacting simultaneously with a continuum. Just the case is considered below.

Let us express the Hamiltonian of a certain quantum system of interacting objects as

$$\hat{H} = \hat{H}_0 + \hat{V}(t) \ , \quad \hat{V}(-\infty) = 0 \ . \tag{9.1}$$

The spectrum of the unperturbed Hamiltonian H_0 runs through a number of discrete values E_n. The discrete states $\{n\}$ lie against the background of the continuous spectrum v, i.e., there exist E_v such that $E_n = E_v$. We assume that the problem of orthogonalization of the wave functions of the continuous spectrum in the absence of an interaction with the discrete spectrum is solved, i.e., we state

$$\langle v | \hat{H} | v' \rangle = E_v \delta(E_v - E_{v'}) \ . \tag{9.2}$$

Then the Hamiltonian (9.1) corresponds to the following system of equations for the amplitudes of the states $|n|$ and $|v|$:

$$i\hbar \dot{a}_n = \sum_{n'} V_{nn'} e^{i\omega_{nn'}t} a_{n'} + \int dv \, V_{nv} e^{i\omega_{nv}t} a_v \ , \tag{9.3}$$

$$i\hbar \dot{a}_v = \sum_{n'} V_{vn'} e^{i\omega_{vn'}t} a_{n'} \ , \tag{9.4}$$

where $\omega_{yy'} = (E_y - E_{y'})/\hbar$ are the frequencies of the transitions $y \to y'$.

Under rather general assumptions it is possible to "lower the order" of the system (9.3, 4). For this it is necessary that the distance from the levels $\{n\}$ to the nearest boundary of the continuum $\tilde{E}$ be large in comparison with the following three quantities: a) the distance between the levels $\hbar\omega_{nn'}$, the resonance of the levels; b) the characteristic frequencies $\hbar/\tau$ of the variation of the operator $V(t)$, the adiabaticity of the perturbation; c) the characteristic widths Γ and shifts Δ of the levels $\{n\}$ (see below). These conditions take the form

$$\hbar\omega_{nn'} \ll \tilde{E} \ , \quad \hbar/\tau \ll \tilde{E} \ , \quad \Gamma, \Delta \ll \tilde{E} \ . \tag{9.5}$$

Integrating (9.4) with respect to time and substituting in (9.3), we can take the functions $V_{nv}(t)$ and $a_n(t)$, which vary slowly in comparison with the oscillating exponentials, outside of the integration with respect to t, after which the integration with respect to t yields the ζ-function $\zeta(\omega_{n'v})$ (cf. [9.8], p. 164 of the English edition). Proceeding then to integration with respect to dv to $dE = (dv/dE)^{-1}v$ (where dv/dE varies little over the interval $\omega_{nn'}$ far from the boundary of the continuum), we reduce the system (9.3) to the form [9.9, 10]:

$$i\hbar \dot{a}_n = \sum_{n'} (V_{nn'} - i/2 \, W_{nn'}) e^{i\omega_{nn'}t} a_{n'} \ , \tag{9.6}$$

where

$$W_{nn'} \equiv \Gamma_{nn'} + i\Delta_{nn'} \ ,$$

$$\Gamma_{nn'} = 2\pi \frac{dv}{dE} V_{nv}(t) V_{vn'}(t)\big|_{E_v = E} \ , \tag{9.7}$$

$$\Delta_{nn'} = 2P \int_{-\infty}^{\infty} \frac{dE}{E_{n'} - E} \frac{dv}{dE} V_{nv}(t) V_{vn'}(t) \ . \tag{9.8}$$

The above expressions (9.6) constitute the fundamental system of equations describing the resonance of the discrete states against the background of the continuum. The derivation of this system is based only on the condition (9.5). On the other hand, the relations between the quantities $V_{nn'}$, $W_{nn'}$, $\hbar\omega_{nn'}$ and $\hbar/\tau$ can in general be arbitrary. The conditions (9.5) have enabled us to determine the amplitudes a_n from the much simpler system (9.6) and substitute the result into (9.4).

The interaction with the continuous spectrum has been reduced in (9.6) to a change of the transition matrix in the equations for the amplitudes a_n. This change, however, is of a fundamental character and becomes manifest, in particular, in the non-Hermitian character of the transition matrix. This leads to the nonconservation of the normalization of the amplitudes of the discrete states as a result of their decay into the continuum. The rate $\dot{w}_n$ of this decay can be easily obtained with the aid of (9.6):

$$\dot{w}_n \equiv \frac{d}{dt} |a_n|^2 = 2\,\mathrm{Re}\{\dot{a}_n a_n^*\} \ ,$$

$$\dot{w}_n = \frac{2}{\hbar}\,\mathrm{Im}\left\{ \sum_{n'} \left(V_{nn'} + \frac{1}{2}\Delta_{nn'} - \frac{i}{2}\Gamma_{nn'} \right) e^{i\omega_{nn'}t} a_n^* a_{n'} \right\} \ . \tag{9.9}$$

Summing (9.9) over n and using the Hermitian character of the matrices $V_{nn'}$ and $\Delta_{nn'}$, we obtain

$$-\dot{w} \equiv \sum_n \dot{w}_n = -\sum_{nn'} \Gamma_{nn'} e^{i\omega_{nn'}t} a_n^* a_{n'} \ . \tag{9.10}$$

The quantity $\dot{w}$ is the total rate of the transition from the aggregate of the discrete states $\{n\}$ into the continuum. This can be verified by directly calculating $\dot{w}$:

$$\dot{w} = \frac{d}{dt} \int |a_v(t)|^2 dv \ . \tag{9.11}$$

From (9.10, 11) it follows that the normalization of the total wave function, with allowance for the continuous states, is conserved.

9.1.2 Several Continua. Scattering Problems

Let us generalize the results of Sect. 9.1 to include the case of several continua $\{v_n, v_m\}$. In this case the initial system of equations differs from (9.3, 4) only in the presence of an additional summation over the continua m. We note that this notation presupposes orthogonalization of the wave function not only of each continuum, but also of the different continua with one another. If the conditions (9.5) are satisfied for $\tilde{E} = \min\{\tilde{E}_m\}$, where $\tilde{E}_m$ is the distance from the levels $\{n\}$ to the boundary of the m-th continuum $\{v_m\}$, then relations (9.6−9) are valid, with

$$\Gamma_{nn'} = \sum_m \Gamma_{nm,mn'} = 2\pi \sum_m \frac{dv_m}{dE} V_{nv_m} V_{v_mn'} \Bigg|_{E_{v_m} = E_{n'}}, \qquad (9.12)$$

$$\Delta_{nn'} = \sum_m \Delta_{nm,mn'} = \sum_m 2 \,\mathscr{P} \int_{-\infty}^{\infty} \frac{dE}{E_{n'}-E} \frac{dv_m}{dE} V_{nv_m} V_{v_mn} . \qquad (9.13)$$

We can also introduce the decay rate $\dot{w}_m$ of the states in each of the continua and the summary rate $\dot{w}$:

$$\dot{w}_m = \sum_{n,n'} \Gamma_{nm,mn'} e^{i\omega_{nn'}t} a_n^* a_{n'} , \qquad \dot{w} = \sum_m \dot{w}_m . \qquad (9.14)$$

The analog of (9.11) is

$$\dot{w}_m = \frac{2\pi}{\hbar} \left| \frac{dv_m}{dE} \sum_n V_{nv_m} e^{iE_nt/\hbar} a_n \right|^2 , \qquad (9.15)$$

$$|\Gamma_{nm,mn'}|^2 = \Gamma_{nm,mn} \Gamma_{n'm,mn'} , \qquad |\Gamma_{nn'}|^2 \leqslant \Gamma_{nn} \Gamma_{n'n'} . \qquad (9.16)$$

The equality in the last formula corresponds to the case of one continuum.

Certain scattering problems can be reduced to the equations obtained above. The scattering problems differ mainly in that the Hamiltonian H_0 does not have "purely" discrete states in this case, and therefore are labeled as states by the two subscripts discrete k and continuous v, and represent the eigenvalues E_{kv}, in the form $E_k + E_v$.

We are interested in a certain limited group of scattering problems connected with the following assumptions. Assume that it is possible to separate from the aggregate of the discrete levels $\{k\}$ as a subgroup $\{n\}$ of resonating states, i.e., such that the frequencies of the transitions between them are much lower than the other frequencies: $\omega_{nn'} \ll \omega_{nm}$. Here $\{m\}$ are the remaining nonresonant states from the aggregate $\{k\}$. We assume that the transition between the states of the group $\{n\}$ are not accompanied by changes in the energy of the continuum. Then, solving the elastic scattering problem, we obtain

$$\langle v_m | \hat{H} | v_{m'} \rangle = E_{vm} \delta_{mm'} \delta(E_{vm} - E_{vm'}) . \qquad (9.17)$$

In this case, the states $\{n\}$ appear as discrete levels, for which equations of the type (9.3) are valid, while the change in index v is connected with transitions to the "remote" levels $\{m\}$.

Let the system be initially in one of the resonant states $\{n\}$, while the wave function of the continuous spectrum corresponds to a definite state v_0. If the matrix elements $V_{nv_0,n'v'}$ and V_{nv_0,mv_m} are small in comparison with transition frequencies $\omega_{nm'}$, we can regard the amplitudes a_{nv_0} as the principal terms, and the remaining amplitudes a_{mv_m} and a_{nv} $(v \neq v_0)$ as correction terms, thus obtaining

$$i\hbar \dot{a}_{mv_m} = \sum_{n'} V_{mv_m,n'v_0} e^{i\omega_{mv_m,nv_0}t} a_{nv_0} . \tag{9.18}$$

Further transformations of (9.18) lead naturally to the expressions (9.6–9, 12–16), where a_n should be taken to mean a_{nv_0}, and V_{nv_m} should be taken to mean V_{nv_0,mv_m}. The conditions for the applicability of these expressions coincide with (9.5) where $\tilde{E}$ stands for the distance to the nearest level of the group $\{m\}$, namely min (ω_{nm}). This means, in particular, that the initial wave packet becomes slightly smeared out in comparison with $\tilde{E}$, thus justifying the formulation (9.18).

9.1.3 Two-Level Problem with a Stationary Perturbation

We consider the system (9.6) for two levels, $n = 1$ and 2, and for a time-independent perturbation V that is instantaneously turned on at $t = 0$:

$$i\dot{a}_1 = -i\gamma_1 a_1 + (V-i\gamma)e^{i\Delta\omega t}a_2 , \tag{9.19}$$

$$i\dot{a}_2 = -i\gamma_2 a_2 + (V-i\gamma)e^{-i\Delta\omega t}a_1 , \tag{9.20}$$

with the initial conditions $a_1(0) = 1$ and $a_2(0) = 0$. Here we have introduced the notation

$$\gamma_1 \equiv \Gamma_{11}/2\hbar, \gamma_2 \equiv \Gamma_{22}/2\hbar, \gamma = \Gamma_{12}/2\hbar ,$$

$$V = V^* = \left(V_{12} + \frac{1}{2}\Delta_{12}\right)\bigg/\hbar ,$$

$$\Delta\omega = \omega_{12} + \left(V_{22} + \frac{1}{2}\Delta_{22} - V_{11} - \frac{1}{2}\Delta_{11}\right)\bigg/\hbar ,$$

where the quantities $\Gamma_{nn'}$ and $\Delta_{nn'}$ are defined by (9.7, 8) or (9.12, 13). We note that V and γ are real quantities; this is the result of the proper choice of the phases of the discrete wave function.

Relations (9.9, 10) are rewritten in the form

$$\dot{w}_1 = -2(\gamma_1 w_1 + \gamma w_{12}) , \quad \dot{w}_2 = -2(\gamma_2 w_2 + \gamma w_{12}) , \tag{9.21}$$

$$\dot{w} = 2(\gamma_1 w_1 + 2\gamma w_{12} + \gamma_2 w_2) \,, \tag{9.22}$$

$$w_{12} = \mathrm{Re}\{e^{i\Delta\omega t} a_1^* a_2\} = \mathrm{Re}\{e^{-i\Delta\omega t} a_1 a_2^*\} \,. \tag{9.23}$$

For the case of the one continuum we have $\gamma = (\gamma_1 \gamma_2)^{1/2}$ and (9.22) takes the form

$$\dot{w} = 2\,|\gamma_1^{1/2} a_1 + \gamma_2^{1/2} e^{i\Delta\omega t} a_2|^2 \,. \tag{9.24}$$

We consider the limiting cases of the solution of the system (9.21–23). We shall show that the Fano result follows from (9.19,20) if perturbation theory is used in the term of the quantity $V - i\gamma$. Putting $a_1 = 1$ in (9.20), we obtain a_2; then putting a_2 in (9.24) and changing over to sufficiently long times $\gamma_2 t \gg 1$, we arrive at the Fano formula [9.6].

$$\dot{w} = 2\,\frac{|\gamma_1^{1/2}\Delta\omega + V\gamma_2^{1/2}|^2}{\Delta\omega^2 + \gamma_2^2} \,. \tag{9.25}$$

It is clear from the derivation that (9.25) is valid under the condition

$$\gamma_2 \gg |V - i\gamma| \,, \quad \text{or} \quad \gamma_2 \gg \gamma_1, V \,. \tag{9.26}$$

In essence, the condition (9.26) expresses the simplest criteria of a quasistationary regime.

Let us consider the limit that is the converse of perturbation theory, when

$$V \gg \gamma_1, \gamma_2 \,. \tag{9.27}$$

In this case we can retain in (9.19,20) only the terms containing V. Then the solutions of (9.19,20) and the probability (9.24) oscillate rapidly with a frequency Ω_0. For such an "oscillatory" regime, physical interest is attached to quantities averaged over the oscillation period $2\pi/\Omega_0$:

$$\bar{w}_1 = \frac{\Delta\omega^2 + 2V^2}{\Delta\omega^2 + 4V^2} \,, \quad \bar{w}_2 = \frac{2V^2}{\Delta\omega^2 + 4V^2} \,, \tag{9.28}$$

$$\bar{\dot{w}} = \gamma_1 + \frac{|\gamma_1^{1/2}\Delta\omega + \gamma_2^{1/2} 2V|^2}{\Delta\omega^2 + 4V^2} \,. \tag{9.29}$$

Expression (9.29) describes, just as the Fano formula (9.25) does, a certain resonant interference process. We note that the form of the resonance for (9.25,29) is determined by the same function

$$f(x) = \frac{(x+q)^2}{x^2+1} \,, \tag{9.30}$$

where in the Fano case we have

$$x = \Delta \omega / \gamma_2 , \quad q = V/\gamma = V/\sqrt{\gamma_1 \gamma_2} , \tag{9.31}$$

and in the case of (9.29) we have

$$x = \Delta \omega / 2V , \quad q = \sqrt{\gamma_2 / \gamma_1} . \tag{9.32}$$

The oscillating decay in a two-level system has been studied in considerable detail (e.g., [9.10]), but the interference effect described by (9.30) has not yet been analyzed, since the considered problems pertained usually to the case $\gamma = 0$ (Sect. 5, example 2).

9.1.4 Certain Examples

The theory developed above has certain applications in atomic and possibly nuclear physics. We shall now discuss several concrete physical examples. We are using the concept of a compound object, namely, to reveal the resonant levels. We do not consider the states of individual objects contained in the quantum system, but rather the levels of the entire compound system as a whole. The advantage of this approach lies in its consistency, a particularly importance factor in complicated cases (see examples 2, 3, and 4).

All of the examples considered here are connected with the transitions due to the interaction with a monochromatic (e.g., laser) electromagnetic field of intensity

$$\mathscr{E}(t) = \mathscr{E}_0 \cos \omega t . \tag{9.33}$$

These examples make it possible (at least from the fundamental point of view), on the one hand, to trace the continuous transition from a weak perturbation to a strong one by increasing $\mathscr{E}_0$; and on the other hand, to investigate the form of the resonance by varying ω.

Adhering to the concept of a compound object, it is convenient in the analysis of the resonant states to regard the field as a system with a definite number of quanta n_ω, and to include it in the unperturbed Hamiltonian $\hat{H}_0$. In the concrete expressions for the matrix elements, on the other hand, going to the limit as $n \to \infty$, we shall use the quantity $\mathscr{E}_0$.

Example 1 (Fig. 9.1). Photoeffect Under the Conditions of Auto-Ionization of an Atom. This illustrates the simplest case, the resonance of two discrete states against the background of one continuum. We assume that the electron shell of the atom X consists of a valence electron e and an atomic residue X^+ with known wave function. These objects, in conjunction with the electromagnetic field, form the compound system "atomic residue X^+ + electromagnetic field $\mathscr{E}$ + valence electron e," the Hamiltonian $\hat{H}$ which is of the form (9.1), where

$$\hat{V} = \hat{V}_{eX^+} + \hat{V}_{\mathscr{E}X^+} + \hat{V}_{e\mathscr{E}} . \tag{9.34}$$

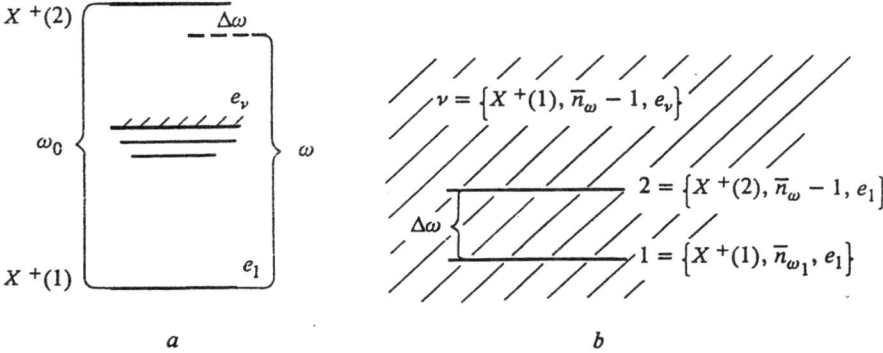

Fig. 9.1a, b. Photoeffect under the conditions of auto-ionization of the atom (Example 1, see text) [9.10]

Let the atomic residue have an isolated excited state X^+ (2) with excitation energy $\hbar\omega_0$ exceeding the detachment energy of the valence electron (see Fig. 9.1). Let also the frequency ω be close to ω_0. Then the Hamiltonian $\hat{H}_0$ corresponds to two discrete states (Fig. 9.1):

$$1 = \{X^+ (1),\ n_\omega,\ e_1\}\ ,\quad 2 = \{X^+ (2),\ n_\omega - 1,\ e_1\}\ ,\tag{9.35}$$

that resonate against the background of one continuum

$$\nu = \{X^+ (1),\ n_\omega - 1,\ e_\nu\}\ .\tag{9.36}$$

Here $X^+ (1)$ is the ground state of the atomic residue, while e_1 and e_ν denote the valence electron in the ground and free states, respectively. Thus, we can apply to this problem directly the analysis of Sects. 4, 5 with the quantities V, γ_1, γ_2 expressed simply in terms of the matrix elements of the operators (9.34). Thus, for example, we have

$$\gamma_2 = \varrho(\omega_0)\,|V_{eX^+}|^2\ .\tag{9.37}$$

Here $\varrho(\omega_0) = 2\pi\hbar^{-1}(d\nu/dE)$, and $E = \hbar\omega_0$ is the density of states of the detached electron. For this problem, the condition for the validity of the Fano theory (9.27) obviously reduces to a limitation on the field intensity:

$$\mathscr{E}_0^2 \ll \frac{\varrho^2(\omega_0)\,|V_{eX^+}|^4}{2d_{12}^2 + \varrho^2(\omega_0)(d_{1\nu})^2\,|V_{eX^+}|^2} = \mathscr{E}_{0\mathrm{cr}}^2\ .\tag{9.38}$$

A deviation from the Fano formula should be observed in fields $\mathscr{E}_0$ that are comparable with (or larger than) the critical field $\mathscr{E}_{0\mathrm{cr}}$.

If the parameter γ is determined by the electrostatic interaction of the valence electron with the atomic residue then usually (e.g., [9.11]) $\gamma_2 \simeq 10^2$ eV and we have $\mathscr{E}_{0cr} \simeq 10^{-2}$ a.u. $\simeq 10^8$ V/cm, i.e., the Fano formula is valid for most real cases. However, auto-ionization states exist with large lifetimes, which decay either radiatively or (slower still by a factor $\hbar c/e^2$) as a result of spin interaction [9.6, 11]. For these states, the case $\mathscr{E}_0 > \mathscr{E}_{0cr}$ can be frequently realized.

It is most convenient to investigate the shape of the resonance curve in the photoeffect with the aid of tunable lasers. So far, no such lasers have been developed for the short wavelength region corresponding to excitation of the auto-ionization levels of atoms. There is, however, a situation that is favorable from this point of view in the case of negative ions, where the auto-ionization levels are already quite low.

Example 2. Induced Transition Between Spontaneously Decaying States of an Atom. The monochromatic field is regarded as a discrete object, and the spontaneous field is continuous (Fig. 9.2).

States 1 and 2 of the atom X, coupled by the field (9.33), decay spontaneously into other states $X(m)$. Here $H_0 = H_X + H_{\mathscr{E}} + H_{vac}$ (where H_{vac} is the Hamiltonian of the zero-point oscillation of the field) has the following discrete states:

$$1 = \{X(1), n_\omega, \dots, n_{\omega_{1m}}, n_{\omega_{2m}}, \dots\} , \quad 2 = \{X(2), n_\omega - 1, \dots, n_{\omega_{1m}}, n_{\omega_{2m}}, \dots\}$$

which resonate against the background of the continua that are characterized by the states of the atom $X(m)$ and of the field (9.33) and by the photon occupation numbers:

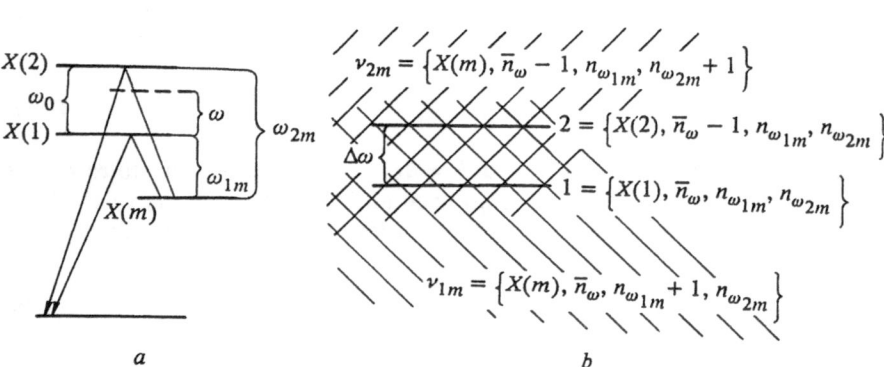

Fig. 9.2a, b. The induced transitions between spontaneously decaying states of an atom (Example 2, see text) [9.10]

$$v_{1m} = \{X(m), n_\omega, \ldots, n_{\omega_{1m}} + 1, n_{\omega_{2m}} \ldots\} \ ,$$

$$v_{2m} = \{X(2), n_\omega - 1, \ldots, n_{\omega_{1m}}, n_{\omega_{2m}} + 1, \ldots\} \ .$$

The transitions $1 \to v_{2m}$ and $2 \to v_{1m}$ are forbidden, since they correspond to a simultaneous change of the states of three pairwise interacting objects. Each state interacts only with "its own" continua, thus corresponding to the case $\gamma_1 \gamma_2 \neq \gamma^2 = 0$. Expressions for γ_1, γ_2 and V are widely known. This problem has been considered in detail for weak and strong fields.

Example 3 (Fig. 9.3). Ionization and Pairwise Excitation of Atoms in the Course of a Collision in a Radiation Field, i.e. a Radiative Collision. This illustrates the more complicated case of resonance against the background of two continua with a time-dependent interaction.

The atoms X and Y collide in the field (9.33), and the energy is close to the sum of the excitation energies of the atoms, $\hbar\omega_0 = E_{21}^X + E_{21}^Y$ and exceeds the ionization energy of each of the atoms. In this case the compound object is "atom X + atom Y + electromagnetic field $\mathscr{E}$." The Hamiltonian is

$$\hat{H}_0 = \hat{H}_X + \hat{H}_Y + \hat{H}_{\mathscr{E}} \ .$$

In this problem, two discrete states of the unperturbed Hamiltonian H_0 (see Fig. 9.3)

$$1 = \{X(1), \ Y(1), \ n_\omega\} \ , \quad 2 = \{X(2), \ Y(2), \ n_\omega - 1\}$$

resonate against the background of two continua

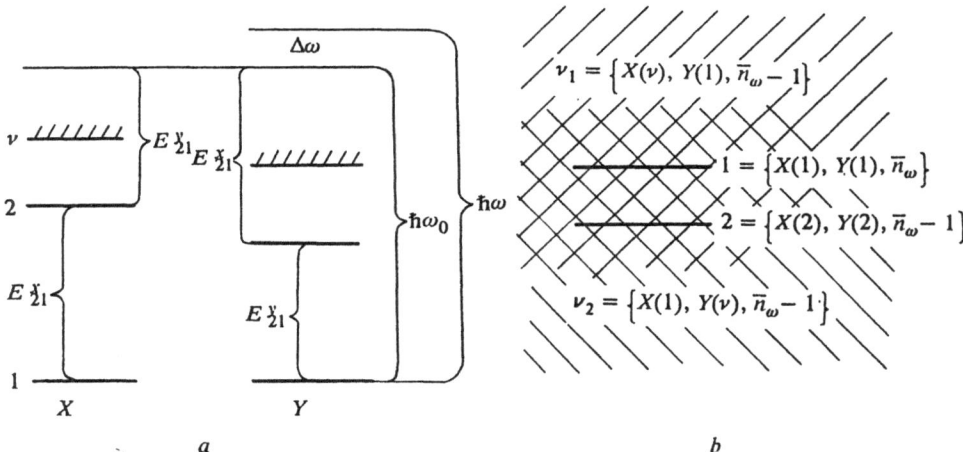

Fig. 9.3 a, b. Pairwise excitation of atoms with ionization by collision in a radiation field (Example 3, see text) [9.10]

$$v_1 = \{X(v),\ Y(1),\ n_\omega - 1\}\ ,$$

$$v_2 = \{X(1),\ Y(v),\ n_\omega - 1\}\ .$$

The interaction of the atoms V_{XY} depends parametrically on the distance between the nuclei and consequently on the time. The transitions $1 \rightarrow v_1$ and $1 \rightarrow v_2$ correspond to atomic photo-ionization reactions; the transition $1 \rightarrow 2$, which correspond to the excitation of a pair of interacting atoms (quasi-molecular) by one photon is a radiative-collision reaction (analogous to that considered in [9.9]) the transitions $2 \rightarrow v_1$ and $2 \rightarrow v_2$ comprise the result of de-excitation of the other (analogous to the Penning effect, see [9.8, 11]).

Example 4 (Fig. 9.4). Scattering of an Electron by Atoms Situated in a Resonant Electromagnetic Field. The application of the theory to scattering problems is illustrated.

The atoms (or molecules) X are situated in a field (9.33) whose frequency is close to the frequency ω_0 of the transition $X(1) \rightarrow X(2)$. An electron is scattered by the atom. Interest attaches to excitation of the atom by an electron with participation of an optical quantum. The chosen compound system is "atom X + electromagnetic field E + free electron".

According to the approach developed in Sect. 3 for scattering problems, we find for the object "atom X + field E", which has discrete levels, an aggregate of state $\{n\}$, from which we separate subgroups of resonating states $\{n\}$ and

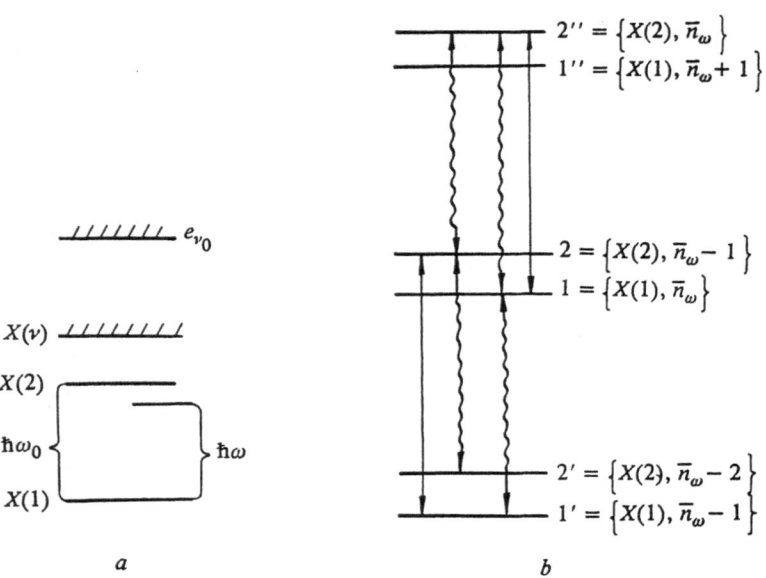

Fig. 9.4a, b. Scattering of an electron by an atom situated in a resonant electromagnetic field (Example 4, see text), [9.10]

states $\{m\}$ that are far from consideration, then we can confine ourselves to the following six discrete states:

$$1 = \{X(1), n_\omega\}, \qquad 2 = \{X(2), n_\omega - 1\},$$

$$1' = \{X(1), n_\omega - 1\}, \qquad 2' = \{X(2), n_\omega - 2\},$$

$$1'' = \{X(1), n_\omega + 1\}, \qquad 2'' = \{X(2), n_\omega\},$$

of which 1 and 2 form the subgroup $\{n\}$. Figure 9.4b shows the possible transitions from the subgroup $\{m\}$. These transitions (to the "remote" levels) are accompanied by corresponding changes in the free electron energy. On the other hand, transitions within the group of states $\{1,2\}$, $\{1',2'\}$, $\{1'',2''\}$ can be regarded as elastic. The latter makes it possible to reduce the solution of this problem to the general results (9.18) of Sect. 3 for the scattering problems. We note that the considered process of excitation of the atom by the electrons in the presence of an optical field is possible, as a result of interference between a bremsstrahlung transition with the collision transition, as a consequence of the virtual transitions $1 \rightarrow 1' \rightarrow 2$ or $1 \rightarrow 2'' \rightarrow 2$ (Fig. 9.4).

9.2 Damping of Atomic States Due to Their Relaxation in Plasmas

9.2.1 Impact Relaxation of Atomic Levels

Atomic states immersed in plasmas do not remain stationary and are under the continuous influence of plasma particles. The condition energy exchange takes place between interatomic electron states and the surrounding plasma. The observed quantized light corresponding to the atomic transition is emitted in essence by the whole particle system of "atom + a plasma". This viewpoint of the radiation goes back, as already mentioned, to Max Born. The action of the surrounding medium on the atom may be under some conditions reduced to a relaxation of atomic states characterized by the damping constant γ. The reduction contains the procedure of averaging over the plasma variables being of a type of a continuum for the transitions considered.

The damping of atomic states due to their collisions with fast plasma electrons is well known in the impact theory of spectral line broadening. The foundation of the theory is the possibility of presentation of the average evolution operator $\{\hat{T}(t, t')\}$ of atomic states in the form of a damping exponent $\exp(-\hat{\Phi}_n t)$, where $\hat{\Phi}_n$ is independent of the time electron impact broadening operator for a given level n. We will consider the structure of the operator $\hat{\Phi}_n$ and some features of atomic spectra connected with the structure. The introduction of the constant operator $\hat{\Phi}_n$ is possible for the difference of two typical times, namely, the "dynamical" time of collision $\tau_c \sim \varrho/v$ (ϱ is the impact parameter, v is the electron velocity) and the "kinetic" time between collisions $\gamma^{-1} = (N v \sigma_c)^{-1} \sim |\Phi_n|^{-1}$, where $\sigma_c \sim \pi \varrho_c^2$ is the collision cross section, and N is the electron density. Under the condition $\gamma \tau_c \ll 1$, expressed the isolated nature of collisions the operator $\hat{\Phi}_n$ is connected with the averaged value of the evolution operator $T_n(t + \Delta t, t)$ of the atomic level at the time interval Δt, obeying the conditions $\tau_c \ll \Delta t \ll \gamma^{-1}$, namely

$$\Phi_n = (\Delta t)^{-1} \exp[-i H_n^0 t/\hbar]\{\hat{T}_n(t + \Delta t, t) - 1\}_{Av} \exp[i H_n^0 t/\hbar] . \tag{9.39}$$

Here H_0 is the stationary atomic Hamiltonian, generally including the interaction with the constant ionic field $\boldsymbol{d} \cdot \boldsymbol{F}_i$; and the symbol $\{ \| \}_{Av}$ defines the averaging over the broadening plasma electron ensemble.

The condition $\Delta t \gg \tau_c$ allows replacement of the evolution operator at the time interval Δt by the scattering matrix $S_n = T_n(+\infty, -\infty)$. As a result the operator $\hat{\Phi}_n$ takes the form:

$$\langle \alpha | \hat{\Phi}_n | \alpha' \rangle = N \int_0^\infty dv f(v) v \int_0^\infty 2 \pi \varrho d\varrho \{\langle \alpha | S_n - 1 | \alpha' \rangle\}_{Av} , \tag{9.40}$$

where $f(v)$ is the electron velocity distribution function, α, α' label the sublevels of a given level n, and the symbol $\{ \| \}_{Av}$ represents an averaging over angular variables.

The introduced operator $\hat{\Phi}_n$ describes the atomic state relaxation due to a collision with plasma electrons. Such a collision not results obligatorily in transitions of the electron in the other atomic levels, that is in its "destination" at the given level. For example, their role may be expressed in the sudden displacement ("shaking") of the phase of the atomic state due to pure elastic electron scattering. Damping of the averaged evolution operator means simply disordering the phase coherence of the radiating state. The broadening mechanism not connected with the inelastic transition of the atomic electron to other states was considered for the first time by V. *Weisskopf*. The operator $\hat{\Phi}_n$ for such a mechanism of phase relaxation is diagonal in α, α' quantum numbers, which obviously leads to a Lorentzian profile of emitted spectral lines with their widths defined by the operator $\Phi_{\alpha\alpha}$ [9.12, 13].

For plasma electrons the inelastic broadening mechanism connected with the transition of the atomic electron on other sublevels is more typical. The calculation of the operator $(\Phi_n)_{\alpha\alpha}$ may be performed in the case with the help of perturbation theory over the interaction $-e r \cdot F_e(t)$ of the atomic dipole moment with the field $F_e(t)$ produced by plasma electron. The possibility of the perturbation theory (Born approximation) application is connected with the distant nature of electron Coulomb field F_e leading, as for a number of other plasma characteristics, to the logarithmic dependence of cross sections reflecting the main role of distant collisions. In the simplest approximation of the classical trajectories of plasma electrons, the calculation of S-matrix in (9.40) using second-order perturbation theory over the interaction $e r \cdot F_e$ leads to the well-known result:

$$-(\Phi_n)_{\alpha\alpha'} = (w_n)\alpha\alpha' + i(d_n)\alpha\alpha' = \gamma_0 \sum_{\alpha''} \left\langle \alpha \left| \frac{r}{a_0} \right| \alpha'' \right\rangle \left\langle \alpha'' \left| \frac{r}{a_0} \right| \alpha' \right\rangle (\Lambda + i) \ . \tag{9.41}$$

Here the real (w_n) and image (d_n) parts of the operator Φ_n define the width and shift of the spectral line, $\gamma_0 \equiv \frac{3}{2}\pi(\hbar/m)^2 N\{v^{-1}\}_{\text{Av}}$ is the typical scale of the broadening collision frequency, a_0 is the Born radius, $\Lambda \equiv \ln(\varrho_{\max}/\varrho_{\min})$ is the typical logarithm defined cross section, and i is the imaginary unit.

We are not interested in numerical values of the width w and the shift d of the spectral line in the following discussion, but we rather investigate the types of their dependencies on the quantum numbers α, α' of atomic states. If the distribution of perturbing plasma electrons is spherical symmetric such a symmetry will be true also for the operator $\hat{\Phi}_n$ which contains the average over the electron distribution. This means that the operator $\hat{\Phi}_n$ may be diagonalized in the spherical basis.

The specifics of the influence of the plasma on the atom is that together with the impact action of the electrons the atom is under the influence of the static ion fields $d \cdot F_i$, as well. This leads to the fact that matrix elements of the operator $\hat{\Phi}_n$ must be considered in the atomic state basis where the interaction with the ion field F_i is diagonal. This basis for hydrogenic states is parabolic and it responds to the atomic quantization along the field F_i. However, in the basis the operator Φ_n is not already diagonal, resulting in definite features of the spectral line shape formation.

Calculations of non zero matrix elements of the operator $r \cdot r$ defining properties of the operator Φ_n in a parabolic basis are conveniently performed with the help of the four-dimensional symmetry properties of hydrogen atom. Accounting for the connection between vector r and the Runge-Lentz vector A for atomic states with a given n, one may reduce the calculation of matrix elements $r \cdot r$ to the corresponding calculation of the operator $A * \cdot A$. Using for the last case the connection of the operator A with operators of generalized orbital momenta, J_1 and J_2 it is easy to obtain [9.12]:

$$\langle n_1 n_2 m | A \cdot A | n_1 n_2 m \rangle = \frac{\hbar^2}{2} [n^2 + (n_1 - n_2)^2 - |m|^2 - 1] \ , \tag{9.42}$$

$$\langle n_1 - 1, n_2 + 1, m | A \cdot A | n_1 n_2 m \rangle = \hbar^2 [n_1(n - n_1)(n_2 + 1)(n - n_2 - 1)]^{1/2} \ , \tag{9.43}$$

$$\langle n_1 + 1, n_2 - 1, m | A \cdot A | n_1 n_2 m \rangle = \hbar^2 [n_2(n - n_2)(n_1 + 1)(n - n_1 - 1)]^{1/2} \ . \tag{9.44}$$

As a result the analogous dependence is realized for the matrix elements of the operator Φ_n as well.

Let us consider the concrete structure of the operator Φ_n for the simplest spectral line $L_\alpha (n = 2)$. The splitting of the level $n = 2$ in an ion field reduced, as is well known, to two shifted lateral components responding to the projection of the magnetic quantum number $m = 0$ and one central (unshifted) component, responding to values $m = \pm 1$. Let us introduce designations for the wave functions φ_i of the parabolic atomic states and matrix elements of the operator Φ:

$$
\left.
\begin{aligned}
&\varphi_1 \equiv |010\rangle, \ \varphi_2 \equiv |100\rangle, \quad |000\rangle \equiv \varphi_0; \ w, \beta, d > 0 \\
&(-\Phi_2)_{11} = w - id, \qquad -(\Phi_2)_{22} = w + id, \\
&(-\Phi_2)_{12} = \beta + id, \qquad -(\Phi_2)_{21} = \beta - id, \\
&(H_2^0)_{11} = \omega_0 - \frac{\Delta}{2}, \qquad (H_2^0)_{22} = \omega_0 + \frac{\Delta}{2}.
\end{aligned}
\right\}
\tag{9.45}
$$

Here ω_0 is the unperturbed transition frequency, and $\Delta/2$ is the component Stark shift in the ion field.

The structure of the operator Φ_2 (9.45) consists in correspondence with (9.42–44) of the diagonal (w) and nondiagonal (β) elements. The appearance of both sect. of elements is easy to recognize from Fig. 9.5. It is particularly clear that the nondiagonal matrix element $-\Phi_{12} \equiv \beta$ is responsible for the connection of the atomic states 1 and 2 through the third atomic state 0. We have already encountered such a nondiagonal relaxation connection between atomic states in Sect. 9.1, where a continuum played the role of the third state. As will be shown, the appearance of such a nondiagonal connection has an effect on the merging of the components.

9.2.2 Features of the Spectral Line Shape Under Impact Relaxation of Atomic Sublevels in an Ion Field

Let us consider some features of the spectral line shapes in ion fields caused by the relaxation operator Φ_n structure mentioned above. The expression for

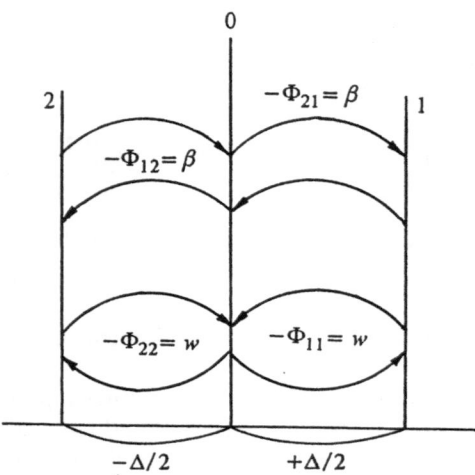

Fig. 9.5. Collisional transitions between Stark components of the L_α line responsible for the formation of a diagonal and nondiagonal matrix elements of the electronic impact broadening operator Φ

the spectral line shape $I_{nn'}(\omega)$ corresponding to the transition from the upper state n to the nondegenerate lower state n' is obtained by the Fourier transform of the evolution operator [9.12, 13]:

$$I_{nn'}(\omega) = \frac{1}{\pi} \mathrm{Re} \left\{ \sum_{\alpha\alpha'} \langle \alpha | d | n' \rangle \langle n' | d | \alpha' \rangle \right.$$

$$\left. \times \left\langle \alpha \left| \left[i \left\{ \omega - \frac{1}{\hbar} H_n^0(F) - \hat{\Phi}_n \right\} \right]^{-1} \right| \alpha' \right\rangle \right. . \tag{9.46}$$

When calculating $I_{nn'}(\omega)$ from (9.46) one can find either matrix elements of the inverse matrix $[i\{\omega - (H_n^0 - H_{n'}^0)/\hbar - \Phi_{nn'}\}]^{-1}$ in a given system of wave functions (for example, the parabolic ones diagonalizing the Hamiltonian H_n^0) or the complete system of eigenfunctions of the matrix $(H_n^0 - H_{n'}^0)/\hbar - i\Phi_{nn'}$ with subsequent calculation of the matrix elements in this system. Naturally, both approaches are closely connected.

Let E_α and Ψ_α be the eigenvalues and eigenfunctions of the operator $\hbar^{-1} H_n^0(F) - i\Phi$ and E_β^0, φ_β are the same for the operator $\hbar^{-1} H_n^0(F)$. Furthermore, the matrix $\hat{C}_{\beta\alpha}$ determines the transition from the basis $|\varphi_\beta\rangle$ to the basis $|\Psi_\alpha\rangle$. The operator $\hbar^{-1} H_n^0 - i\Phi_n$ is non-Hermitian, so for the determination of the complete system of wave functions in (9.46) it is necessary to supplement the system of ket vectors $|\Psi_\alpha\rangle$ by the system of bra vectors $\langle \chi_\alpha |$ orthogonal to it. The former system is obtained from $\langle \varphi_\beta |$ with the help of the matrix $\hat{C}_{\alpha\beta}^{-1}$. Note that the non-Hermitian nature of the operator $H_n^0 - i\Phi_n$ results in the fact that the matrix $\hat{C}$ is not unitary $C^{-1} \neq C^+$. After that the resolvent operator $[i\omega - \Phi_{nn'}]^{-1}$ in (9.46) will contain only diagonal elements, so the problem reduces to the determination of its eigenvalues and concrete expressions of the matrix C, C^{-1}. When determining the matrix $\hat{C}$ for the L_α line it is enough to consider only the shifted line components, corresponding to $m = 0$. At the same time the matrix block $H_n^0 - i\Phi_n$ is diagonal for central components $m = \pm 1$, so the corresponding matrix C is equivalent to a unity matrix. Introducing the destinations (9.45), we write down the solutions of the secular equation

$$|\delta_{\alpha\beta}(E_\beta^0 - E_\alpha) - i\Phi_{\alpha\beta}| = 0 \tag{9.47}$$

in the following form

$$E_{12} = \omega_0 \pm \Omega - iw , \tag{9.48}$$

where

$$\Omega = \sqrt{\frac{\Delta^2}{4} - \Delta d - \beta^2} .$$

The role of the nondiagonal matrix element β is seen from (9.48). Indeed, in contrast with the real nondiagonal matrix element V_{12} (Sect. 9.2.1) causing the repulsion of atomic terms, the "relaxation" matrix element β being the nondiagonal part of the operator $\hat{\Phi}_n$ causes the effective attraction of the Stark sublevels. It is seen that for a decrease in the ion splittings of the atomic levels ($\Delta \to 0$) the magnitude Ω becomes purely imaginary and its contribution to the energy eigenvalue (9.48) reduces to a renormalization of the component width w. So the relaxation interaction of atomic terms through an intermediate state results in the effect of their merging ("collapse" [9.13]).

Let us find the specific form of the transformation matrix C and C^{-1}. Taking into account (9.47) we obtain

$$
C = \begin{bmatrix} 1 & \dfrac{i(\Phi_2)_{12}}{\hbar^{-1}(H_2^{00})_{11} - E_2 - i(\Phi_2)_{11}} \\[2ex] \dfrac{i(\Phi_2)_{21}}{\hbar^{-1}(H_2^{0})_{22} - E - i(\Phi_2)_{22}} & 1 \end{bmatrix} ,
$$

$$
C^{-1} = \frac{1}{\det C} \begin{bmatrix} 1 & \dfrac{-i(\Phi_2)_{12}}{(\hbar^{-1}H_2^{0})_{11} - E_2 - i(\Phi_2)_{11}} \\[2ex] \dfrac{-i(\Phi_2)_{21}}{(\hbar^{-1}H_2^{0})_{22} - E_1 - i(\Phi_2)_{22}} & 1 \end{bmatrix} .
$$

$$(9.49)$$

The line intensity $I(\omega)$ is expressed as the sum of two terms $I(\omega) = I_1(\omega) + I_2(\omega)$, corresponding to two eigenvalues (9.48) transforming for large values of Δ to shifted Stark components of the spectral line. In such a way the contribution of component 1 takes the form:

$$
I_1(\omega) = \frac{1}{\pi} \operatorname{Re} \left\{ \sum_{a,a'} \frac{(\hat{C}^{-1})_{1a}(\hat{C})_{a'1}}{i(\omega - E_1)} d_{a0} \cdot d_{0a'} \right\} . \tag{9.50}
$$

In the domain of real values of $\Omega (\Delta \geq 2\beta)$ the spectral line shape takes the form [9.12, 13]:

$$
I(\omega) = \frac{d^2}{\pi} \left[\frac{w + \beta(\omega - \omega_0 + \Omega)/\Omega}{(\omega - \omega_0 + \Omega)^2 + w^2} + \frac{w - \beta(\omega - \omega_0 - \Omega)/\Omega}{(\omega - \omega_0 - \Omega)^2 + w^2} \right] . \tag{9.51}
$$

It is seen from (9.51) that the appearance of the nondiagonal matrix element $\beta \neq 0$ results in the difference of the shape emitted from the composition of two Lorentzian line shapes. The transition to this former case takes place only for well isolated components corresponding to the large splitting Δ in the ion field: $\beta/\Delta \ll 1$. In the domain $\Delta \leq \beta$ these components are effectively attractive and the Lorentzian profiles are deformed.

9.3 Emission of Forbidden Spectral Lines and Decay of Metastable Levels in Plasmas

9.3.1 The Polarization Mechanism for Forbidden Transitions in an Atom

The atomic state polarization due to collisions with charged particles may result in the radiation of the atomic states which have not possessed any previous radiative transitions. Thus, the spectral lines arise in atomic spectra at frequencies of transitions which have been forbidden by selection rules for dipolar radiation. The striking example of the effect is the forbidden helium lines $4F-2P$ arising for radiation emission of the atom in a plasma. One of the most interesting examples of such forbidden transitions is the effect of the destruction of a metastable level due to collisions with the charged particles. Here the classic examples are the $2p_{1/2}$ and $2s_{1/2}$ atomic levels of a hydrogen or hydrogen-like ion. The level $2p_{1/2}$ is radiative with a large decay rate constant γ, but the level $2s_{1/2}$ is metastable and the decay rate to the $1s$ state (due to the two photon transition) may be neglected. Under the collisions with a slow charged particle (e.g. a plasma ion) the dipolar moment appears at the transition $2s \rightarrow 1s$ due to polarization in the system of the $2s-2p$ levels, leading to a possibility of quantum emission at the frequency of the $2s \rightarrow 1s$ transition. That results in the finiteness of the lifetime τ of the metastable level.

Two main mechanisms of metastable level destruction must be mentioned [9.14–21]. The first one is due to direct collisional transitions from the $2s$ to the $2p$ state, and it is determined by the inelastic cross section of the $2s \rightarrow 2p$ transitions accompanied by further emission of a $2p$ level. The second mechanism stipulated by polarization under slow collisions doesn't lead to inelastic transitions from the $2s$ to the $2p$ level, but causes only polarization mixing of these states. This results in the appearance of the nonzero dipole moment at the $2s \rightarrow 1s$ transition. So, namely this second (polarization) mechanism of the metastable level emission is responsible for forbidden line formation. These lines may be considered as some type of polarization emission of atoms. Note that the forbidden lines arise even from a static external influence, e.g., for static atomic state polarization.

The investigation of the forbidden line spectra is complicated enough [9.15–17], so we will limit our discussion by the consideration of the integral (over spectrum) characteristics, namely the lifetime τ of the metastable level. The time determines the total intensity of the forbidden transition [9.17].

The initial system of Schrödinger equations for the amplitudes a_0 and a_1 of the stationary (or metastable) and radiative states has the form

$$\left. \begin{array}{l} i\dot{a}_0 = V_{01}(t)e^{i\omega t}a_1 \ , \\[2mm] i\dot{a}_1 = -i\gamma a_1 + V_{10}(t)e^{-i\omega t}a_0 \ . \end{array} \right\} \tag{9.52}$$

Here ω is distance between the levels, V_{10} is the matrix element of the dipole interaction of the atom with the electric microfield F of the system of N charged particles:

$$\hbar V_{01}(t) = d_{10} \cdot F(t) \; ,$$

$$F(t) = e \sum_i (r_{0i} + v_i t) |r_{0i} + v_i t|^{-3} \; , \tag{9.53}$$

where r_{0i} and v_i are the initial coordinates and velocities of the perturbing particles with charge e, and d is the dipole moment of the atom.

The problem concerning the lifetime τ of the stationary level, that is, the time it takes for the value of $|a_0|^2$ to decrease by a factor of e, breaks up into two stages. The first stage consists in the solution of the dynamic problem, that is, the solution of the system (9.52) with the appropriate initial conditions. The second stage consists in averaging the result with respect to the statistical ensemble of the random variables r_{0i} and v_i.

The conventional approach to the calculation of τ, developed by *Purcell* [9.14] and improved in [9.15], − the calculation of the cross section for isolated fast collisions − corresponds to the solution of the system (9.52) for $\gamma \equiv 0$ according to perturbation theory. The position of this method (henceforth called the Purcell method) within the framework of the general theory will be clarified below. Incidentally, without such a clarification it is not completely obvious that such an approach is internally consistent since it combines the neglect of γ (i.e., it assumes $\gamma = 0$) with a non-allowance for the possibility of the inverse transition from a_1 to a_0 even before the emission of the photon (which is equivalent, on the other hand, to the assumption $\gamma \to \infty$).

The most realistic case $\gamma \ll \omega$ is considered below (since the ratio $\gamma/\omega \approx 0.1$ for the hydrogen levels $2s - 2p_{1/2}$). In this connection it turns out that the finiteness of γ itself leads to such interesting physical consequences as an unusual "intermingling" of the concepts of the elasticity and inelasticity of scattering, an "intermingling" of the concepts of adiabaticity and nonadiabaticity, and the appearance of effective cross sections of a more complicated (in comparison with an ordinary cross section) nature containing γ. Furthermore, sometimes it even leads to the result that τ cannot be reduced to a cross section in general.

We start with the case of the decay of a metastable atomic level due to the presence of the constant electric field F created by the surrounding charged particles in the plasma. In the absence of radiative damping ($\gamma = 0$) a constant perturbation would simply lead to a periodic oscillation of the amplitudes of both atomic states with frequencies $\omega^{(1,2)}$, which are determined by the roots of the corresponding secular equation [9.11]. Taking the attenuation into account ($\gamma \neq 0$) leads to the appearance of the imaginary parts in the frequencies $\omega^{(1,2)}$ and, thereby, leads to an attenuation of the amplitude of the metastable state. In the case $\gamma \ll \omega$ under consideration, the probability $|a_0(t)|^2$ that the system remains in the metastable state is given by the expression

$$|a_0(t)|^2 = \exp\left[-\gamma t \left(1 - \frac{\omega}{\sqrt{\omega^2 + 4|V_{01}|^2}} \right) \right] . \tag{9.54}$$

This formula is obtained from the exact expression, which is much more cumbersome, by making certain simplifications which are permissible in the lifetime problem of interest to us. The simplifications are the terms which are most slowly damped with time are kept, and the time independent, pre-exponential factor, which is close to unity, has been replaced by unity.

For sufficiently strong fields ($|V_{01}|_{\text{eff}} \gg \omega$) the states a_0 and a_1 are strongly "intermixed" and the lifetime τ of both states turns out to be of the order of $1/\gamma$. Therefore, from the point of view of visualizating the dependence of τ on the plasma parameters, the most interesting case is $V_{01} \ll \omega$ when [compare with (9.54)]

$$\text{Im}\{\omega^{(1,2)}\} = -\frac{\gamma}{\omega^2}|V_{10}|^2 = -\frac{\gamma}{\omega^2}\frac{|d_{01}|^2}{\hbar^2}F^2 \equiv \frac{C_4}{e^2}F^2 \ . \tag{9.55}$$

It follows from (9.55) that in this case we are dealing with a distinctive quadratic Stark effect and the value of the constant C_4 depends on the value of γ.

To determine τ it is necessary to average (9.54) with respect to the distribution of the plasma microfield, which we assume to be a Holtsmark distribution [9.15]. Then the equation for the determination of τ takes the form

$$\int_0^\infty \exp\{-\gamma\tau[1-1/((1+2\lambda\mu z)^2)^{1/2}]\}H(z)dz = e^{-1} \ . \tag{9.56}$$

Here $H(z)$ is the Holtsmark function, $\lambda \approx 2.603$, $\mu = (\hbar N^{2/3}/m\omega)$ (N denotes the density of the perturbing particles) is the characteristic dimensionless parameter which determines the order of magnitude of the ratio of the value of the splitting in the field for an average interparticle distance ($\approx eN^{2/3}$) to the distance between the levels.

For $\mu \gg 1$ the lifetime τ determined from (9.56) is of the order of $1/\gamma$, in agreement with what was said earlier. For $\mu \ll 1$ we obtain

$$\tau \approx \gamma^{-1}0.13/(\lambda\mu)^2 \equiv \tau^{st} \ . \tag{9.57}$$

It is interesting to determine what value F^* of the *constant* field F corresponds to τ from (9.57). Using the expression for the lifetime in a constant field (obtained from (9.54) or (9.56) for $\mu \ll 1$) and equating it to τ as given by (9.57), we have: $F \approx 5.11\,eN^{2/3}$.

9.3.2 Interrelation Between the Nonelastic and Polarization Mechanisms. The Weisskopf Mechanism for Inelastic Transitions

Let us consider the solution of the system (9.52) in the approximation of perturbation theory with regard to the magnitude of the interaction V. We seek the solution for a_0 in the form $a_0 = e^{-i\varphi(t)}$, where in second-order perturbation theory we write down the following expression for the phase $\varphi(t)$ (assuming $a_0(0) = 1$):

$$\varphi(t) = -i \int_0^t dt' \, V_{01}(t) e^{(i\omega-\gamma)t'} \int_0^{t'} dt'' \, V_{10}(t'') e^{(-i\omega+\gamma)t''} . \tag{9.58}$$

Since the phase shifts are small in the case under consideration, we assume that the resulting shift $\varphi(t)$ is the summation of the shifts from individual collisions: $\varphi(t) = \sum_{k=1}^{\mathcal{N}} \varphi_k(t)$. This enables us to reduce the averaging over the phase space of $\mathcal{N}$ particles (denoted below by $\langle \rangle_{\mathcal{N}}$) to an averaging over the phase space of a single particle. Introducing the normalization volume V containing $\mathcal{N}$ particles and going to the limit $\mathcal{N}, V \to \infty$ in the usual manner, we obtain

$$\langle |a_0(t)|^2 \rangle_{\mathcal{N}} = \langle e^{2\,\mathrm{Im}\,\{\varphi(t)\}} \rangle_{\mathcal{N}} = \left\langle \prod_{k=1}^{\mathcal{N}} e^{2\,\mathrm{Im}\,\{\varphi_k(t)\}} \right\rangle_{\mathcal{N}} = (\langle e^{2\,\mathrm{Im}\,\{\varphi_1(t)\}} \rangle_1)^{\mathcal{N}}$$

$$= \left[1 - \frac{1}{V} \int dr (1 - e^{2\,\mathrm{Im}\,\{\varphi_1(t)\}}) \right]^{N \cdot V} = \exp\left[-NV(t) \right] , \tag{9.59}$$

where the "collision volume" $V(t)$ is given by

$$V(t) = \int dr (1 - e^{2\,\mathrm{Im}\,\{\varphi_1(t)\}}) . \tag{9.60}$$

Expressions (9.59, 60) resemble the expressions for the correlation function in the adiabatic theory of line broadening [9.19] with the difference that the phase $\varphi(t)$ is complex in the general case.

Let us now consider the static limit, which is connected to the present case with the perturbation of the atom by individual particles. Calculating (9.60) with (9.55, 58) (where $F = e/r_0^2$) taken into consideration, we find

$$V(t) = 4\pi \int_0^\infty dr_0 r_0^2 [1 - \exp(-2C_4 t/r_0^4)] = \frac{4\pi}{3} \Gamma(1/4)(2C_4 t)^{3/4} . \tag{9.61}$$

According to (9.59, 61) we obtain the following result for the lifetime

$$\tau \simeq \gamma^{-1} \mu^{-2} \frac{1}{2} (3/4\pi \Gamma(1/4))^{4/3} \simeq 0.013 \, \gamma^{-1} \mu^{-2} \equiv \tilde{\tau}_{st} . \tag{9.62}$$

This "binary" $\tilde{\tau}_{st}$ differs from the exact expression (9.57) for τ_{st} by a numerical factor ≈ 0.7 which obviously characterizes the accuracy of the binary scheme (the additivity of the phases φ_k) in the static limit. The noted difference is (besides, for example, the Holtsmark broadening of the lines) one of a few examples of the nonbinary nature of the effect microfield on the atom.

As one moves away from the static limit, the role of the nonbinary effects continuously decreases. As the static limit is approached, the impact limit and the effects due to the action of the microfield (line broadening, Coulomb collisions and others) are completely converted into binary effects.

Now let us take the time dependence of the perturbation V_{01} into account. We shall consider $V_{01}(t)$ in the so called rotating coordinate system, where at any instant of time the axis of quantization z is directed toward the perturbing particle; this permits us to greatly simplify the investigation while preserving all of the characteristic dynamic features of the problem. For the dipole interaction under consideration, we have

$$V_{10}(t) = \alpha R^{-2}(t), \ \alpha = \frac{e}{\hbar} d_{01}, \ R^2(t) = \varrho^2 + v^2 t^2 \ , \tag{9.63}$$

(ϱ is the impact parameter and v is the velocity of the perturbing particle).

A rigorous calculation of the time dependence of V_{01} would require utilization of the quantity $v^2(t-t_0)^2$ in expression (9.63), where t_0 denotes the time of closest approach. The customary simplification $t_0 \equiv 0$ is obviously equivalent to taking account of only one of the completed trajectories, which is completely analogous to the *impact* approximation in the theory of broadening. By using the standard arguments corresponding to this approximation, we obtain the following result (9.58, 60, 63)

$$NV(t) = \Gamma t \ , \tag{9.64}$$

$$\Gamma = N v \int\limits_0^\infty 2\pi \varrho \, d\varrho \ [1 - e^{2 \operatorname{Im} \{\varphi_1(\infty)\}}] \equiv N v \sigma(v) \ , \tag{9.65}$$

$$\operatorname{Im} \{\varphi_1(\infty)\} = -\operatorname{Re} \left\{ \int\limits_0^\infty e^{(i\omega - \gamma)\tau} d\tau \int\limits_{-\infty}^\infty V_{01}(t) V_{10}(t - \tau) dt \right\} \ . \tag{9.66}$$

As is clear from (9.64, 65) in the approximation under consideration the quantity τ is expressed in terms of a certain cross section $\sigma(v)$, which corresponds to the binary nature of the collisions in this approximation. Since the connection between the quantity τ and $\sigma(v)$ is well known in this case, we shall often write out only one of these quantities.

Let us evaluate expression (9.66) by using (9.63) and the condition $\gamma \ll \omega$. Direct integration yields

$$\operatorname{Im} \{\varphi(\infty)\} = -\frac{\pi}{2} \left(\frac{\alpha}{\varrho v} \right)^2 \left[e^{-2\varrho \omega/v} + \frac{\gamma}{\omega} F \left(2 \frac{\varrho}{v} \omega \right) \right] \ , \tag{9.67}$$

where the function $F(x)$ has the form

$$F(x) = x[e^{-x}\overline{\text{Ei}}(x) + e^{x}\overline{\text{Ei}}(-x)] \simeq \begin{cases} 2x^{-1}, & x \gg 1 \\ 2x(\ln x + C), & x \ll 1 \end{cases} . \tag{9.68}$$

(Ei denotes the exponential integral and $C = 0.577$ is Euler's constant).

It follows from (9.67, 68) that the first term in (9.67) gives the major contribution to the cross section (9.65) for large velocities, and the second term gives a major contribution for small velocities. According to (9.58), expression (9.67) is the inelastic scattering phase, where the first term corresponds to the usual Born approximation of the type used by *Purcell* [9.14]. With regard to its structure, the second term resembles the elastic scattering phase in the theory of broadening associated with the presence of a perturbing level, [9.15, 19], which is related to the usual adiabatic (V. Weisskopf) mechanism for line broadening. The appearance of such a phase shift in (9.67), characterizing the inelastic scattering means that the presence of an attenuation γ leads to an "intermingling" of the elastic and inelastic scattering amplitude. Thus, here we encounter a new, unique inelastic transition mechanism, resulting from the presence (to a "sufficient degree") of the elastic phase in the ratio γ/ω. In analogy with the theory of line broadening in what follows we shall call this mechanism the Weisskopf mechanism.

Now we trace the transition between the Purcell and the Weisskopf mechanisms for inelastic scattering. Changing to the dimensionless variables $x = 2\varrho\omega/v$ and $\beta = \omega a/v^2$ in (9.65, 67), we obtain

$$\Gamma = \pi^{5/3}\Gamma(1/3)N\frac{v^{1/2}a^{4/3}}{\omega^{2/3}}I_{\gamma/\omega}(\beta) , \tag{9.69}$$

$$I_{\gamma/\omega}(\beta) = \frac{\beta^{-4/3}}{2\pi^{2/3}\Gamma(1/3)}\int_0^\infty dx x\{1 - \exp[-\beta^2\chi_{\gamma/\omega}(x)]\} , \tag{9.70}$$

$$\chi_{\gamma/\omega}(x) = 4\pi\left(\frac{\pi e^{-x}}{x^2} + \frac{\gamma}{\omega}\frac{F(x)}{x^2}\right) \simeq 4\pi\begin{cases} \dfrac{\gamma}{\omega}2x^{-3}, & x \gg 1 \\ \pi x^{-2}, & x \ll 1 \end{cases} . \tag{9.71}$$

As $\beta \to \infty$, $I_{\gamma/\omega}(\beta)$ approaches $(\gamma/\omega)^{2/3}$, and we arrive at the Weisskopf limit where according to (9.69) the lifetime and the cross section are given by

$$\tau_w^{-1} \equiv \Gamma_w = \pi^{5/3}\Gamma(1/3)Nv^{1/3}(\gamma a^2/\omega^2)^{2/3} . \tag{9.72}$$

For $\beta \to 0$, we have $I_{\gamma/\omega}(\beta) = \pi^{4/3}\beta^{2/3}\ln(\beta^{-1})/\Gamma(1/3)$

which gives the Purcell limit (within logarithmic accuracy) after substitution into expression (9.69)

$$\tau_P^{-1} \equiv \Gamma_P = \pi^3 N \frac{\alpha^2}{v} \ln(\beta^{-1}) \ . \tag{9.73}$$

Expression (9.73) differs from Purcell's result [9.14] by the numerical factor $\pi^2/4$, which is typical for the difference between the results obtained in rotating and fixed coordinate systems [9.15].

The transition from the "collisional" case (9.64) under consideration (more precisely, from its Weisskopf limit) to the static case (9.61) is given by the general formula (9.60). Such a transition is quite analogous to the transition from the impact limit to the static limit in the theory of broadening associated with the quadratic Stark effect. It is possible to evaluate the integral (9.60) analytically. We are primarily interested in the boundary between the regions of the applicability of (9.61, 64). It is well known from the theory of broadening that the static and impact results are valid for times $t \ll \varrho_0/v$ and $t \gg \varrho_0/v$, respectively, where ϱ_0 is the characteristic radius of the collision. In our case it is related to the "Weisskopf" cross section (9.72). We note that the quantity ϱ_0 is, in a sense, the analog of the Weisskopf radius in the theory of broadening; here it has a more complicated nature (which appears in the explicit dependence of ϱ_0 on γ). The "transition" value can also be obtained by directly setting the "collisional" (9.64) (in its Weisskopf limit) and the static (9.61) expression for τ equal to each other.

9.3.3 The Adiabatic Approximation for Polarization Radiation

Now let us consider another approximation which is valid for perturbations which vary sufficiently slowly, the adiabatic approximation. Such an approximation is applicable in the region $\beta \gg 1$, where $\beta = \omega \alpha/v^2$ (9.69). In this case the solution is usually based on the static result (9.54), where the perturbation V_{01} is assumed to parametrically depend on the time [9.15, 19, 20]. On the other hand, one can also approach the solution of this problem in the spirit of [9.15]. Starting from (9.54), let us determine the steady state value of the transition probability per unit time, $W(t)$. If the perturbation varies adiabatically, then one can assume that the change of the total (at the moment t) probability $P(t)$ for a transition from the state a_0 to a_1 is equal at each moment of time to the product of the probability $1 - P$ of observing the system in the state a_0 by probability W, which depends parametrically on the time:

$$dP/dt = -W(t)(1-P); \quad W(t) = -d \ln |a_0(t)|^2/dt$$

$$= \gamma [1 - \omega/(\omega^2 + 4V_{01}^2)^{1/2}] \ . \tag{9.74}$$

Integrating (9.74) with the initial condition $P(-\infty) = 0$, we obtain the following result with (9.54) taken into account:

$$P(\infty) = 1 - \exp\left[-\gamma \int_{-\infty}^{\infty} \left(1 - \frac{\omega}{\sqrt{\omega^2 + 4\,|V_{01}(t)|^2}} \right) dt \right] \,. \tag{9.75}$$

Defining on the basis of (9.75) the "adiabatic" cross section σ_{ad} in the usual manner (compare with (9.65)) and introducing the dimensionless parameter ξ, we obtain:

$$\sigma(\xi) = 4\pi \frac{\alpha}{\omega} \Lambda \left(\frac{\gamma}{\omega} \sqrt{2\omega\alpha/v^2} \right) \,, \tag{9.76}$$

where

$$\Lambda(\xi) = \int_0^{\infty} \frac{dz}{z^3} [1 - \exp(-\xi P(z))] \,, \tag{9.77}$$

$$p(z) = \frac{1}{z} \int_{-\infty}^{\infty} dx \left(1 - \frac{1}{\sqrt{1 + z^4/(1+x^2)^3}} \right) \approx \begin{cases} \pi z^3/4 \,, & z \ll 1 \\ B\left(\dfrac{3}{4}, \dfrac{3}{4} \right) \approx 1.70 \,, & z \gg 1 \end{cases} \tag{9.78}$$

(B is the beta function). For $\xi \gg 1$ we have $z_{eff} \ll 1$ and $\Lambda \propto \xi^{2/3}$, which now gives for (9.76) the well known Weisskopf cross section (9.72). For $\xi \ll 1$ we have $z_{eff} \gg 1$ and $\Lambda \sim \xi$ which gives

$$\sigma \approx \frac{8\pi}{3} B\left(\frac{1}{4}, \frac{5}{4} \right) \frac{\gamma\alpha^{3/2}}{v\omega^{3/2}} \approx 31.1 \frac{\gamma\alpha^{3/2}}{v\omega^{3/2}} \equiv \sigma_{ad} \,. \tag{9.79}$$

The lifetime corresponding to this cross section is given by

$$\tau \approx 0.032\,\omega^{3/2}/N\gamma\alpha^{3/2} \equiv \tau_{ad} \,. \tag{9.80}$$

We note that this "adiabatic" lifetime does not depend on the velocity, just like the static τ^{st} given by expression (9.57). It is interesting to compare these two lifetimes:

$$\tau_{ad}/\tau_{st} \sim (\alpha N^{2/3}/\omega)^{1/2} \sim \mu^{1/2} \ll 1 \,. \tag{9.81}$$

The value $\xi \approx 1$ corresponds to the value $\beta \sim (\omega/\gamma)^2 \gg 1$.

9.4 Decay of Atomic States and Some Elementary Processes in Plasmas

9.4.1 Transition Discrete Spectrum − Continuum in Hydrogenic Plasmas

The problem of merging a discrete spectrum and a continuum is considered in numerous articles [9.22, 23]. One of the possible methods of such a merging is founded on taking into account the decays of highly excited atomic states in an ion electric field F. The ionization of excited states in the field F leads to a weakening of bound-bound transitions from these levels. Simultaneously such a "loss of oscillator strengths" in the discrete spectrum is compensated by their increase of free-bound (photorecombination) transitions [9.23].

To take into account the weakening of bound-bound transitions, one may introduce the weakening factor $j(F)$ defined by the ratio of the radiation decay probability A of the level to the probability $\Gamma(F)$ of its ionization in an electric field. This ratio strongly depends on the parabolic quantum numbers n, n_1, n_2 of the Stark sublevels belonging to the given atomic level n split in an electric field. In a simplified version one can introduce some average for a given level "k" decay rate $\Gamma_k(F)$. Then, for the transition from the level "k" to a given level "n" accompanied by the radiation of a quantum with frequency

$$\omega = \frac{1}{2n^2} - \frac{1}{2k^2} , \tag{9.82}$$

the weakening factor $j_{kn}(F)$ takes the form [9.23]

$$j_{kn}(F) = A_{kn}/[A_{kn} + \Gamma_k(F)] , \tag{9.83}$$

where A_{kn} is the average rate of radiation transition $k \rightarrow n$.

The intensity $I_{kn}(\omega)$ of the discrete transition $k \rightarrow n$ considered is schematically written in the form:

$$I_{kn}(\omega) = \int dF W(F) j_{kn}(F) I_{kn}^0(\omega, F) , \tag{9.84}$$

where I_{kn}^0 is the intensity distribution in the spectral line for a fixed ion field F with a distribution function $W(F)$. The corresponding contribution of free-bound transitions at the frequency ω (9.82) is determined by the relation:

$$I_n(\omega) = I_n^0(\omega) \int_0^\infty dF W(F) [1 - j_{kn}(F)] , \tag{9.85}$$

where $I_n^0(\omega)$ is the photorecombination intensity continued analytically through the boundary of the discrete spectrum.

The calculation of a smooth transition between the discrete and continuous spectra performed using the scheme mentioned leads to a satisfactory agreement with the experiment, as illustrated by Fig. 9.6 from [9.23].

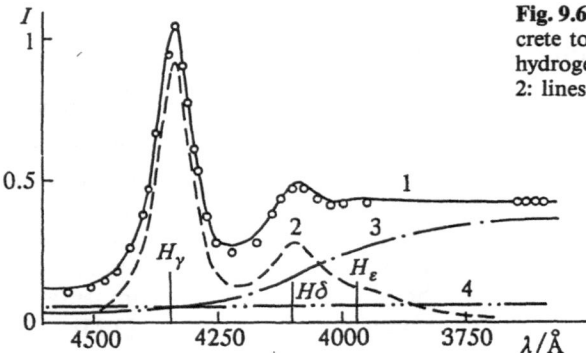

Fig. 9.6. Smooth transitions from discrete to continuum spectra for atomic hydrogen in plasmas [9.23]. 1: sum, 2: lines, 3: continuum, ○: experiment

9.4.2 Charge Exchange of Atoms at Multicharged Ions as a Decay Process

The concept of atomic state decay may be successfully applied for an estimation of charge exchange cross sections of neutral atoms at multicharged ions [9.24–26]. Let us consider for simplicity the charge exchange of an atomic hydrogen nucleus with charge $Z \gg 1$. The nucleus with the charge Z being at the distance R from the hydrogen atom produces the electrical field $F = ZR^{-2}$. The probability of atom's decay w in the field F is defined by the expressions [9.24, 26] (a.u.):

$$w(F) = \frac{4}{F}\exp\left(-2/3F\right) = \frac{4}{Z}R^2\exp\left(-2R^2/3Z\right) \equiv \Gamma(R) \ . \tag{9.86}$$

It is seen that the probability values $w(R) \approx 1$ are achieved at distances $R \approx \sqrt{Z}$ at which the effective decay of the atom in the field of multicharged ion takes place. It is clear that for slow enough ion-atom collisions, the corresponding cross sections are of the order of $\sigma \sim R_{\text{eff}}^2 \sim Z$. For small relative collision velocities $v \ll Z^{1/2}$ the ejected electron proves to be captured at a level of a multicharged ion. The corresponding cross section σ determines the rate of the charge exchange process:

$$H^0 + A^{+Z} \rightarrow H^+ + A^{+(Z-1)} \ . \tag{9.87}$$

More detailed calculations of the charge exchange cross section may be performed using the adiabatic solution of the Schrödinger equation [9.26]. The transition probability $w(\varrho)$ for the collision with the impact parameter ϱ is connected with the decay rate Γ during the collision by the obvious relation [9.26]

$$w(\varrho) = 1 - \exp\left(-\int_{-\infty}^{\infty} \Gamma[R(t)]\,dt\right) \ , \tag{9.88}$$

where $R(t)$ is the straight particle trajectory.

Using in the place of $\Gamma(R)$ in (9.88) the expression (9.86) and making the transition to the charge exchange cross section σ_{cx} it is easy to obtain the following estimation of the cross section [9.24] (a.u.):

$$\sigma_{cx} = \frac{3\pi}{2} a_0^2 Z \ln \left[\frac{\sqrt{Z}}{v} \ln \left(\frac{\sqrt{Z}}{v} \right) + \text{const.} \right] . \tag{9.89}$$

The general dependence on the ion charge Z is linear, as expected. A more detailed account of the decay model is given in [9.25].

9.4.3 Auto-ionization Decays and Dielectronic Recombination in Plasmas

The influence of the electrical field on atomic states in plasmas may change sharply the rates of auto-ionization decays Γ_A as well as the probabilities of processes connected with these rates, particularly the dielectronic recombination.

Here we briefly discuss the evidence of these effects for $2p^2 {}^3P$ levels of helium-like ions in a dense plasma, following [9.27–29]. The states $a \equiv 2p^2 {}^3P$ are metastable relative to dipole auto-ionization transitions to the $1s$ level. On the contrary, the auto-ionization decays of the near lying level $d \equiv 2s2p$ into the state $c \equiv 1s2s$ are allowed. After applying an external field F the mixing of the state wave functions a and d takes place, resulting in a sharp increase of the auto-ionization decay probability of the level a. The effect is analogous to the decrease of the lifetime of a metastable level in the plasma, as considered in Sect. 9.3.

It is easy to obtain the expression for the auto-ionization decay rate $\Gamma_A(a \to c)$ with the help of second-order perturbation theory:

$$\Gamma_A(a \to c) = \frac{2\pi}{\hbar} \left| \sum_d \frac{\langle a | d \cdot F | d \rangle \langle d | V | c \rangle}{(E_a - E_d) + i\Gamma_d/2} \right|^2 \delta(E_a - E_c) , \tag{9.90}$$

where V is the inter-electron interaction, corresponding to the auto-ionization decay of the level d. Allowing the field F to be spherically symmetric and averaging (9.90) over all directions of F and polarizations of d, one obtains [9.27]

$$\Gamma_A(a \to c) = \frac{A(a \to d)\Gamma_A(d \to c)}{4\omega_{ad}^3 \alpha^3 [(E_a - E_d)^2 + \Gamma_d^2/4]} \langle |F^2| \rangle , \tag{9.91}$$

where $\alpha = e^2/\hbar c$, and the symbol $\langle || \rangle$ represents the averaging over values of the field F. The change of level auto-ionization widths exerts a noticeable influence on the intensity distribution of dielectronic satellites in dense plasmas [9.27].

The atomic state mixing in plasmas results in particular in the fact that the true atomic states for auto-ionization width determination becomes parabolic

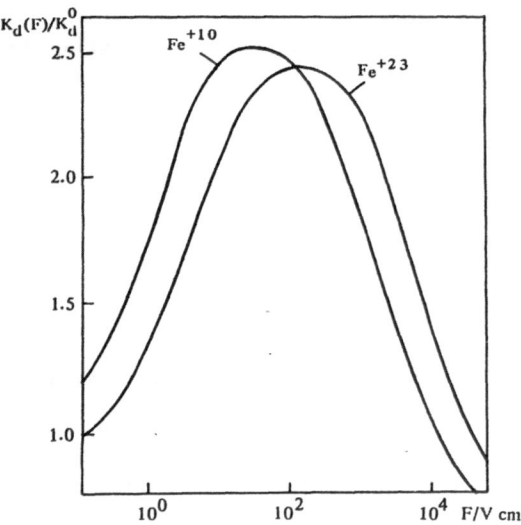

Fig. 9.7. The dependence of the dielectronic recombination rates K_d on the electric field strength F for ions Fe^{+10} and Fe^{+23} [9.28]

states, but not spherical states. The circumstance leads to an expansion of the atomic quantum number domain taking part in the electron capture processes on auto-ionization levels. The expansion of phase space may result in an increase of probabilities of processes with the participation of auto-ionization states, particularly dielectronic recombination [9.28, 29]. In Fig. 9.7 from [9.28], the results of the dielectronic recombination rate calculation for the ions Fe^{+10}, Fe^{+23} versus the electric field strength F are shown. It is seen that even for moderate fields $F \approx 10^2$ V/cm, the noticable increase of the recombination rate takes place. The further increase of values of F is accompanied by the decay of atomic levels in an electric field leading to the decrease of total number of atomic states taking part in the process and to the corresponding decrease of the dielectronic recombination rate.

10. Excited Hydrogen-Like Atom in Electrical Fields of Charged Particles

The problem of charged particle interaction with an excited hydrogen-like level n of an atom or an ion is of great interest both in its fundamental and applied aspects. The fundamental meaning of the problem is connected with the fact that it is one of the realistic examples of a precise solution of quantum mechanical problems both for classical and quantum motion of perturbing charged particle. The possibility of such a solution is connected with the dipole interaction potential $V = -deR^{-2}$ between dipole moment d of the excited hydrogen-like state n and the electrical field $F = -eR^{-2}$ of the charged particle. It is seen that the dependence of the interaction on the coordinate R is the same as for the centrifugal potential l^2R^{-2}. Therefore, it is clear that the radial wave functions of the scattered charge must be of an analytical form, it being even simpler than in Coulomb field $-ZeR^{-1}$, which decreases with the distance by a law other than the centrifugal one. Moreover it is principally clear that the possibility of an analytical solution of the problem for a particle under the simultaneous influence of Coulomb and dipole potentials.

The problem considered is also of a large practical interest for a number of problems connected with the description of the evolution of the excited hydrogen state. Here we will focus on two such problems. First, the broadening of hydrogen or hydrogen-like spectral lines due to collisions with charged particles. The second problem we will consider is the charge exchange of atoms at a multicharged ion while taking into account the evolution of degenerated hydrogen-like states of the former.

10.1 The Atomic State Evolution in the Electric Field of a Classically Moving Charged Particle

Let us consider the evolution of degenerated hydrogen-like states belonging to an atomic level with a given n in the electrical field $F(t)$ produced by a classically moving charged particle along the trajectory $R(t)$. Let us make a transition to the rotating coordinate system $X'Y'Z'$, in which the $0x'$ axis lies along $R(t)$ of any moment of time. The transformation of the wave function from the fixed to the rotating coordinate system takes the form

$$\psi' = \exp\left[i l_z \chi(t)\right] \psi \;, \tag{10.1}$$

where l_z is the z component of the angular momentum of the atomic electron, and $\chi(t)$ is the angle of the rotation of the vector $R(t)$ in the collision plane.

The substitution of (10.1) into the Schrödinger equation leads to the result:

$$i\hbar \frac{\partial \psi'}{\partial t} = [H_0 + d_x |F(t)| + L_z \dot\chi(t)] \psi' = [H_0 + V(t)] \psi'(t) , \qquad (10.2)$$

where H_0 is the Hamiltonian of the free atom, and $F(t) \equiv |F(t)|$.

From (10.2) it follows that both the electrostatic $(d_x F)$ and "magnetic" $(L_z \dot\chi)$ interactions exist in the rotating system. The latter designation is justified by the fact that there is a complete analogy between the interaction $L_z \dot\chi$ and the interaction of an atom with a magnetic field in the absence of spin: $\mu_0 L_z H (\mu_0 = e\hbar/2mc)$. Thus, the third term in the Hamiltonian (10.2) can be regarded as the interaction with a certain effective magnetic field $H_{eff} = \hbar \dot\chi/\mu_0$, which appears in the rotating system. Therefore, the problem has been reduced to finding the energy levels and wave functions of the hydrogen atom in the mutually perpendicular (variable) electric and magnetic fields.

The possibility of an exact solution of this problem is based on the utilization of the degeneracy, specific for hydrogen, with respect to the orbital quantum number l. The said degeneracy is closely related to the presence in a Coulomb field of an additional integral of the motion, namely, the Runge-Lenz vector A (Sect. 1.3). States pertaining to a fixed principal quantum number n are responsible for the effects under consideration. However, as is well known, it is precisely for such effects that it is possible to use the symmetry properties (corresponding to the rotation group O4) of the hydrogen atom discovered by V. Fock. The mutual disposition of the vectors $(a/e)F(t)$, $(\mu_0/h)H_{eff}(t)$, $\omega_1(t)$, and $\omega_2(t)$ in the rotating coordinate system is shown in Fig. 10.1.

Therefore, following the usual procedure, we introduce the new "angular momentum operators" J_1 and J_2:

$$J_1 = \frac{1}{2}(L+A) , \quad J_2 = \frac{1}{2}(L-A) . \qquad (10.3)$$

Then $V(t)$ in (10.2) can be represented in the form

$$V(t) = ex \cdot F(t) + \hbar L \cdot \dot\chi(t) = J_1 \cdot \omega_1(t) + J_2 \cdot \omega_2(t) , \qquad (10.4)$$

$$\omega_{1,2}(t) = \mu_0 H_{eff}(t) \pm \frac{\alpha}{e} F(t) , \qquad (10.5)$$

and $\alpha \equiv (3/2)ne^2 a_0/\hbar$ (a_0 denotes the Bohr radius).

In (10.4, 5) it is assumed that the vectors F and H_{eff} are directed, respectively, along the x and z axes of the rotating coordinate system (see Fig. 10.1).

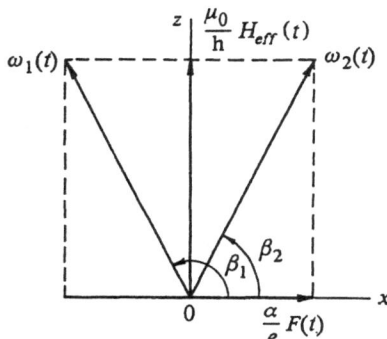

Fig. 10.1. The geometry of the effective electrical (F) and magnetic (H) fields and the directions of the atom quantization $\omega_{1,2}$

The subsequent solution consists in the construction of wave functions $u_{nn'n''}$ which diagonalize the Hamiltonian (10.4). These wave functions correspond to a definite projection of J_1 on ω_1 (characterized by the quantum number n') and a definite projection of J_2 on ω_2 (characterized by the quantum number n''). In the case of constant ω_1 and ω_2, the functions $u_{nn'n''}$ can be obtained from the usual parabolic wave functions $u_{n i_1 i_2}$ (where i_1 and i_2 are the quantum numbers corresponding to the projections of J_1 and J_2 on the x axis) by means of rotations through the angles β_1 and β_2, determining the axis of quantization of the atom (see Fig. 10.1).

In our case there is an essential complication due to the time dependence of the vectors ω_1 and ω_2. However, one can show that the direction of $\omega_{1,2}(t)$ does not change during the collision process. In fact, from a direct investigation of the geometry of the trajectory in the collision plane it follows that

$$\tan \beta_2 = \frac{\mu_0 H_{\text{eff}}(t)}{(\hbar\, a/e)\alpha F(t)} = \frac{\dot{\chi}(t)}{(a(e)F(t)} = \frac{\varrho v}{\alpha} \equiv \frac{1}{\delta}\ , \tag{10.6}$$

where ϱ is the impact parameter characterizing the flight path of the particle.

Thus, in the process of collision on the atom there are selected "directions of quantization", determined by the angle β_2, which depends on a single characteristic dimensionless parameter δ. Relation (10.6) shows that in the case of close and slow collisions ($\delta \gg 1$) the direction of the axis of quantization coincides with the direction of the electric field. However, in the case of fast and distant collisions ($\delta \ll 1$), the direction of the axis of quantization coincides with the direction of the "magnetic field". The boundary value of the impact parameter corresponds to the Weisskopf radius a/v [10.1]. Conservation of the direction of quantization during the collision process means the absence of transitions between states having different values of n' and n''. The preceeding discussion immediately permits us to generalize to our case the results which were obtained for constant F and H, by treating the dependence on the time as dependence of a parameter, since the wave functions $u_{nn'n''}$ diagonalize the Hamiltonian. Thus, for the energy $E_{nn'n''}$ we obtain

$$V_{nn'n''}(t) = \hbar(n'+n'') \, |\omega_{1,2}(t)| = \hbar(n'+n'') dv/dt \ , \tag{10.7}$$

$$v(t) = \chi(t)[1+(a/\varrho v)^2]^{1/2} \ , \tag{10.8}$$

where the $u_{nn'n''}$ are obtained, as indicated, from $u_{n i_1 i_2}$ by means of simple rotations as discussed in [10.1−5].

The evolution operator $\hat{T}(t_2, t_1)$ of the atomic states is obviously diagonal in the n', n'' representation:

$$\langle n', n''| \, \hat{T}(t_2, t_1) \, | n', n'' \rangle = \exp [i\Delta v(t_2, t_1)(n'+n'')] \ , \tag{10.9}$$

$$\Delta v(t_2, t_1) = [1+(a/\varrho v)^2]^{1/2} \Delta\chi \ , \quad \Delta\chi = \chi(t_2) - \chi(t_1) \ . \tag{10.10}$$

Accounting for the mentioned connection (10.10) of the wave functions $u_{nn'n''}$ with parabolic functions $u_{n i_1 i_2}$ it is easy to express the evolution operator $\hat{T}$ in terms of a parabolic quantum number [10.2]

$$\langle i_1' i_2'| \, \hat{T} | i_1 i_2 \rangle = \sum_{n', n'' = -j}^{j} D_{i_1' n'}^{j}(\beta_2) D_{i_2' n''}^{j}(\pi - \beta_2)$$

$$\times D_{i_1 n'}^{j}(\beta_2) D_{i_2 n''}^{j}(\pi - \beta_2) \exp [-iv(n'+n'')] \ , \tag{10.11}$$

where D_{in}^{j} are *Wigner* functions [10.3] and the quantities β, α, v are defined in (10.6, 11).

The state $|0\rangle$, to which the charge transfer is made, has the form $|0\rangle = |i_1 i_2\rangle = |J, -J\rangle$. Equation (10.11) contains a sum of n^2 terms, and is therefore too complex for specific calculations when $n \gg 1$. However, the calculation can be significantly simplified by means of the following transformations. Let us write the double sum over $n'n''$ in (10.11) in the form of a product of two sums

$$\sum_{n'n''} = \left(\sum_{n'} \cdots \right) \left(\sum_{n''} \cdots \right) \ . \tag{10.12}$$

It is easy to see that each of the sums is a product of three rotations, which can be replaced by one resultant rotation described by one D-function (for example, [10.4]). As a result, we obtain

$$\langle i_1' i_2'| \, \hat{T} | i_1 i_2 \rangle = D_{i_1' i_1}^{j}(\alpha_1 \beta_0 \gamma_1) D_{i_2' i_2}^{j}(\alpha_2 \beta_0 \gamma_2) \ . \tag{10.13}$$

The angles β_0, $\alpha_{1,2}$, which define the resultant rotation, are connected with the original angles β and $\Delta\chi$ by the relations

$$\cot \alpha_{1,2} = \cos \beta \, (\cot v + 1/\sin v) \ , \quad \cot \gamma_{1,2} = -\cot \alpha_{1,2} \ ,$$

$$\cos \beta_0 = 1 - 2\sin^2 \beta \sin^2 (v/2) \ . \tag{10.14}$$

Since into the expressions for the transition probability enters the modulus, $|\hat{T}|$, of the $\hat{T}$ operator, with allowance for the relation

$$D^j_{pq}(\alpha, \beta, \gamma) = e^{ip\alpha} D^j_{pq}(\beta) e^{iq\gamma} \ ,$$

we have

$$|\langle i'_1 i'_2 | \hat{T} | i_1 i_2 \rangle|^2 = D^{j^2}_{i'_1 i_1}(\beta_0) D^{j^2}_{i'_2 i_2}(\beta_0) \ . \tag{10.15}$$

Let us consider the form of the evolution operator in the limiting cases of low and high speeds. For $a/\varrho v \gg 1$, it follows from (10.15) that $\beta \to 0$, $\beta_0 \to 0$, and the evolution operator is diagonal in the rotating coordinate system (since $D_{i_i i'_i}(0) = (\delta)_{i'_i i_i}$). This means that the excited state of the ion has time to "follow" the electric field of the proton flying by. (More trivial is the diagonality of $\hat{T}$ for $\Delta \chi \to 0$.)

In the other limiting case, $a/\varrho v \ll 1 (\beta \to \pi/2)$, from (10.14) we have $\cos \beta_0 \simeq \cos v \simeq \cos \Delta \chi$. In this case the effect of the evolution operator amounts to the rotation of the wave function through the angle $\Delta \chi$, which is equal to the angle of rotation of the internuclear axis. The aforementioned properties of the $\hat{T}$ operator are in accord with the qualitative arguments discussed. The obvious analytical form (10.13) for the evolution operator is of fundamental importance because it describes the multichannel problem of atomic state evolution accounting for strong coupling between scattering channels.

10.2 Effect of Hydrogenic State Mixing During Charge Exchange of an Atom at the Multicharged Ion

Accounting for the strong coupling of different scattering channels arises in a large number of problems in the physics of atomic collisions. We will dwell on the processes of charge exchange of a neutral hydrogen atom at the nucleus of a multicharged ion X^{+Z} with the charge $Z \gg 1$. The principle distinctive features of the charge transfer process consist of the following. The initial term, $E_0(r)$, of the quasimolecule $H(1\,S) + X^{+Z}$ for an internuclear distance $R \to \infty$ coincides with the ground level of hydrogen and varies slowly as the particles approach each other due to the weak polarization interaction ($\propto 1/R^4$). The final $H^+ + X^{+(Z-1)}$ term, $E_n(R)$, for $R \to \infty$ coincides with the level of the multicharged ion with energy $E_n = -Z^2/2n^2$, and for finite R, contains a strong Coulomb interaction between the proton H^+ and the multicharged ion $X^{+(Z-1)}$, so that

$$E_n(R) = -Z^2/2n^2 + (Z-1)/R \ . \tag{10.16}$$

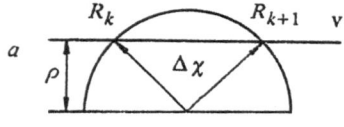

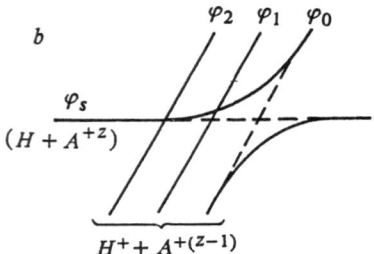

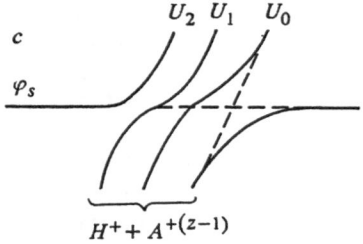

Fig. 10.2a–c. The geometry of the charge exchange process of an hydrogen atom at a multicharge ion [10.8]: the domains near term crossing points R_k and R_{k+1} are pointed out

It is easy to see that the initial term $E_0(R)$ has a set of points of intersection, R_n, with the hydrogen-like system of levels in the final term. These points are given by the relation

$$R_n = 2(Z-1)n^2/(Z^2-n^2) , \quad n < Z .$$ (10.17)

At relative velocities of the nuclei lower than the atomic velocities (i.e., for $v \ll 1$), the points of intersection make the dominant contribution to the charge-transfer probability. As usual, the system passes through the points of intersection twice: once when the nuclei are approaching each other and once when they are separating (Fig. 10.2).

One of the consequences of the symmetry of the one-electron system in the field of two Coulomb centers is the fact that only one out of all of the n^2 ion states belonging to the level with the principal quantum number n gets occupied in the charge transfer made from the initial term. This state corresponds to the maximum component of the dipole moment of the hydrogen-like level of the ion along the direction of the electric field F produced by the proton. Consequently, the parabolic quantum numbers $n_2 = n-1$, $n_1 = 0$, and $m = 0$ characterizes the state. For ions of low multiplicity ($Z \leq 10$), the number of levels that contribute to the charge transfer is quite limited (two to three); therefore, the replacement of these levels by an equivalent continuum becomes problematic.

The previous discussion indicates first and foremost two distinctive features of the charge transfer between a hydrogen atom and a bare nucleus. First, the high degree (n^2-fold) of degeneracy of the final states of the hydrogen-like ion. Second, the existence of selection rules according to which there can be charge transfer to only one of these degenerate states, namely, the state with the quantum numbers $n_2 = n-1$, $n_1 = 0$, $m = 0$, which we shall hereafter denote by the symbol $|0\rangle$. These two circumstances lead to the appearance of new effects. Basically, they amount to the following. The total transition probability is made up of contributions from transitions at two points: during the approach (R_k) and during the separation (R_{k+1}) as in Fig. 10.2a. During the motion of the system from the point R_k to the point R_{k+1} there can occur transitions between the populated state $|0\rangle$ and the remaining n^2-1 degenerate states. In particular, this can lead to the depopulation of the state $|0\rangle$ as the system approaches the point R_{k+1}. Thus, the total transition probability is determined not only by the population of the states $|0\rangle$ at the points R_k and R_{k+1}, but also by the nature of the interaction of the n^2 degenerate states between these points.

The specific properties of hydrogen-like states are connected with the fact that the effects of mixing of the states in the process of charge transfer can be taken into account exactly in spite of the high degree (n^2-fold) of level degeneracy. It is to these problems that the following discussion is devoted. Let us consider the effect of state mixing on the charge exchange, allowing the charge transfer domains (near the points R_k, R_{k+1}) to be independent from the mixing domains (between the points R_k and R_{k+1}). Let the transition probabilities, p, at the points of intersection R_k and R_{k+1} (Fig. 10.2a) be known. Let the operator, $\mathbb{T}(\Delta\chi)$, of evolution of the system of n^2 degenerate sublevels between the points R_k and R_{k+1} separated by an angle $\Delta\chi$ of rotation of the internuclear axis be also known. Since the exchange interaction $V(R)$ (and along with it the transition probability p) has a simple form in the coordinate system oriented along the internuclear axis, it is convenient to carry out all of the computations in this rotating system. Within the framework of the model under consideration, it is not difficult to derive an expression for the total transition probability, W, with allowance for the operator $\mathbb{T}$:

$$W = 2p(1-p)+(1-p)^2[1-|\mathbb{T}_{00}(\Delta\chi)|^2] \ , \tag{10.18}$$

where $\mathbb{T}_{00}$ is the matrix element of the operator $\mathbb{T}$, taken with respect to the states $|0\rangle$, to which the charge is transferred. The effects connected with the rotation of the internuclear axis will be called rotation effects.

For the transition probability p, we have the Landau-Zener formula [10.6–9]

$$p = \exp\left[-2\pi V^2(R_n)/v_R F\right] \ , \tag{10.19}$$

where the magnitude of the exchange interaction V, the force F, acting on the electron, and the radial velocity $v_R = v \cdot (1-\varrho^2/R^2)^{1/2}$ are evaluated at the point of intersection of the terms R_n (10.16). We shall assume the motion of the nuclei is rectilinear: $R^2 = \varrho^2 + v^2 t^2$ (ϱ is the impact parameter and v is the relative velocity at infinity). The evolution operator $\mathbb{T}(\Delta\chi)$ of the degenerate hydrogen-like states with a given n in the field of the passing charge is known from (10.10)

$$T_{00}^J(\Delta\chi) = [1 - \sin^2\beta\sin^2(v/2)]^{4J} , \quad J = (n-1)/2 . \tag{10.20}$$

Substituting (10.20) into (10.18), and defining the cross section in the usual manner

$$\sigma = 2\pi \int_0^{R_n} \varrho\, d\varrho\, W(\varrho) ,$$

we find

$$\sigma_n = 2\pi R_n^2 [f_{L-Z}(\Delta\chi) + f_{rot}(\delta,\Delta)] , \tag{10.21}$$

$$f_{L-Z} = \int_0^1 dx \exp(-\Delta/\sqrt{x})[1 - \exp(-\Delta/\sqrt{x})] , \tag{10.22}$$

$$f_{rot}(\delta,\Delta) = \int_0^1 x\,dx[1 - \exp(-\Delta/\sqrt{1-x^2})]^2[1 - |T_{00}|^2] . \tag{10.23}$$

Here $\Delta = 2\pi V^2/Fv$ and $\delta = a/R_n v$, $f_{L-Z}(\Delta)$, describing the usual velocity dependence of the cross section in the Landau-Zener model [10.9] and the function $f_{rot}(\delta,\Delta)$ takes account of the contribution of the rotation effects.

The influence of the rotation effects can be easily elucidated directly from the general formulas (10.18, 20). Let us assume that $p\to0$, so that the system moves along adiabatically. Then the charge transfer probability is wholly determined according to (10.18) by the probability $1 - |T_{00}|^2$ for the depopulation of the state $|0\rangle$. The depopulation of the state $|0\rangle$ is caused, as has been noted, by transitions into other degenerate states not interacting with the initial term (Fig. 10.2b).

One can perform more precise accounting for rotation effects using the "true" basis of wave functions $U_{n'n''}$ which diagonalizes the evolution operator, as in (10.10). The states may be called "dynamic" terms [10.2] because they depend parametrically on the collision parameters ϱ and v. The main purpose of the introduction of the dynamic terms is to have the rotation effects automatically taken into account, treating the charge transfer as occurring directly between the initial term and the system of (mutually independent) dynamic terms. With allowance for (10.11), we obtain the value of the effective exchange matrix element, $V_{n'n''}$, coupling the state $u_{n'n''}$ to the initial state φ_s (we have taken account of the fact that the only nonzero matrix element couples φ_s to $|0\rangle = |i_1 = +J, i_2 = -J\rangle$):

$$V_{n'n''} = D_{n'J}^J(\beta)D_{n''-J}^J(\pi-\beta)V . \tag{10.24}$$

It can be seen that this interaction also depends parametrically on the velocity.

In Fig. 10.2c we show the scheme of the intersections of the dynamic terms $u_{n'n''}$ with the initial term φ_s. Let us emphasize that the states $u_{n'n''}$ no longer interact with each other. Thus, the problem of a two level charge exchange with a subsèquent mixing of the states has been reduced to the problem of transitions upon the intersection of the initial term with a system of levels. To solve this latter problem, we can use the results for the transitions from a given term when it intersects a system of parallel terms [10.7].

The described approach makes it possible to precisely take into account the rotation effects in the dipole approximation. Here we will not concentrate on the concrete results of the calculations (for details, [10.8]). Note only that accounting for the rotation effects results, as follows from (10.24), in the population of a broad spectrum of excited states n_1, n_2 of the multicharged ion, that is in a change of the distribution over orbital momenta at the given excited level n. The effect is confirmed by the experimental observations [10.5].

10.3 Quantum Motion of an Electron in an Electric Field of an Hydrogen-Like Atom or Ion. Connection with the Line-Broadening Problem

10.3.1 Classical and Quantum Formulations of the Problem of Electron Interaction with an Excited Atom

We now make a transition to the case of the quantum motion of a perturbing particle in the field of an excited hydrogen atom. The problem is the quantum generalization of the classical results of Sect. 10.1. In the classical approach the motion of a perturbing particle is assumed to be determined and the atomic wave function $\psi(r_A)$ evolution is considered to be in the electric field of a perturbing particle. In the quantum approach the energy exchange between the atom and the particle results in a deformation of its motion. Therefore in this case the information on the interaction is contained in the equations for the wave function $\psi(r_i)$ of the perturbing particle. Dividing the total wave function $\psi(r_A, r_i)$ into the multiplication $\psi(r_A)\,\psi(r_i)$ is possible for the processes with large cross sections when the domains of the atomic (r_A) and perturbing (r_i) coordinates are essentially divided. The considered lower processes of the particle scattering at the excited hydrogen-like levels as well as the broadening of the hydrogen spectral lines are examples of such processes.

10.3.2 The System of Wave Functions of an Excited Hydrogen Atom and a Broadening Particle

As frequently mentioned above, the particular properties of hydrogen are connected with the degeneracy of levels a and b in the angular momentum l. Therefore, the main difficulty in the solution of the problem is in taking into account the interaction of all of the degenerate states of the atom during the scattering of the broadening ion. We must now find the wave function $\psi(r_i)$ of the ion interacting with the excited hydrogen atom. The Hamiltonian for the system is

$$H = H_A + H_i + \hat{V}_{Ai} \tag{10.25}$$

where H_A, H_i are the Hamiltonians for the free atom and free ion, respectively, and $\hat{V}_{Ai}$ is the operator for their dipole interaction: $\hat{V}_{Ai} = -r_i \cdot r_A r_i^{-3}$. We note that the potential V_{Ai} is noncentral and, in contrast to the above case, the orbital angular momentum l_i of the ion is therefore not conserved.

The conserved quantity is only the resultant angular momentum $L = l_i + l_A$, where l_A is the orbital angular momentum of the atom. We now write the wave function $\psi(r_A, r_i)$ of the ion+atom system in the form of an expansion over the products of the unperturbed wave functions corresponding to the level n of the atom, i.e., $\psi^{(n)}(r_A)$, and the wave functions of the ion, $\psi(r_i)$. After substitution into the Schrödinger equation with the Hamiltonian given by (10.25), we obtain a system of coupled equations for the functions $\psi(r_i)$. This system is analogous to the system of equations with a strong coupling in the theory of scattering [10.9], the solution of which is a relatively difficult problem. In the present case, however, we can avoid having to perform a direct solution of this problem by using the fact that for the dipole potential defined by (10.25), there is an additional constant of motion being verified by direct commutation with the Hamiltonian:

$$\Lambda = l_i^2 - 2M \cdot r_A (n_i \cdot n_A), \quad \hat{V}_{Ai} = \frac{1}{r_i} - \frac{1}{|r_A - r_i|} \approx -\frac{r_i \cdot r_A}{r_i^3}, \quad \Lambda \psi = \lambda \psi \tag{10.26}$$

where $n = r/r$ is a unit vector in the direction of r (henceforth, we shall use the atomic system of units in which $e = \hbar = m = 1$).

The use of wave functions with a definite eigenvalue λ is convenient because we know the solutions $R_{\lambda L}(r)$ of the equation describing the radial motion of the ion. In fact, the dipole potential V_{Ai} falls off as r_i^{-2}, i.e., in the same way as the centrifugal potential $V_{cf} = -l_i^2 r_i^{-2}$. The sum of the two potentials then contains the constant of motion Λ given by (10.26) as a coefficient of r_i^{-2}. In view of this, the radial equation for the ion has the form (the subscript i will be omitted henceforth)

$$\frac{d^2R}{dr^2} + \frac{2}{r}\frac{dR}{dr} + \left(q^2 - \frac{\lambda}{r^2}\right)R = 0 . \tag{10.27}$$

The solution of this equation can be expressed [10.10] in terms of the Bessel functions:

$$R^{\lambda L} = \left(\frac{q}{r}\right)^{1/2} J_{(\lambda + 1/4)^{1/2}}(qr) . \tag{10.28}$$

The solution thus has the same form as for the free motion of an ion except that, instead of the orbital angular momentum $l_i + 1/2$ the subscript on the Bessel function now contains the quantity λ which contains information on the dipole interaction between the ion and the atom.

Despite the apparently simple form of the solution given by (10.28), we must recall that, before we can use it, we must find the value of λ and the wave functions $\psi_{\lambda LM}$, i.e., we must solve the secular equation given by (10.26). The functions $\psi_{\lambda LM}$ corresponding to definite values of the total angular momentum L, its projection M, and the constant of motion λ, can be conveniently determined in the form of the expansion

$$\Psi_{\lambda LM} = R^{\lambda L}(q) | \lambda LM \rangle , \quad | \lambda LM \rangle = \sum_{l_A l_i} \langle l_A l_i | \lambda L \rangle | l_A l_i LM \rangle , \qquad (1.29)$$

where the functions $| LM l_A l_i \rangle$ can be expressed in terms of the spherical harmonics Y_{lm} in the usual way (with the aid of the Clebsch-Gordan coefficients [10.9]).

Substitution of (10.29) into (10.26) leads to a system of algebraic equations for the coefficients $\langle l_A l_i | \lambda L \rangle$, the determinant of which can be used to find the value of λ. The problem of determining λ and the coefficients $\langle l_A l_i | \lambda L \rangle$, for the $n = 2$ level (Ly$_\alpha$ line) has been considered by *Seaton* [10.13] in connection with scattering problems. We shall not reproduce the relevant and rather unwieldy formulae, and will suppose henceforth that the coefficients $\langle l_A l_i | \lambda L$ and the values of λ are known. We must relate the function $\psi_{\lambda LM}$ to the functions $| \pm q l_A m_A \rangle$ which contain the incident plane wave and diverging (and converging) spherical waves:

$$| \pm q l_A m_A \rangle \underset{r \to \infty}{\approx} | l_A m_A \rangle e^{iqr} + \frac{f^{(\pm)}}{r} e^{\pm iqr} , \qquad (10.30)$$

where $| l_A m_A \rangle$ is the wave function of the atom in the state $l_A m_A$.

This relationship can be established with the aid of the expansion

$$| \pm q l_A m_A \rangle = \sum_{\lambda LM} \langle \lambda LM | \pm q l_A m_A \rangle \Psi_{\lambda LM} , \qquad (10.31)$$

where the coefficients $\langle \lambda LM | \pm q l_A m_A \rangle$ can be obtained by demanding that the functions given (10.30, 31) must be equal as $r \to \infty$.

Equating corresponding coefficients we find:

$$\langle \lambda LM | \pm q l_A m_A \rangle = \frac{(2\pi)^{3/2}}{q} \exp \left[\pm i \frac{\pi}{2} \left(v_{\lambda L} - \frac{1}{2} \right) \right]$$

$$\times \sum_{lm} (\pm 1)^l \langle m_A m | LM \rangle \langle l_A l | \lambda L \rangle Y_{lm}^*(q) , \qquad (10.32)$$

where $Y_{lm}(q)$ are the spherical functions, $v_{\lambda L} = (\lambda + 1/4)^{1/2}$.

The coefficients (10.32) together with the functions (10.29) give the total solution of the problem. These wave functions of a perturbing particle in the field of an excited atom will be used for cross section and line-shape calculations.

10.3.3 The Hydrogen Line Shape and the Overlap Integral of the Wave Functions of a Broadening Particle

Let us begin by deriving the quantum mechanical expression for the line shape $I_{ab}(\omega)$ due to a transition between level a and b of the atom. The levels are assumed to be nondegenerate in l, and the interaction between the atom and ion in states a and b is represented by the spherically symmetric potentials $U_a(r)$ and $U_b(r)$.

We shall suppose that the quantum $\hbar\omega$ is emitted by a single system consisting of the atom and the broadening ion. The wave function of this system in the initial and final states is given by the product of the wave functions of the atom, φ_a, φ_b, and the wave function $\psi^{\pm}_{q_{a,b}}$ of the ion. The scattering of the particle of momentum q by the potentials U_a and U_b is described by the wave function. The two signs correspond, respectively, to asymptotically converging and diverging spherical waves at infinity.

The probability of a transition of the system from state a to state b with the emission of a photon of frequency ω and momentum k is given by the following well-known expression:

$$W = \frac{2\pi}{\hbar} |\langle a|V|b\rangle|^2 \delta(\varepsilon_b - \varepsilon_a) \frac{dq_b dk}{(2\pi)^3 (2\pi)^3} , \qquad (10.33)$$

where V is the operator representing the interaction between the atom and the radiation field, and the wave functions $\langle a|$, $|b\rangle$ and the energy difference $\varepsilon_b - \varepsilon_a$ of the atom+ion system are given by

$$\varepsilon_b - \varepsilon_a = (\hbar^2/2M)(q_b^2 - q_a^2) - \hbar\Delta\omega ;$$

$$|a\rangle = \varphi_a q_a^{-1/2} \psi_{q_a}(r_a^+), \ |b\rangle = \varphi_b \psi_{q_b}(r_b^-) . \qquad (10.34)$$

The wave function $|a\rangle$ is assumed to be normalized to a unit flux of the broadening particles (ions). After averaging over the initial states and summing over the final states, the expression given by (10.33) will therefore give the differential cross section $d\sigma/d\omega$ for the emission of a photon of frequency ω in the range $d\omega$. The cross section $d\sigma/d\omega$ is completely analogous to the bremsstrahlung emission cross section. However, there is one important difference: in the bremsstrahlung case, the interaction V with the electromagnetic field includes the dipole moment of the scattering particle, whereas, in the case of broadening, it includes the dipole moment of the atom. Therefore, the total radiation intensity in the case of broadening will not depend on whether the atom scatters an ion or an electron.

It is clear from the preceeding discussion that to evaluate the cross section $d\sigma/d\omega$ for the emission of a photon of frequency ω by the atom, we can use the expressions for the bremsstrahlung emission cross section (for example, Sobel'man [10.14], Chap. 34, Part 3) with the dipole moment $d = er$ of the electron replaced by the dipole moment of the atom, and the wave function of the atom+ion system given by (10.34). The final result is

$$\frac{d\sigma}{d\omega} = \frac{4\omega^3 |d_{ab}|^2 \pi^2}{3c^3 v_a E_a q_{bl}} \sum (2l+1)|A_l|^2 \; , \tag{10.35}$$

where $E_a = \hbar^2 q_a^2/2M$ and v_a are the energy and velocity of the electron in state a, respectively. The matrix element of the dipole moment d of the atom over the wave functions of the atom+ion system is thus split into the product of the matrix element of d between states a and b of the atom and the overlap integral A_l for the wave functions of the ion with angular momentum l:

$$A_l = \int\limits_0^\infty dr\, r^2 R_l(q_a r) R_l(q_b r) \; , \tag{10.36}$$

where $R(q_i r)$ are the solutions of the radial Schrödinger equation with an angular momentum l in the potentials $U_a(r)$ and $U_b(r)$.

The expression for the radiated power per unit volume per unit frequency can be expressed in terms of the cross section $d\sigma/d\omega$, as in the case of the bremsstrahlung radiation, and is given by

$$Q(\omega) = N_A N_i \hbar\omega \int\limits_0^\infty v_a f(v_a) \frac{d\sigma}{d\omega} dv_a \; , \tag{10.37}$$

where N_A, N_i are the concentrations of atoms and ions, and $f(v_a)$ is the distribution over the initial relative velocities. As before, we shall not average over v_a, i.e., we shall suppose that $f(v_a) \sim \delta(v_a - v_0)$ (whenever necessary, this averaging can be carried out in the last stage). Dividing (10.37) by the total intensity of radiation emitted by the atoms, $N_A 4\omega^4 |d_{ab}|^2/3c^3$, we obtain the following expression for the line shape or intensity distribution emitted by an individual atom:

$$I(\omega) = N_i \frac{\pi^2 \hbar}{E_a q_b} \sum (2l+1)|A_l|^2 \; . \tag{10.38}$$

Thus, in the quantum mechanical formulation, the determination of the line shape reduces to the evaluation of the overlap integral A_l, whose dependence on the frequency shift is given by the relation

$$\frac{\hbar^2 q_a^2}{2M} - \frac{\hbar^2 q_b^2}{2M} = \hbar \Delta\omega \; . \tag{10.39}$$

Now we make the transition to the case of hydrogen-like atom. Proceeding as in the derivation of (10.35, 38) we substitute the functions $|\pm q\,l_A\,m_A\rangle$ into the general formulae for the cross section $d\sigma/d\omega$ for the emission of a photon of frequency ω, and obtain the following expression for the line shape connected with the transition from level n to a nondegenerate lower level:

$$I(\omega) = \frac{N_i \pi^2 \hbar}{3 E_a q_b} \sum_{\lambda L} (2L+1) [(\langle 1\,L-1\,|\,\lambda L\rangle A_{\lambda L}^{L-1})^2 + (\langle 1\,L+1\,|\,\lambda L\rangle A_{\lambda L}^{L+1})^2] \ . \tag{10.40}$$

Comparison of (10.40) with the expression for the two level approximation given by (10.38) shows that transitions with a change in the angular momentum l_i of the ion by ± 1 contribute to the line shape with weights given by the coefficients $\langle l_A l_i | \lambda L \rangle$ (10.29). Moreover, the line shape includes contributions of all scattering channels corresponding to different values of the constant of motion λ.

Further investigation of (10.40) involves an analysis of the radial overlap integrals $A_{\nu\nu'}$, which, according to (10.28), have the form

$$A_{\nu\nu'} = (q_a q_b)^{1/2} \int_0^\infty r J_\nu(q_a r) J_{\nu'}(q_b r) dr \ , \tag{10.41}$$

where ν_a, ν_b are related to λ in a simple fashion (10.28). The overlap integral given by (10.41) can be expressed in terms of the complete hypergeometric function [10.15 − 17]

$$_2F_1 = F\left[\frac{\nu_a + \nu_b}{2}, \frac{\nu_a - \nu_b}{2}, \nu_a + 1, \left(\frac{q_a}{q_b} \right)^2 \right] \ ,$$

where, according to (10.39), the dependence on the frequency determining $\Delta\omega$ is given by

$$(q_a/q_b)^2 = (1 + 2M\Delta\omega/q_a^2)^{-1} \ . \tag{10.42}$$

The line shape given by (10.40) is thus seen to admit a direct analytic evaluation. The question arises as to what is the relationship between the quantum mechanical solution and the exact classical solution given in Chap. 10.1. This can be established by the direct limiting transition in (10.40, 41) to high angular momenta L:

$$L \gg [3n(n-1)M/2]^{1/2} \ . \tag{10.43}$$

If we recall that $L = Mv\varrho$, we find that the quantum mechanical expression (10.40) becomes identical to the classical one for large values of angular momenta [10.10].

10.3.4 Generalization onto the Case of Hydrogen-Like Ions

The introduction of an additional interaction of a perturbing particle with a Coulomb field of a hydrogen-like ion with a charge Z makes the form of the spectrum determined by the overlap integral (10.50) more complicated. Nevertheless, such a problem also has an analytical solution. The overlap integrals arising here are analogous to ones in quantum relativistic bremsstrahlung theory [10.18]. Therefore, it is of interest to look for a direct transition of analytical results in the ion field to results for the neutral atom and further onto classic results. The example under consideration is, as mentioned, one of only a few quantum mechanical problems solved in analytical form.

The Hamiltonian of the interaction between the ion and the broadening electron of the dipole term (on top of the Coulomb term):

$$H = H_0 - \frac{Ze^2}{r} + \hat{\Lambda}/2Mr^2 \ , \quad \hat{\Lambda} = l^2 - 2Mr_A \cdot r_e e^2/r_e \ , \tag{10.44}$$

where H_0 is the Hamiltonian of the unperturbed ion. The operator takes into account the sum of the centrifugal and the dipole potentials (l, M, r, and r_a are the orbital momentum, the mass, the coordinate of the broadening electron, and the coordinate of the atomic electron, respectively). It is seen from (10.44) that since the centrifugal and the dipole potentials decrease in identical fashion with distance, it is possible to separate the radial and the angular motions. Indeed, using the wave functions corresponding to definite values λ of the operator $\hat{\Lambda}$ and of the total angular momentum of the system L, we easily see that the radial Schrödinger equation that follows from (10.44) differs from the equation for the free motion in the Coulomb field only in that the eigenvalues of the orbital momentum l are replaced by the eigenvalues λ of the operator $\hat{\Lambda}$. As a result, we can immediately write down the expression for the radial function $R_q(r)$ of the broadening electron:

$$R_q(r) = \left(\frac{2}{\pi}\right)^{1/2} q \, e^{(\pi\eta)/2} \frac{|\Gamma(a)|}{\Gamma(b)} (2qr)^{\nu-1/2} e^{-iqr} {}_1F_1(a,b,2iqr) \tag{10.45}$$

$$\nu = \sqrt{\lambda+1/4} \ ; \quad a = \nu+1/2+i\eta \ ; \quad \eta = Ze^2/\hbar v \ ; \quad b = 2\nu+1 \ ,$$

where $F(a,b,x)$ is a confluent hypergeometric function that is regular at zero; $\Gamma(Z)$ is the gamma function. The radial function $R_q(r)$ is normalized to $\delta(q-q')$.

The angular part of the problem, which reduces to finding the eigenvalues λ and the eigenvectors $|l_a L' \lambda L\rangle$ of the operator $\hat{\Lambda}$ is the same for an ion as for a neutral atom. The basic problem is to calculate the overlap integrals of the radial wave functions A, expressed in the form

$$A_{L'}^{\lambda L} = \int_0^\infty dr\, r^2 R_{q_i}(r) R_{q_f}(r) = \frac{v_i^2 - v_f^2}{q_i^2 - q_f^2} \int_0^\infty R_{q_i}(r) R_{q_f}(r) \equiv \frac{v_i^2 - v_f^2}{q_i^2 - q_f^2} B \ . \quad (10.46)$$

Using the explicit form (10.45) of the functions $R_q(r)$ we obtain the formal solution of the problem:

$$B = \frac{1}{\pi} (q_i q_f)^{1/2} x^{v_1} (-y)^{v_2} e^{i(\pi/2)\alpha - \pi/2(\eta_i + \eta_f)}$$

$$\times \frac{|\Gamma(a_i)\Gamma(a_f)\Gamma(\alpha)|}{\Gamma(b_i)\Gamma(b_f)} F_2(\alpha, a_i, a_f, b_i, b_f, x, y) \ ,$$

$$(10.47)$$

$$\alpha = \frac{b_i + b_f}{2} - 1 \ , \quad a_{i,f} = v_{i,f} + \frac{1}{2} \pm i\eta_{i,f} \ ; \quad \eta_{i,f} = Ze^2/\hbar v_{i,f} \ ,$$

$$x = \frac{2q_i}{q_i - q_f} \ ; \quad y = -\frac{2q_f}{q_i - q_f} \ , \quad q_i^2 - q_f^2 = 2M\hbar\Delta\omega \ ,$$

where F_2 is a hypergeometric function of two variables, the Appel function [10.17] defined by the series

$$F_2(\alpha, a_1, a_2, b_1, b_2; x, y) = \sum_{m,l=0}^\infty \frac{(\alpha)_{l+m}(a_1)_l(a_2)_m}{(b_1)_l(b_2)_m} \frac{x^l y^m}{l!\, m!} \ ,$$

$$(\alpha)_m = \Gamma(\alpha + m)/\Gamma(\alpha) \ ,$$

which converges if $|x+y| < 1$.

In our case $x + y \equiv 2$ in (10.47), so that the solution obtained remains formal until we obtain its analytic continuation in the variable range of interest to us. The indicated analytic continuation can be obtained in analogy with the theory of relativistic bremsstrahlung [10.18]. The analytic continuation formulae are given in [10.19], where the method used to obtain them is briefly described. The results are expressed in the form of a rapidly converging double hypergeometric series and can be used for numerical calculations.

We now trace in the general formula (10.47) the transition to straight trajectories, $Z/q \to 0$, including the known results [10.11] for neutral hydrogen ($Z \equiv 0$). The radial integral (10.47) is expressed in terms of an analytic function and can therefore in principle be investigated also without an analytic continuation. Under the condition $Z/q \to 0$, we have in formula (10.47) $q_{1,2} = v_{1,2} + 1/2$ and $b_{1,2} = 2a_{1,2}$. Using the quadratic transformations of the hypergeometric function ([10.17], Chaps. 2, 11), and taking into account the condition $x + y = 2$, we obtain

$$F_2(\alpha, a_1, a_2, 2a_1, 2a_2; x, y) = e^{-i(\pi/2)\alpha} \left(\frac{y}{2}\right)^{-\alpha} \frac{\Gamma(a_2+1/2)\Gamma(1/2)}{\Gamma(a_2+1/2-\alpha)\Gamma(\alpha/2+1/2)}$$

$$\times {}_2F_1 \left[\frac{\alpha}{2}, \frac{\alpha+1-2a_2}{2}, a_1+1/2, \left(\frac{x}{y}\right)^2\right].$$

$$(10.48)$$

Substitution of (10.48) into (10.40) directly yields for the contour the expression previously obtained for the neutral hydrogen atom.

It is of interest to consider the case of a small frequency shift $(\Delta\omega \to 0)$, when the overlap integrals are expressed in terms of scattering phases in Coulomb and dipole potentials. In the impact limit, corresponding to small frequency deviations $(\Delta\omega \to 0)$, the line shape $I(\omega)$ is known [10.11] to be Lorentzian with a width γ expressed in terms of the difference between the phase shifts for the scattering of the broadening electron by the levels of the radiating atom. In the considered case of ion lines, a peculiar relation appears between the scattering phase shifts in the Coulomb and in the dipole potentials.

The scattering phases Δ in the Coulomb potential are expressed as known [10.9] in terms of a Γ-function argument depending on the parameters $Ze^2/\hbar v \equiv \eta$, namely $\Delta = \arg\Gamma(i\eta)$. In the case under consideration, it is obvious that the asymptotic behavior of the wave functions will be also connected with phases scattering of Coulomb types. This is due to the case of H-like ions being of interest, so the motion of the broadening electron differs from one in a pure Coulomb field by replacing the orbital momentum l by the nonintegral parameter λ only. To meet the criteria, it is necessary that the main part of overlap integrals (10.46) be expressed in terms of Γ-functions. This is really the case, and it can be verified using the formulae of analytical continuation [10.19].

After straightforward but rather laborious transformations taking (10.47) into account, we obtain

$$|A|^2 = \frac{4q^2}{\pi^2(v_i^2 - v_f^2)^2} \sin^2\left[\frac{\pi}{2}(v_i^2 - v_f^2) + \Delta_i - \Delta_f\right],$$

$$(10.49)$$

$$\Delta_{i,f} = \arg\Gamma\left(v_{i,f} + \frac{1}{2} + i\eta\right).$$

In the limit as $Z/v \to 0$ (the phase shifts $\Delta_{i,f} \to 0$) we get from (10.49) the known result (10.38) for the impact width of a neutral hydrogen atom (e.g., [10.19]). At $Z \neq 0$ the Coulomb scattering leads, as is seen from (10.49), to the appearance of an additional phase difference $\Delta_i - \Delta_f$. We note that the Coulomb phase shifts $\Delta_{i,f}$ depend also on the dipole interaction constant that enters in $v_{i,f}$ (this is precisely why their difference in the upper and lower states does not vanish). The more detailed relation with line broadening problems is described in [10.19, 20].

10.4 Differential Cross Sections for Electron and Ion Scattering at the Excited Hydrogen Atom. Precise Solutions

The wave functions constructed in Sect. 10.3.2 for the particle in the field of an excited hydrogen atom make it possible to find the precise cross sections for the scattering of these particles at the atom without using either Born or quasi-classical approximations. The mass of a scattered charged particle will be designated by m. Using the wave functions from (10.30–32), we find the scattering amplitude for the transition of the atom from the state s to the state $l_A m_A$

$$f = \sum_\lambda f_\lambda, f_\lambda = \sum_{Ll} \left(\frac{2\pi}{iq}\right) \langle \lambda L|0L\rangle \langle \lambda L|l_A l\rangle C^{L_0}_{l_A m_A, l - m_A}$$

$$\times \left(\frac{2L+1}{4\pi}\right)^{1/2} (S^{\lambda L} - 1) Y_{lm_A}(qr) , \quad S^{\lambda L} = e^{2i\delta_{\lambda L}} ,$$

$$\delta_{\lambda L} = \frac{\pi}{2}\left(L + \frac{1}{2} - \sqrt{\lambda + 1/4}\right) .$$

(10.50)

Utilization of the integral of the motion makes it possible to divide the motion in λ channels and represent the total amplitude as the sum of independent λ items. For differential cross sections corresponding to the definite atomic transitions we have

$$\frac{d\sigma}{d\Omega}(l_A - l_{A'}) = \frac{1}{2l_A + 1} \sum_{m_A, m_{A'}} |f(l_A m_A \to l_{A'} m_{A'})|^2 .$$

(10.51)

The scattering at the level $n = 2$ may be investigated analytically because in the case the integral of the motion Λ is diagonalized in an analytical form. The eigenfunctions are placed above. For corresponding amplitudes we have

$$f(s \to s) = \frac{1}{2iq} \sum_{\lambda L} (2L+1)\langle \lambda L|0L\rangle^2 (S^{\lambda L} - 1) P_L(\cos\theta) ,$$

$$f^{m = 1,0}_{(s \to p)} = \frac{e^{im\varphi}}{2iq} \sum \frac{2L+1}{[L(L+1)]^{m/2}} \langle \lambda L|0L\rangle \langle \lambda L|1l\rangle$$

(10.52)

$$\times \left(\frac{2l+1}{2L+1}\right)^{1/2} C^{LM}_{1Ml0} (S^{\lambda L} - 1) P^m_L(\cos\theta) .$$

The expressions (10.52) are precise under the limits of a dipole approximation and may be used for numerical calculations of corresponding cross sections. In the limit of large angular momenta it is easy to obtain the following simple analytical expressions

$$|f(s \to s)| = \frac{(ma)^2}{q} \int x \, dx \, \frac{z}{(1+z)^2} [1 - \cos(\pi\sqrt{1+x^{-2}})] J_0(Ax) ,$$

$$|f^{(m=1)}(s \to p)| = \frac{(ma)^2}{2^{3/2} q} \int x \, dx \, \frac{z^{1/2}(1-z)}{(1+z)^2} [1 - \cos(\pi\sqrt{1+x^{-2}})] J_1(Ax) ,$$

(10.53)

$$|f^{(m=0)}(s \to p)| = \frac{(ma)^2}{2q} \int x \, dx \, \frac{z^{1/2}}{1+z} \sin(\pi\sqrt{1+x^{-2}}) J_0(Ax) ,$$

$$a = 6 \text{ a.u.}, \quad A = ma\theta/2, \quad z = (x + \sqrt{1+x^2})^{-2} .$$

The condition $L \gg m^{1/2}$ for electrons coincides with the condition $L \gg 1$, but for ions it is more rigorous, however, quite enough for our consideration. It may be shown that the condition $L \ll m$ together with the condition $L \gg m^{1/2}$ is necessary and enough for the quasiclassical nature of external particle motion. Because for electrons $m = 1$ their scattering at the level $n = 2$ cannot be taken into account in the quasiclassical approximation.

In the quasiclassical domain ($x = (2L+1)/ma \ll 1$, $Z \approx 1$) we find from (10.53)

$$\frac{d\sigma}{d\Omega}(s \to s) = 3\pi \cos^2 \sqrt{12\pi m\theta}/4E\theta^2 \sin\theta ,$$

$$\frac{d\sigma}{d\Omega}(s \to p) = 3\pi \sin^2 \sqrt{12\pi m\theta}/4E\theta^2 \sin\theta ,$$

(10.54)

$$12\pi m\theta \gg 1 ,$$

where $E = q^2/2m = mv^2/2$. Expression (10.54) coincides with the quasiclassical cross sections obtained in [10.2].

Let us further investigate the general expression (10.53) in the limit $A \ll 1$ (here $x \gg 1$, $z = x^{-2}/4 \ll 1$). The condition $L \gg m$ makes it possible, in contrast with the quasiclassical condition $m^{1/2} \ll L \ll m$, to account for the potential of the dipole interaction of the atom with a particle as a small perturbation. From this, it is clear that results obtained in the limit must correspond to the Born approximation. In the limit $m\theta \ll 1$ we obtain from (10.53):

$$\frac{d\sigma}{d\Omega}(s \to s) = \frac{m^3 a^4}{8E} \ln^2\left(\frac{1}{ma\theta}\right) , \quad \frac{d\sigma}{d\Omega}(s \to p) = \frac{ma^2}{2E\theta^2} .$$

(10.55)

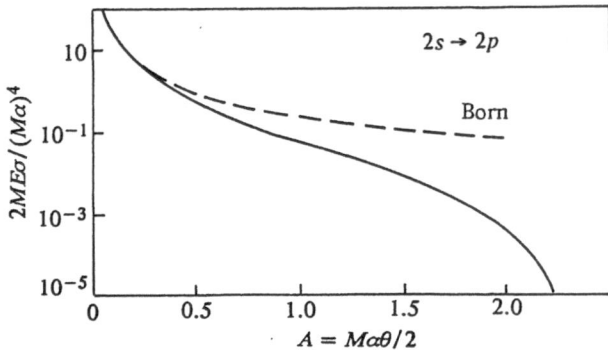

Fig. 10.3. Differential cross section $d\sigma/d\theta$ of an electron scattered by an excited hydrogen atom at the energy level $n = 2$ [10.21] (the Born approximation is also shown)

Results of cross section calculations in the intermediate angle domain performed with the help of (10.50, 53) are placed in Fig. 10.3. The typical peculiarity of the cross sections is the transition of the slow dependence σ on θ in the Born domain to rapid oscillations in the quasiclassical domain, as compared with [10.21].

Let us use Born differential cross sections for determining the total cross sections for transitions between fine structure components. The result of the calculations may be considered to confirm the fact that the Born angle domain is dominant for a total cross section $\sigma = 2\pi \int d (\cos \theta) \, d\sigma/d\Omega$:

$$\sigma(s\rightarrow p_{1/2}) = \frac{1}{3}\sigma(s\rightarrow p) = \frac{24\pi}{v^2} \ln \frac{\theta_{\max}}{\theta_{\min}} \, . \tag{10.56}$$

The cross section (10.56) diverges logarithmically without taking into account the Lamb shift. Let $\theta_{\min} \sim \Omega_{1/2}/E$ (here and below, $\Omega_{1/2} = 4.34 \times 10^{-6}$ eV is Lamb shift, $\Omega_{3/2} = 4.1 \times 10^{-5}$ eV is the fine structure). Defining $\theta_{\max}$ as the boundary of Born domain $\theta_{\max} \, m^{-1}$ and making analogous procedure for $\sigma(s)\rightarrow p_{3/2})$ we find with logarithmic accuracy $(v^2 \gg \Omega)$:

$$\sigma(s\rightarrow p_{1/2}) = \frac{24\pi}{v^2} \ln \frac{v^2}{2\Omega_{1/2}} \, , \quad \sigma(s\rightarrow p_{3/2}) = \frac{48\pi}{v^2} \ln \frac{v^2}{2\Omega_{3/2}} \, ,$$

$$\tag{10.57}$$

$$\sigma(s\rightarrow p_{1/2}) = \frac{12\pi}{v^2} \ln \frac{v^2}{2\Omega_0} \, , \quad \Omega_0 = [\Omega_{1/2}\Omega_{3/2}^2]^{1/3} \, .$$

The cross sections (10.57) coincide with those obtained in [10.13].

Now we consider the problem of the applicability of the optical theorem [10.6] for calculations of total cross sections for corresponding processes. Let us consider the case of scattering at the s-atom state, which is of the most practical interest. According to the optical theorem, the imaginary part of the scattering amplitude $s-s$ at the angle $\theta = 0$ must coincide with the total cross section for elastic and inelastic transitions from the s-state: $\sigma = (ns \rightarrow ns, np \dots)$. Using the general expression (10.52) we obtain

$$\sigma = \sigma(ns \rightarrow ns, np \dots) = \frac{4\pi}{q} \operatorname{Im} \{f_{s \rightarrow s}(0)\}$$

$$= -\frac{2\pi}{q^2} \operatorname{Re} \left\{ \sum_{\lambda L} (2L+1) \langle \lambda L | 0L \rangle^2 (S^{\lambda L} - 1) \right\} . \tag{10.58}$$

Considering the state with $n = 2$ and taking into account that the main contribution to the scattering at zero angle is due to impacts with large angular momenta we obtain

$$\frac{4\pi}{q} \operatorname{Im} \{f_{s \rightarrow s}(0)\} = \frac{12\pi}{v^2} \ln \frac{L_{max}}{L_{min}} . \tag{10.59}$$

The angular momenta are limited in formulae (10.59) connected with the scattering at the zero angle in contrast to (10.56), where for arbitrary angular momenta the angles are limited due to relativistic effects. Defining L_{min} as the boundary of the Born domain and L_{max} from the condition $\Omega L_{max}/E \sim 1$ one becomes sure that (10.59) coincides with (10.57) (under the logarithmic accuracy the cross section of $s \rightarrow s$ scattering is of no importance). The analogous consideration may be applied to the case of electron scattering at dipolar molecules.

11. Collisions of an Atom with Atomic Particles in External Fields

In the following sections, we will consider the problems connected with collisions of a radiating atom with atomic particles in external fields that have an influence on the elementary collision event itself. The influence of external fields is of importance for resonance processes where the external electromagnetic (E.M.) field may result in detuning the resonance or, oppositely, coming into resonance. Of principal importance are the collisions of atoms with particles in a resonant laser field $\mathscr{E}_0 \cos \omega t$, whose frequency ω is close to the frequency ω_0 of the atomic transition. These conditions respond to the standard formulation of nonlinear spectroscopy problems with the essential difference that the field $\mathscr{E}_0$ influences not only relaxation kinetics, but the collision event itself. Accounting for multiparticle atomic collisions with perturbing particles in a resonant E.M. field is also of great importance because it is the case when the time of the collision and the time between collisions cannot be separated. The consideration of the problems mentioned has its foundation on [11.1–8].

11.1 Collisional Transitions Between Fine Structure Sublevels of a Hydrogen Atom in a Magnetic Field

The simplest example of the influence of external fields on collision transition processes is the transition between the fine structure sublevels of a hydrogen atom in a magnetic field B [11.1]. The energy structure $\varepsilon_{l \cdot j_z}$ of the hydrogen atomic level $n = 2$ in a magnetic field is shown in Fig. 11.1. The values of the terms are marked by the orbital momentum magnitudes l and projections j_z of the total angular momentum on the direction B of the magnetic field, the magnitudes Δ_1 and Δ_2 designating the Lamb shift and the fine splitting of the levels correspondingly. The characteristics of the variation of the system term are connected with the appearance of the crossing points $1, 2, 3, 4$ depicted in the figure. The crossings obviously represent a definite spin polarization of the states. The magnitudes of magnetic fields in the crossing points are equal to following: $B_1 = 572\,\mathrm{G}$, $B_2 = 1184\,\mathrm{G}$, $B_3 = 5165\,\mathrm{G}$, $B_4 = 7835\,\mathrm{G}$.

Let us find the cross sections of the transitions between the s and p sublevels in a magnetic field due to the collisions of the atom with heavy particles, e.g., ions (protons). The cross sections determine the finiteness of the

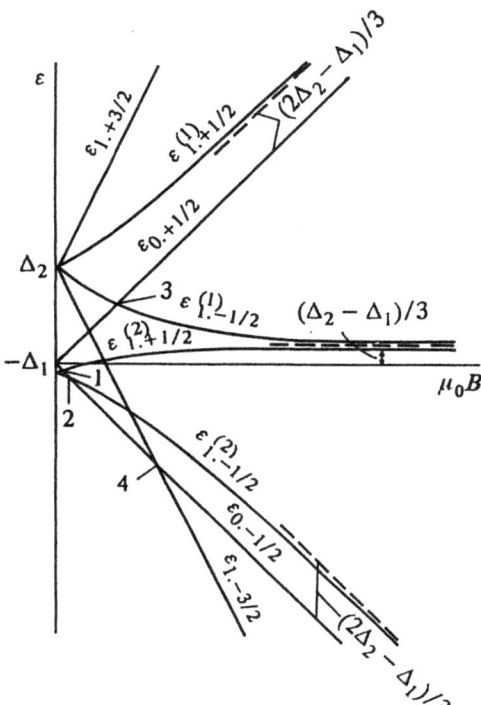

Fig. 11.1. The energy terms of the hydrogen atomic level $n = 2$ in a magnetic field B [11.1]: the crossing points of the terms are shown

lifetime of the metastable s-state, caused by the transitions into the p-state decaying by radiative emission. The corresponding cross sections clearly depend on the angle between the particle velocity and the magnetic field. It is convenient to determine three types of such cross sections ($k = 1, 2, 3$) corresponding to $\theta = 0$ and $\theta = \pi/2$, as well as to one averaged over θ:

$$\sigma_1(v) \equiv \sigma(v, 0); \quad \sigma_2(v) \equiv \sigma\left(v, \frac{\pi}{2}\right); \quad \sigma_3(v) \equiv \frac{1}{2}\int_0^\pi d(\cos\theta)\,\sigma(v, \theta) \ . \quad (11.1)$$

The cross section calculations are performed with the help of matching the solutions neglected the level splittings with those determined by perturbation theory [11.1]. The structure of corresponding cross sections is of the Bethe-Born formulae form [11.9]:

$$\sigma_k(E) = \frac{A}{E}\left(\ln\frac{E}{\Delta_2 F_k(B, \mathscr{P})} - 8.57\right) \quad (11.2)$$

where $A = 1.58 \times 10^{-10}$ eV cm^2, and $E(\text{eV}) = Mv^2/2$ is the kinetic energy of colliding particle. The function $F_k(B)$ is determined by the dependence of atomic terms on the magnetic field B as well as on the spin polarization $\mathscr{P}$ of the initial atomic s-state. The variation of the function $F_k(B)$ near the crossing points B_i of the term is of the most interest, it being determined by the power law

$$F_i(B) = C(\Delta B)^b , \quad \Delta B = B - B_i . \tag{11.3}$$

Substitution of (11.3) into (11.2) gives the following dependence of the cross sections near the crossing points:

$$\sigma_k(E) = \frac{A}{E}(\ln E + a_k - b_k \ln |\Delta B|) . \tag{11.4}$$

Constants a_k and b_k depend on the type of cross section considered as well as on the initial polarization $\mathscr{P}$ of the hydrogen atom. Values of the constants a_k and b_k for three types of cross sections and four crossing points are listed in Table 11.1, taken from [11.1] (values "a_k" are given for E expressed in electron volts, B in Gauss).

It is seen that near the distinguished points 1, 2, 4, the coefficient "b_k" is proportional to $(1 - \mathscr{P})$, and near point 3, to the magnitude $(1 + \mathscr{P})$. It is due to the fact that in points 1, 2, 4 one of the $2p$ energetic terms crosses the $2s$ term which corresponds to the atomic state with a spin projection at the z axis equal to $s_z = -1/2$ as well as in the point 3 to the projection $+1/2$.

So, the peculiarity of the cross sections calculated consists first in its essence arising near the feature points $(\Delta B \rightarrow 0)$ and, second, in the strong dependence on atomic spin polarization $(1 \pm \mathscr{P})$. This characteristic leads to an interesting possibility regarding the creation of the polarized hydrogen atom. If one introduces some quantity of ions into an atomic hydrogen gas in the state $2s$ near critical value of magnetic field then the atomic $2s$ states with one of two spin projections on the magnetic field direction will "burn" more quickly than ones with opposite projections. As a result the gas is polarized. One may also obtain the spin-polarized atomic beams by means of their passage through the plasma. Note that the destruction of the metastable $2s$ state may be under the influence of a Lorentzian electric field, arising in the coordinate system of the atom moving across magnetic field. For estimations of the effect, consult [11.1].

11.2 Collisions of a Two-Level Atom with Particles in a Strong Resonant Electromagnetic Field

The theory of atomic collisions for optical problems [11.10–16] was primarily developed for the problems of radiation transfer and diagnostics. Usually the light field $\mathscr{E}_0$ in these problems can be considered to be weak. More precisely,

Table 11.1.

$B[G]$	$\sigma_1(E)$		$\sigma_2(E)$		$\sigma_3(E)$	
	a_k	b_k	a_k	b_k	a_k	b_k
$B_1 = 572$	$3.42 - 1.27\,\mathscr{P}$	$0.174\,(1 - \mathscr{P})$	$3.10 - 0.61\,\mathscr{P}$	$0.087\,(1 - \mathscr{P})$	$3.21 - 0.83\,\mathscr{P}$	$0.116\,(1 - \mathscr{P})$
$B_2 = 1184$	$2.25 - 0.27\,\mathscr{P}$	0	$2.97 - 0.985\,\mathscr{P}$	$0.1\,(1 - \mathscr{P})$	$2.76 - 0.71\,\mathscr{P}$	$0.066\,(1 - \mathscr{P})$
$B_3 = 5165$	$3.32 + 0.81\,\mathscr{P}$	$0.159\,(1 + \mathscr{P})$	$2.785 + 0.31\,\mathscr{P}$	$0.08\,(1 + \mathscr{P})$	$2.94 + 0.55\,\mathscr{P}$	$0.106\,(1 + \mathscr{P})$
$B_4 = 7835$	$4.12 - 2.24\,\mathscr{P}$	$0.25\,(1 - \mathscr{P})$	$2.91 - 1.27\,\mathscr{P}$	$0.125\,(1 - \mathscr{P})$	$3.22 - 1.59\,\mathscr{P}$	$1/6\,(1 - \mathscr{P})$

it can be assumed that the field has no effect on a broadening-collision event. Therefore, in the theory of broadening the effect of the field $\mathscr{E}_0$ on the shape of the spectral line is neglected, and light absorption is treated as a set of events of photon absorption by atoms broadened "beforehand" by collisions.

We will attempt to construct a nonlinear broadening theory free from the limitations, i.e., a theory that generalizes the results of the standard broadening theory to the case of arbitrary fields and frequency shifts $\Delta\omega$. The analysis is based on a consistent use of the idea that the atom and the field constitute a compound system such that the transitions occurring in it are responsible for the absorption of light by the medium. We are able to derive for the total dissipatable light power an expression that coincides in form with the *Karplus-Schwinger* result [11.15, 16]. However, the fundamental parameters are, in the general case, not relaxation constants, but complicated functions of the frequency and the field strength. In the weak-field limit the obtained expression yields the results of the standard (linear) theory of broadening; for the frequency region near the line center (i.e., in the impact region), the Karplus-Schwinger result; in the general case it describes complex nonlinear effects.

The analysis is carried out in the framework of the scheme usually adopted in broadening theory [11.10−16]: the motion of the broadening particles is assumed to be classical and rectilinear. Additionally, the condition for the collisions to be binary encounters, i.e., the condition $N\varrho_{\mathrm{eff}} \ll 1$, where N is the density of the broadening particles and ϱ_{eff} is the effective impact parameter, which determines the contribution to the corresponding collision cross section, is assumed to be satisfied.

11.2.1 Optical Collisions. The Basic System of Equations

Let us consider the absorption of light energy by the atomic particles X and Y colliding in an external electromagnetic field

$$\mathscr{E}(t) = \mathscr{E}_0 \cos \omega t \ . \tag{11.5}$$

The frequency ω is close to the natural frequency ω_0 of the transition between the states 1 and 2 of the X atom (Fig. 11.2). We shall henceforth call such collisions optical collisions (OC). This term is borrowed from the standard broadening theory, where it was introduced by V. Weisskopf for collisions that destroy the phase of the oscillations of the atom.

In the present consideration we regard optical collisions as isolated events, in the course of which the atoms interact only with each other and with the electromagnetic field. The other interactions that occur in the course of an OC are assumed to be unimportant. In particular, this means that we neglect the spontaneous relaxation of the states $X(1)$ and $X(2)$ during an OC, i.e., we assume that in the absence of interactions with the atom Y and the field (11.5) these states are strictly discrete states.

In the framework of the assumed approximations we can speak of the discrete state of some compound system composed of the atom X and the field

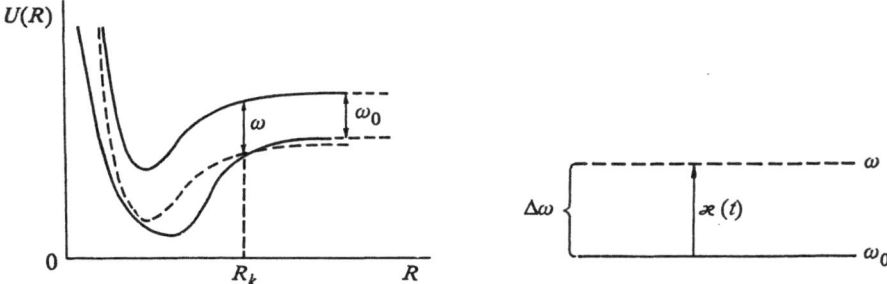

Fig. 11.2. The energy term crossing in the system "atom + E.M. field" arising after shifting of the atomic level at the frequency $\hbar\omega$ of the light quantum

$\mathscr{E}$. The states of the "atom X + field $\mathscr{E}$" compound system are more convenient for the description, since it is precisely the transition induced between these states by the collisions with the Y atoms that are responsible for light absorption in the course of an OC.

Let the total Hamiltonian describing the states of the colliding atoms and the field have the form

$$\hat{H} = \hat{H}_X + \hat{H}_{\mathscr{E}} + \hat{V}_{X\mathscr{E}} + \hat{V}_{XY} . \tag{11.6}$$

Here H_X and $H_{\mathscr{E}}$ are the Hamiltonians of the free X atom and the field $\mathscr{E}$ respectively, $V_{\mathscr{E}}$ is their interaction operator; V_{XY} is the $X-Y$ interaction operator, the motion of the X and Y nuclei is, as has already been indicated, assumed to be classical and rectilinear: $R^2 = \varrho^2 + v^2 t^2$ (where ϱ is the impact parameter and v is the relative velocity of the atoms). For convenience, we shall henceforth treat the electromagnetic field (11.5) as an ensemble of n quanta of frequency ω; its Hamiltonian is included in H. When, however, we go over in the specific expressions for the matrix elements to the classical limit $n \to \infty$, we shall use the quantity $\mathscr{E}_0$.

Let us first consider the standard treatment of photon absorption in the course of an OC. Since in such a transition the number of photons before and after the collision is assumed to be fixed, it is natural to use the system of eigenfunctions $|1\rangle = |X(1), n_\omega\rangle$ and $|2\rangle = |X(2), n_\omega - 1\rangle$ of the Hamiltonian $H_0 = H_X + H_{\mathscr{E}}$. For the amplitudes a_1 and a_2 of these states, we have the equations

$$i\dot{a}_1 = U_1 a_1 + V e^{i\Delta\omega t} , \quad i\dot{a}_2 = U_2 a_2 + V e^{-i\Delta\omega t} a_1 ,$$

$$U_k(t) = \langle \varphi_k | V_{XY} | \varphi_k \rangle , \quad V = |\langle \varphi_1 | V_{X\mathscr{E}} | \varphi_2 \rangle| , \quad \Delta\omega = \omega - \omega_0 , \tag{11.7}$$

in a dipole transition $V = D \cdot \mathscr{E}_0$, where $D = \mathscr{E}_0^{-1} |(d \cdot \mathscr{E}_0)_{12}|$.

The system (11.7) is the basic system of equations for standard broadening theory, which is valid for a sufficiently small field strength $\mathscr{E}_0$ (i.e., for a sufficiently small V). Indeed, for an arbitrary V it is difficult to even establish the initial conditions for (11.7), since V does not go to zero as $t \to \pm \infty$ ($V = \text{const.}$). Therefore, V is assumed to be small, so as to make the probability of a transition occurring in the absence of the broadening interaction (when $U_1(-\infty) = U_2(-\infty) = 0$) negligibly small in comparison with the transition probability in the course of an OC. Then setting one of the amplitudes equal to unity as $t \to -\infty$ and the other equal to zero (to be specific, $a_1(-\infty) = 1$ and $a_2(-\infty) = 0$), and solving (11.7) with the aid of perturbation theory, we find

$$|a_2(\infty)|^2 = V^2 \left| \int\limits_{-\infty}^{\infty} dt \exp \left\{ i \left[\Delta \omega t - \int\limits^t \kappa(t') dt' \right] \right\} \right|^2 , \qquad (11.8)$$

where $\kappa = U_2 - U_1$ is the term shift due to the interaction.

The expression (11.8) determines the probability of emission (absorption) in one OC of a photon with a frequency shift. The resulting shape arising from many OC is, generally speaking, not equal to the sum of the independent terms. This is due to the finiteness of the mean free time of the atom, which leads to the correlation of the individual events in which photons of frequency close to the line center (i.e., for which $\Delta \omega = 0$) are emitted. But this fact is important only in the narrow frequency region $\Delta \omega \lesssim \gamma_{im}$, where γ_{im} is the impact width of the line (for example, [11.10]). However, in the region $\Delta \omega \gg \gamma_{im}$ (which may be called the "single-particle" region), the line shape $I(\Delta \omega)$ is completely determined by the probability of a transition occurring in an OC with one Y particle and is proportional to the number of such OC that occur in unit time. Notice that the region of values $\Delta \omega \lesssim \gamma_{im}$ is, in the case of binary collisions, fairly narrow. It is located deep inside of the impact region of the spectrum, so that practically all of the characteristics of the line shape — in particular, the transition between the impact and static broadening mechanisms [11.11] — can be described by the single-particle approximation.

For the single-particle region it is convenient to transform with the aid of integration by parts, which corresponds to the exclusion of the region $\Delta \omega \lesssim \gamma_{im}$. The first term is proportional to the delta function $-\delta(\Delta \omega)$, and vanishes at $\Delta \omega \neq 0$, while the second yields

$$|a_2(\infty)|^2 = \left| \int\limits_{-\infty}^{\infty} dt \frac{V}{\Delta \omega} \kappa(t) \exp \left\{ i \left[\Delta \omega t - \int\limits^t \kappa(t') dt' \right] \right\} \right|^2 . \qquad (11.9)$$

It can be seen from a comparison of (11.9) and (11.8) that an effective replacement has occurred in the single-particle region of the "transition potential" V by $V\kappa/\Delta \omega$. This seemingly formal circumstance has a profound physical meaning in connection with the investigation of the transitions in the compound system (see below). Notice that the expression (11.9) is not quite correct. It diverges upon passage to the limit, i.e., $\lim (V/\Delta \omega) = \infty$ for $\Delta \omega \to 0$ and $\Delta \omega \gg \gamma_{im}$, which is a direct indication of the limited applicability of the conventional conception of strong fields.

Let us proceed to the case of an arbitrary V. In this case we cannot speak of the absorption of an individual photon: to describe the dissipation of the light energy in the course of an OC, we must consider the transitions between the states of the Hamiltonian $H_{X\mathscr{E}} = H_X + H_{\mathscr{E}} + V_{X\mathscr{E}}$ of the compound system, states which are characterized by the eigenfunctions ([11.9], p. 135):

$$\Psi_I = b_1 \varphi_1 + b_2 \varphi_2 , \quad \psi_{II} = b_2 \varphi_1 - b_1 \varphi_2 , \tag{11.10}$$

where

$$b_{1,2} = \frac{1}{\sqrt{2}} \left(1 \pm \frac{\Delta\omega}{\Omega_0} \right)^{1/2} , \quad \Omega_0 = \sqrt{\Delta\omega^2 + 4V^2} .$$

Everywhere below we shall use the capital letters (K, K') and the Roman numerals (I, II) to denote quantities pertaining to the compound system; the small letters (k, k') and the Arabic numerals $(1, 2)$ will be used to denote quantities characterizing the states φ_1 and φ_2. For the amplitudes of the states I and II of the compound system we obtain from (11.7) with allowance for (11.10) the expressions

$$i\dot{a}_I = U_I a_I + V_{I,II} e^{i\Omega_0 t} a_{II} , \quad i\dot{a}_{II} = U_{II} a_{II} + V_{II,I} e^{-i\Omega_0 t} a_I ,$$

where

$$U_I = b_1^2 U_1 + b_2^2 U_2; \quad U_{II} = b_2^2 U_1 + b_1^2 U_2; \quad V_{I,II} = V_{II,I} = b_1 b_2 (U_2 - U_1) .$$

After the substitution $b_K = a_K \exp\left(i \int^t U_K dt' \right)$ we have

$$i\dot{b}_I = b_{II} \kappa \frac{V}{\Omega_0} \exp\left\{ i \left[\Omega_0 t - \frac{\Delta\omega}{\Omega_0} \int^t \kappa(t') dt' \right] \right\} ,$$

$$i\dot{b}_{II} = b_I \kappa \frac{V}{\Omega_0} \exp\left\{ -i \left[\Omega_0 t - \frac{\Delta\omega}{\Omega_0} \int^t \kappa(t') dt' \right] \right\} . \tag{11.11}$$

The system (11.11) is the basic system of equations describing the transitions connected with the dissipation of the energy of the field in the course of an OC.

The role of the potential including the transition in the compound system is, as can be seen from (11.11), played by the quantity $\kappa V/\Omega_0$, which, for $\Delta\omega \gg 2V$ goes over into the effective potential $\kappa V/\Delta\omega$ of the formula (11.9). It is clear from this that the Spitzer transformation, which separates out the term responsible for light absorption in the course of an OC, corresponds to a transition to the states of the compound system in the weak-field case when

$V \ll \Delta\omega$. The aforementioned incorrectness of the formula (11.9) is connected with its inapplicability when $V \gtrsim \Delta\omega$.

If we are interested in the single-particle region $\Omega_0 \gg \gamma_{im}$, then the initial conditions for the basic system (11.11) can be formulated for $t \to -\infty$. Choosing them in the form $b_I(-\infty) = 1$ and $b_{II}(-\infty) = 0$ for the $I \to II$ transition probability w, we have: $w = |b_{II}(\infty)|^2$.

It is clear from (11.11) that, in contrast to the results (11.8,9) of the standard broadening theory, the probability w of a transition occurring in an OC is, generally speaking, not proportional to the light intensity $\mathscr{E}_0^2 \sim V^2$. This is a direct indication of the inseparability in the general case of the two elementary events: the broadening-collision and the light-absorption events. It is clear from the foregoing that nonlinear effects should appear when $V > \Delta\omega$. As will be shown below, however, they can appear in fairly weak fields $V \ll \Delta\omega$.

11.2.2 Optical Collisions and Characteristics of Light Absorption in Media

Let us derive an expression for the energy dissipated in one OC event. The computation of the energy dissipated in any transition (including OC) amounts to finding the mean energy of the field in the states of the compound system before and after the transition. If some state of the compound system is characterized by the wave function $\psi(t) = b_I(t)\psi_I + b_{II}(t)\psi_{II}$, then the mean (the quantum mechanical average taken over the arbitrary initial phases of the statistical ensemble) value $\langle H_\mathscr{E} \rangle$ of the field energy is given by the relation

$$\langle H_\mathscr{E} \rangle = |b_I|^2 \langle H_\mathscr{E} \rangle_I + |b_{II}|^2 \langle H_\mathscr{E} \rangle_{II} \ . \tag{11.12}$$

In the optical collision the compound system was initially in the state $b_I(-\infty) = 1$, $b_{II}(-\infty) = 0$. After the collision the system turned out to be in the state $b_I(+\infty)$, $b_{II}(+\infty)$. For the change in the mean energy of the field occurring in one OC we have

$$\langle H_\mathscr{E}(-\infty) \rangle - \langle H_\mathscr{E}(\infty) \rangle = -\hbar\omega(b_2^2 - b_1^2)|b_{II}(\infty)|^2 = \hbar\omega\frac{\Delta\omega}{\Omega_0} w \ . \tag{11.13}$$

Thus, the energy dissipated in one OC event is proportional to the probability w of the $I \to II$ transition in the compound system. Therefore, to compute the energy absorbed by the medium as a result of an OC, we simply need to find the number of OC occurring per unit time. For this purpose it is convenient to introduce the concept of an "optical collision cross section":

$$\sigma(v, \mathscr{E}_0, \Delta\omega) = 2\pi \int_0^\infty d\varrho\, \varrho\, w(\varrho, v, \mathscr{E}_0, \Delta\omega) \ . \tag{11.14}$$

These cross sections are generalizations of the OC cross sections introduced by V. Weisskopf in the conventional theory of broadening ([11.10, 11], p. 465).

In the single-particle region $\Delta\omega \gg \gamma_{im}$ we can speak of populations N_I and N_{II} of the states I and II of the "$X + \mathscr{E}$" compound system. Then, introducing with the aid of the optical-collision rate Γ_{OC} defined by

$$\Gamma_{OC} = N_Y \langle v\sigma(\Delta\omega, v, \mathscr{E}_0)\rangle \ , \tag{11.15}$$

where N_Y is the density of the broadening atoms Y and $\langle \ldots \rangle$ denotes averaging over the velocity, we can easily find the light power Q_{OC} absorbed as a result of OC by a unit volume of the medium

$$Q_{OC} = \hbar\omega \frac{\Delta\omega}{\Omega_0} \Gamma_{OC}(N_I - N_{II}) \ . \tag{11.16}$$

In computing the total light energy absorbed by the medium, it is necessary to bear in mind, besides the OC, another dissipation channel connected with the mixing of the states of the atom and the field [11.3, 4].

11.3 Landau-Zener Mechanism of Strong Electromagnetic (E.M.) Radiation Absorption in the Wings of a Spectral Line

11.3.1 Landau-Zener Model for Optical Phenomena

As mentioned earlier, the equation system (11.11) describes nonlinear effects of a new type associated with light absorption in the process of an atomic collision with a broadening particle. The investigation of a general solution of the system is analogous to the investigation of a transition in a two level system in the theory of atomic collisions [11.14]. The nontrivial nature of the results obtained is connected with the dependence of the effective potentials of the system (11.11) on the parameters of an electromagnetic field. It is clear that in the weak field case $V \to 0$ the results of linear broadening theory [11.10] are followed from (11.11). In the case of fast collisions $\varrho\sqrt{\Delta\omega^2 + 4V^2}/v \ll 1$ the solution of (11.11) reduces to the model of relaxation constants not dependent on electromagnetic field parameters. This is the standard approach of nonlinear laser spectroscopy [11.15].

The greatest interest from the point of view of the realization of new nonlinear effects is the case of slow collisions when the oscillation frequency Ω of the atom in a laser field (Rabi frequency) becomes large as compared with the frequency of collisions $\Omega_W = \varrho_W/v$ (Weisskopf frequency):

$$\Omega = \sqrt{\Delta\omega^2 + 4V^2} \gg \Omega_W \ . \tag{11.17}$$

The case of comparable weak laser fields $\Delta\omega \gg V$ is of essential interest in this domain. In this case the main contribution to absorption is due to the stationary phase point [11.11]:

$$\Delta\omega = U[R(t_\kappa)] \; ; \quad R(t_\kappa) \equiv R_{\Delta\omega} \; . \tag{11.18}$$

The condition (11.18) is the condition of some type of "terms crossing" in the compound system "atom+E.M. field". The crossing arises when the laser frequency ω is in response with the transition frequency ω_0 shifted by the broadening particle at the value $U(R_K)$ (Fig. 11.2). The light absorption is determined by the inelastic transition probability in the compound system (11.11). For slow collisions the main contribution to the probability of such a transition is due to the crossing point of the terms (11.18), where the probability is defined by the Landau-Zener formulae [11.14]. The probability strongly depends on the magnitude of the repulsion of the term at the crossing point. In the case under consideration the repulsion is due to the interaction with the electromagnetic field $\mathscr{E}_0$. Therefore, the corresponding cross section σ_{OC} of the inelastic transition depends on the ratio of the field $\mathscr{E}_0$ to some critical Landau-Zener field $\mathscr{E}_{L-Z}$, depending on the dynamics of the crossing point passing by a colliding system. In agreement with the scheme mentioned it follows that [11.3, 4]

$$\sigma_{OC} = \pi R_\omega^2 \bar{w}(\mathscr{E}_0^2 / \mathscr{E}_{L-Z}^2) \; , \tag{11.19}$$

$$\mathscr{E}_{L-Z} = [V|dU/dR|_{R_\omega} / 2\pi d^2]^{1/2} \; , \tag{11.20}$$

where $\bar{w}(\alpha)$ is the averaged transition probability, determined by the Landau-Zener mechanism of the transitions in crossing points 1 and 2 corresponding to the approaching and scattering directions of the atoms

$$\bar{w}(\alpha) = \left\langle 2\int_0^1 dx \exp\left[-(\alpha/\sqrt{x})\cos^2\theta_1\right]\{1 - \exp\left[-(\alpha/\sqrt{x})(\cos^2\theta_1 - \cos^2\theta_2)\right]\}\right\rangle \; , \tag{11.21}$$

where the symbol $\langle \ldots \rangle$ designates an averaging over angles θ_1 and θ_2 between the atomic dipole moment d and laser field strength $\mathscr{E}_0$ at the points 1 and 2. Formulae (11.21) respond to transitions without changing the magnetic quantum number $\Delta m = 0$. For transitions with $\Delta m = \pm 1$, the cosine terms in (11.21) must be replaced by sines. In the limiting cases of large and small field strengths, the probability $\bar{w}$ takes the form:

$$\bar{w}(\alpha)_{\Delta\tilde{m}=0} = \begin{cases} \dfrac{4}{3}\alpha \; , & \alpha \ll 1 \\[2mm] \dfrac{4}{3}\dfrac{1}{\sqrt{\alpha}} \; , & \alpha \gg 1 \end{cases} \; ; \quad \bar{w}(\alpha)_{\Delta m=\pm 1} = \begin{cases} \dfrac{8}{3}\alpha \; , & \alpha \ll 1 \\[2mm] \dfrac{8}{3}\dfrac{1}{\alpha} \; , & \alpha \gg 1 \end{cases} \; . \tag{11.22}$$

It is seen that for $\alpha \gg 1$ ($\mathscr{E}_0 \gg \mathscr{E}_{L-Z}$) the qualitative change of the absorption probability dependence on the field strength $\mathscr{E}_0$ occurs. Instead of the linear growth of the absorption with the intensity ($\sim \mathscr{E}_0^2$), the drop of the absorption ($\sim \mathscr{E}_0^{-1}$) takes place corresponding to some kind of medium "brightening". The drop is due to the decrease of the absorption due to the repulsion of the atomic terms at their crossing point in the process of optical collision.

11.3.2 Nonlinear Effects in Absorption for the Collision of Identical Atoms

Let us consider in more detail the mechanism of Landau-Zener transitions as a concrete example of identical atom collisions [11.17, 18]. As is known, in the system of two identical atoms X and Y there is the degeneration connected with the identical nature of the states $\psi_{1X}\psi_{0Y}$ and $\psi_{0Y}\psi_{1X}$, representing the combination of atoms X and Y in the ground (ψ_0) and the excited (ψ_1) states, respectively. The total wave function $\psi(t)$ of the system in a two level approximation takes the form:

$$\psi(t) = C_1\psi_{0X}\psi_{0Y} + C_2\psi_{1X}\psi_{0Y} + C_3(t)\psi_{0X}\psi_{1Y} + C_4(t)\psi_{1X}\psi_{1Y} . \quad (11.23)$$

For real atoms the expansion (11.23) is more complicated and accounts for the orbital momentum projections of excited states. The interaction $V_{XY}(t)$ of identical atoms is of a dipole-dipole nature:

$$\hat{V}_{XY}(t) = \{d_x \cdot d_y R^2(t) - 3[d_x \cdot R(t)][d_y \cdot R(t)]\}R^{-5}(t) \equiv V(t) , \quad (11.24)$$

where $R^2 = R_{XY}^2(t) = \varrho^2 + v^2 t^2$ is the distance between atoms, and d_x, d_y are the operators of the atomic dipole momenta.

The corresponding interactions of the atoms with the laser field are of the form:

$$V_{FX} = -d_X \cdot \mathscr{E}_0 \cos \omega t , \quad V_{FY} = -d_Y \cdot \mathscr{E}_0 \cos \omega t . \quad (11.25)$$

Substituting the expansion (11.23) into the Schrödinger equation with interactions (11.24, 25) we obtain the system of equations for the coefficients $C_i(t)$. Let us account for the degeneration of the states in the orbital momentum projections. Let the orbital momentum L of an upper level be equal to unity ($L = 1$) and the lower level equal to zero ($L = 0$). Then the system of equations for $C_i(t)$ will take the form

$$\left.\begin{array}{l}
\dot{C}_1 + i\Delta\omega C_1 = i V_0^m (C_2^m + C_3^m) , \\[2mm]
\dot{C}_2^m = i(V_0^{m*}C_1 + V_0^k C_4^{mk}) - i V_{mm'}(t)C_3^{m'} , \\[2mm]
\dot{C}_3^m = i(V_0^{m*}C_1 + V_0^k C_4^{km}) - i V_{m'm}(t)C_2^{m'} , \\[2mm]
\dot{C}_4^{mk} - i\Delta\omega C_4^{mk} = i V_0^{k*}C_2^m + i V_0^{m*}C_3^k ,
\end{array}\right\} \quad (11.26)$$

where, for example, the coefficient C_4^{mk} designates the state amplitude in which the atom X possesses the orbital projection m and the atom Y contains the projection k, with the summation performed over repeated indices.

Now we introduce further symmetric and antisymmetric combinations of amplitudes:

$$C_\pm^m = \frac{C_2^m \pm C_3^m}{\sqrt{2}} \; ; \quad C_{4\pm}^{mk} = \frac{C_4^{mk} \pm C_4^{km}}{\sqrt{2}} \; . \tag{11.27}$$

After the substitution of (11.27) into (11.26) the system decays into two systems, namely a symmetrical

$$\left. \begin{array}{l} \dot{C}_+^m = i\sqrt{2}\,V_0^{m*} C_1 + i V_0^k C_{4+}^{mk} - i V_{mm'} C_+^{m'} \; , \\[2mm] \dot{C}_1 + i\Delta\omega\,C_1 = i\sqrt{2}\,V_0^m C_+^m \; , \\[2mm] \dot{C}_{4+}^{mk} - i\Delta\omega\,C_{4+}^{mk} = i V_0^{k*} C_+^m + i V_0^{m*} C_+^k \; . \end{array} \right\} \tag{11.28}$$

and an antisymmetrical system

$$\left. \begin{array}{l} \dot{C}_-^m = i V_0^k C_{4-}^{mk} + i V_{mm'} C_-^{m'} \; , \\[2mm] \dot{C}_{4-}^{mk} - i\Delta\omega\,C_{4-}^{mk} = i V_0^{k*} C_-^m - i V_0^{m*} C_-^k \; . \end{array} \right\} \tag{11.29}$$

The matrix of the interatomic interaction in the coordinate system with axis $0Z \parallel R$ takes the form:

$$V_{mm'}(t) = \frac{|d_{01}|}{R^3(t)} \begin{vmatrix} -2 & 0 & 0 \\ 0 & 1 & 0 \\ 0 & 0 & 1 \end{vmatrix} . \tag{11.30}$$

Under the condition $\Delta\omega \gg V_0$ for the determination of the energetic terms of a quasimolecule consisting of identical atoms, let us set the interaction with the E.M. field V_0 equal to zero in (11.28, 29). Then the energetic terms of the compound system versus V take the form shown in Fig. 11.3. For $\Delta\omega < 0$ terms C_1 and C_{4+}^{mk} switch the roles. It is seen that there are crossing points with the term C^{m+} both the term C_{4+}^{mk} and C_1, the picture of the crossing point of the term is not symmetrical, in general. It is important to emphasize that the both line wings $\Delta\omega > 0$ and $\Delta\omega < 0$ have analogous pictures of crossings (even not identical). This means that there is no strong asymmetry in the absorption line (as is the case for the broadening of nonhydrogenic lines by particles without an internal structure [11.10]).

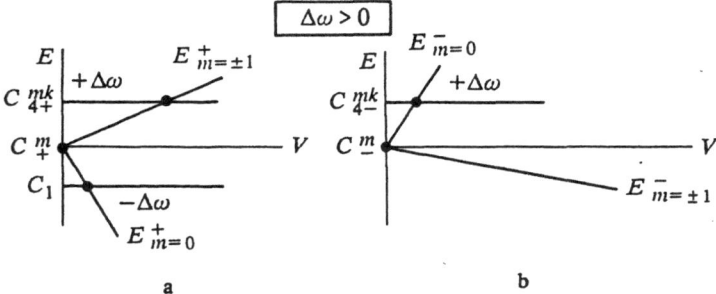

Fig. 11.3a, b. The crossing of the term in the system of identical atoms [11.17]

The transition probabilities we will find in the crossing points with the help of the Landau-Zener formulae [11.14]

$$w = e^{-P} \; ; \quad P = 2\pi |V_0|^2 / |\dot{V}(t_k)| \; , \tag{11.31}$$

where V_0 is the nondiagonal matrix element of the transition between the terms under consideration, and $\dot{V}(t_k)$ is the derivation of the potential in crossing point t_k. According to (11.30) the potentials $V(t)$ are equal to

$$V(t) = d^m / R^3(t) \; , \quad d^0 = -2|d_{10}|^2 \; , \quad d^{\pm 1} = |d_{10}|^2 \; . \tag{11.32}$$

The crossing point t_k is defined from the condition

$$V(t_k) = d^m / R_k^3 = \varDelta \omega \; , \quad \text{with } R_k = (d^m / \varDelta \omega)^{1/3} \; , \quad t_k = \frac{1}{v}\sqrt{R_k^2 - \varrho^2} \; . \tag{11.33}$$

From this we obtain the derivation

$$\dot{V}(t_k) = \frac{3 d^m v}{R_k^4} \sqrt{1 - (\varrho/R_k)^2} = \frac{3|\varDelta \omega|^{4/3} v}{(d^m)^{1/3}} \sqrt{1 - (\varrho/R_k)^2} \; . \tag{11.34}$$

We find the values of the nondiagonal matrix element V_0 with the help of the general expression for dipole moment matrix elements

$$d_{01} = \frac{|d_{01}|}{\sqrt{3}} \left[n_z \delta_{m0} + \frac{n_x - i n_y}{\sqrt{2}} \delta_{m1} + \frac{n_x + i n_y}{\sqrt{2}} \delta_{m-1} \right] \; , \tag{11.35}$$

where n_z, n_x, n_y are the unit vectors of the coordinate system. Accounting for $0Z \parallel R$ we find

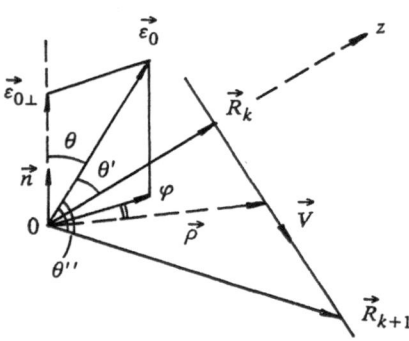

Fig. 11.4. The geometry of the collision of identical atoms in the external laser field $\mathscr{E}_0$ [11.17]

$$|V_0^{m=0}|^2 = \frac{|d_{01}|^2}{3}|\mathscr{E}_{0z}|^2 ; \quad |V_0^{m=\pm1}|^2 = \frac{|d_{01}|^2}{3}\frac{\mathscr{E}_{0x}^2 + \mathscr{E}_{0y}^2}{2} . \tag{11.36}$$

It is easy to show for the crossing of the term $\hat{C}$ with a system of degenerated terms that the nondiagonal matrix element must be replaced by the effective matrix element $|(V_0)_{\mathrm{eff}}|^2 = \sum_k |V_0^k|^2$.

Let us further consider the geometry of an atomic collision in an E.M. field (Fig. 11.4). The field $\mathscr{E}_0$ constitutes an arbitrary angle θ with the vector n normal to collision plane, formed by the vectors ϱ and v. The crossing condition of the terms (11.33) is realized in two points, namely, R_k for the atoms as they approach a collision and R_{k+1} for the atoms after a collision when they are "flying away". The angles θ' and θ'' formed by the vector $\mathscr{E}_0$ with R_k and R_{k+1} are different, so the nondiagonal matrix elements (as well as the transition probabilities) in both points are different, too. Therefore the total transition probability after the passing two crossing points is equal to [11.14]

$$w = w_k(1 - w_{k+1}) + w_{k+1}(1 - w_k) , \tag{11.37}$$

where w is determined by (11.31).

Taking into account the collision geometry we write down the matrix element (11.36) in the form

$$|V_0^{m=0}|^2 = \frac{|d_{01}|^2}{3}\mathscr{E}_0^2 \cos^2\left({}^{\theta'}_{\theta''}\right) = \frac{|d_{01}|^2}{3}\mathscr{E}_0^2 \sin^2\theta \frac{[\varrho\cos\varphi \pm vt_k\sin\varphi]^2}{R_k^2} , \tag{11.38}$$

$$|V_0^{m=\pm1}|^2 = \frac{|d_{01}|^2}{3}\frac{\mathscr{E}_0^2 \sin^2\left({}^{\theta'}_{\theta''}\right)}{2} = \frac{|d_{01}|^2}{3}\frac{\mathscr{E}_0^2}{2}$$

$$\times \left[1 - \sin^2\theta\frac{(\varrho\cos\varphi \pm vt_k\sin\varphi)^2}{R_k^2}\right] , \tag{11.39}$$

where the + and − signs correspond to points R_k and R_{k+1} (Fig. 11.4).

To obtain the total transition probability it is necessary to average (11.37) over all of the angles θ', θ'' (or θ, φ) as well as over the parameters ϱ. Thus the averaged magnitudes obviously define the cross sections of inelastic transitions between the sublevels of the compound system. For the transition between the terms $E_{m=0}^+$ and C_1 shown in Fig. 11.3 we have

$$
\sigma_{m=0}^+ = \int_0^{R_k^{m=0}} 2\pi\varrho\, d\varrho \left[\int_{-1}^{+1} d\cos\theta' \exp\left(-\frac{2\pi \mathscr{E}_0^2 R_k^4 \cos^2\theta'}{9v\sqrt{1-(\varrho/R_k)^2}} \right) \right.
$$

$$
-\frac{1}{2\pi} \int_0^{2\pi} d\varphi \int_{-1}^{+1} d\cos\theta \exp
$$

$$
\left. \times \left(-\frac{4\pi \mathscr{E}_0^2 R_k^4 \sin^2\theta}{9v\sqrt{1-(\varrho/R_k)^2}} \cdot \frac{\varrho^2(\cos^2\varphi - \sin^2\varphi) + R_k^2\sin^2\varphi}{R_k^2} \right) \right] , \quad (11.40)
$$

where the relations (11.32−34) are taken into account as well as the circumstance that in averaging the probabilities $\langle w_k \rangle$, $\langle w_{k+1} \rangle$ in (11.37) it is convenient to use angles θ', θ'', but for multiplication $\langle w_k w_{k+1} \rangle$, the angles θ, φ are more suitable.

By introducing the dimensionless variables $t = (\varrho/R_k)^2$, $x = \cos\theta$ we obtain

$$
\sigma_{m=0}^+ = \pi (R_k^{m=0})^2 \Lambda_1 \left(\frac{2\pi \mathscr{E}_0^2 R_k^4}{9v} \right) , \quad (11.41)
$$

$$
\Lambda_1(z) \equiv \int_0^1 dt \int_{-1}^{+1} dx \left(\exp\left(-\frac{zx^2}{\sqrt{1-t}} \right) - \frac{1}{2\pi} \int_0^{2\pi} d\varphi \exp\left\{ -z\frac{2(1-x^2)}{\sqrt{1-t}} \right. \right.
$$

$$
\left. \left. \times [t\cos^2\varphi + (1-t)\sin^2\varphi] \right\} \right) . \quad (11.42)
$$

The picture of the interaction between the terms C_{mk}^{4-} and $C_{m=0}$ (Fig. 11.3) is seen directly from (11.29). The nondiagonal matrix element between the terms mentioned is equal to V_0^1 (because for $k = m = 0$ the element tends to zero). Analogously to (11.41, 42) we find for $\sigma_{m=0}^-$:

$$
\sigma_{m=0}^- = \pi (R_k^{m=0})^2 \Lambda_2 \left(\frac{2\pi \mathscr{E}_0^2 R_k^4}{9v} \right) , \quad (11.43)
$$

$$\Lambda_2(z) = \int\limits_0^1 dt \int\limits_{-1}^{+1} dx \left[\exp\left(-z\frac{1-x^2}{\sqrt{1-t}}\right) - \frac{1}{2\pi} \int\limits_0^{2\pi} d\varphi \exp\right.$$

$$\left. \times \left(-\frac{2z}{\sqrt{1-t}}\{1-(1-x^2)[t\cos^2\varphi+(1-t)\sin^2\varphi]\}\right) \right]. \qquad (11.44)$$

The interaction between the "vector" term $C_4^{m=\pm1}$ and the "tensor" term C_{mk}^{4+} is more complicated. Because $V_0^{+1} = V_0^{-1}$ and $d^{+1} = d^{-1}$ we will let $C_4^{m=+1} = C_4^{m=-1}$. Then the calculations become simpler, however, both matrix elements V_0^0 and $V_0^{\pm1}$ enter to the interaction of C_{4+}^{mk} with C_1^+. For the cross sections $\sigma_m^+ = 1$ we have taken (11.34) into account:

$$\sigma_{m=\pm1}^+ = \pi(R_k^{m=\pm1})^2 \Lambda_3 \left(\frac{4\pi\mathscr{E}_0^2 R_k^4}{9v}\right), \qquad (11.45)$$

$$\Lambda_3(z) = \int\limits_0^1 dt \int\limits_{-1}^{+1} dx \left[\exp\left(-z\frac{2-x^2}{\sqrt{1-t}}\right) - \frac{1}{2\pi} \int\limits_0^{2\pi} d\varphi \exp\right.$$

$$\left. \times \left(-\frac{2z}{\sqrt{1-t}}\{2-(1-x^2)[t\cos^2\varphi+(1-t)\sin^2\varphi]\}\right) \right]. \qquad (11.46)$$

For $z \ll 1$ the functions (11.42, 44, 46) are of the same type:

$$\Lambda_1(z) \approx \frac{4}{3}z, \quad \Lambda_2(z) \approx \frac{8}{3}z, \quad \Lambda_3(z) \approx \frac{20}{3}z. \qquad (11.47)$$

For $z \gg 1$ the main role is played by the function

$$\Lambda_1(z) \approx 4/\sqrt{z\pi}, \quad z \gg 1, \qquad (11.48)$$

whereas functions $\Lambda_2(z)$ and $\Lambda_3(z)$ are of a higher order of smallness $[\Lambda_2(z) \sim z^{-1}, \Lambda_3(z) \sim \exp(-z)]$.

It follows from (11.41−48) that the cross section of an inelastic transition between the levels of a compound system (as well as light absorption) possesses a qualitative change at the field strength $\mathscr{E}_0$ more than some critical value $\mathscr{E}_0^*$, according to (11.43) equal to

$$\mathscr{E}_0^* \sim \frac{v^{1/2}}{R_k^2} \sim \frac{v^{1/2}\Delta\omega^{2/3}}{|d_{01}|^{4/3}}. \qquad (11.49)$$

Letting $\Delta\omega \sim v/\varrho_w \sim \tau^{-1}$ we obtain the estimation: $\mathscr{E}_0^* \sim v^{3/2}/d^2$. For weak fields $\mathscr{E}_0 \ll \mathscr{E}_0^*$ the light absorption is the usual type which can be obtained quickly from (11.47)

$$\sigma \propto \mathscr{E}_0^2 \frac{R_k^6}{v} \propto |d_{01}|^2 \; \mathscr{E}_0^2 \frac{|d_{01}|^2}{v\Delta\omega^2} \; . \tag{11.50}$$

In other words, the absorption is proportional to the light intensity ($\mathscr{E}_0^2$) and has the standard spectral distribution for resonant broadening in the wings ($\sim \Delta\omega^{-2}$) of the spectrum.

For strong fields $\mathscr{E}_0 \gg \mathscr{E}_0^*$ the absorption decreases with an increase of $\mathscr{E}_0$. With the help of (11.48), we obtain

$$\sigma_{m=0}^+ \propto v^{1/2}/\mathscr{E}_0 \; . \tag{11.51}$$

It is of interest to note that (11.51) does not depend on frequency detuning.

11.3.3 Experimental Aspects

Among different experimental results [11.7, 19, 20] devoted to the investigation of light absorption during slow collisions of heavy atoms, we will concentrate on the data from [11.7, 19] whose authors directly connect the effects observed with the Landau-Zener mechanism of the absorption. In [11.7] the fluorescence from alkali metal vapors was investigated under excitation by the radiation with a frequency being very different from the frequency of the atomic transition $\omega_0(\Delta\omega = \omega - \omega_0 \simeq 2000 \; \text{cm}^{-1})$.

The investigation of the dependence of the effect on density under the addition of a noble gas (Ne) makes it possible to determine that the excitation of the atomic electronic transition takes place in the process of a pair of atomic collisions. The dependence of the fluorescence S of the atomic lines of an alkali metal on excitation power $I \sim \mathscr{E}_0^2$ is displayed in Fig. 11.5. The deflection from the linear dependence is clearly seen for the powers $10^9 - 10^{10} \; \text{W cm}^{-2}$. The value is well correlated with the estimations of the critical field value obtained with the help of the Landau-Zener model mentioned above. Note that the linear dependence of the fluorescence in the system Cs + Ne, also shown in Fig. 11.5, excludes the possibility of any other interpretations of nonlinear effects connected, for example, with multiphoton ionization.

The most detailed measurements of nonlinear effects in the absorption during collisions have been performed in [11.19] for the system Tl + Ar. The peculiarity of the absorption in the case is connected with the parallel nature of the atomic terms $7S_{1/2} - 6P_{1/2,3/2}$ of the system in the frequency domain of the second harmonic of a Nd laser used in the experiment. The peculiarity of atomic terms behavior connected with their touching at a certain point requires a generalization of Landau-Zener model accounting for higher potential derivations. As a result the critical field takes the form [11.19]

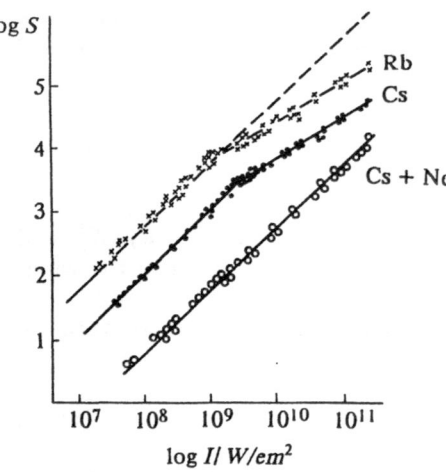

Fig. 11.5. The absorption of strong electromagnetic laser radiation by the system of Rb+Cs atoms [11.7]

$$V_{\rm c} = \left(\hbar\, V^2 \frac{d^2 V}{dr^2} \Big|_{R=R_\omega} \right)^{1/3} . \tag{11.52}$$

The estimation is in good agreement with the experimental data for the fluorescence power Q and its dependence on the thallium line at the transition $7S_{1/2}-6P_{1/2}$ on the intensity I of the exiting radiation. For the light intensity $I \gtrsim 10^9$ W cm^{-2} ($\mathscr{E}_0 > 10^6$ V cm^{-1}) the medium becomes more transparent due to the influence of the electromagnetic field on the broadening collision event. More details are contained in [11.21, 22].

11.4 Multiparticle Effects. The Change of the Atom's Quantization Direction in a Laser Field

11.4.1 Multiparticle Approach to the Powerful Radiation Absorption by the Atom in a Plasma

The previous consideration is founded on the picture of pair (binary) collisions of an absorbing atom with broadening particles. Our main attention is devoted to light absorption processes during a separated collision event. These collision events itself are determined to be separated in time in correspondence with the criterium $N\varrho_{\rm eff}^3 \ll 1$.

Here we will consider the generalization of the theory connected with the simultaneous participation of many broadening particles in the interaction with the radiating (or absorbing) atom. The situation is typical for hydrogen-like atoms in a plasma for which it is also typical the long-range nature of the atomic state interaction with surrounding plasma ions leading to the linear Stark effect. So, it is necessary to generalize the theory to the case of overlapping collisions, i.e., for the condition $N\varrho_{eff}^3 \gg 1$.

The case of multiparticle collisions is obviously realized for a slow variation of the perturbation when the Stark splitting κ of the level exceeds the frequency $N^{1/3} v_{0i}$ of the ion field variation. So, the case under consideration responds to adiabatic collision theory. Under the condition $\Delta\omega \gg V_0 = d_{if}\mathcal{E}_0$ the mean contribution the absorption is due to the crossing points of the term.

$$U_i(t_0) - U_f(t_0) \equiv \kappa(t_0) = \Delta\omega \ , \quad \kappa(t_0) = C_{2k}F_\omega^k(t) = \Delta\omega \ , \tag{11.53}$$

where $U_{i,f}(t)$ are the Stark shifts of the upper and lower atomic levels, and $\kappa(t)$ is the frequency shift of the transition $i \to f$ due to linear $(k = 1)$ or quadratic $(k = 2)$ Stark effects with the Stark constants C_2 and C_4 respectively.

In the every point t_0 the transition probability may be found with the help of the Landau-Zener theory; however, the statistics of such crossing points is of a very sophisticated nature due to the multiparticle character of the interaction $\kappa(t)$. The difference from binary collisions is primarily connected with circumstance, because the only two crossing points are in the binary case. For the transition probability calculation in the multiparticle case one can use an ensemble picture instead of the time crossing picture at the points t_0, the first being characterized by the probability $\kappa = \Delta\omega$ of the appearance of the given value obviously connected with the probability $W(F)$ of the given electric field F realization. The calculations listed below are founded on the elementary model of the oscillator with a varying frequency $\kappa(t)$. However, the model results from the general theory of inelastic transitions in a two level system [11.23 − 26].

11.4.2 Calculation of Spectra in a Laser Field

Let us write down the formula for the power of laser radiation absorbed by an atom in a plasma. For this purpose, simple qualitative arguments are used, and accordingly the atom is considered as a classical oscillator with a varying frequency $\kappa(t) = C_{2k}F^k(t)$. In a weak field $\mathcal{E}_0$ the absorption spectrum Q is [11.11]:

$$Q(\Delta\omega, \mathcal{E}_0) \propto (d_{if} \cdot \mathcal{E}_0)^2 \frac{1}{\pi} \operatorname{Re} \left\{ \int_0^\infty d\tau\, e^{-i\Delta\omega\tau} \left\langle \exp\left[i\int_0^\tau \kappa(t)\,dt\right] \right\rangle \right\} \ . \tag{11.54}$$

Here d_{if} is a matrix element of the atom dipolar momentum between the considered state i and f; $\mathcal{E}_0$, ω are the field strength and laser frequency; $\omega - \omega_0 \equiv \Delta\omega$ is a frequency detuning ($\omega_0 = (E_f - E_i)/\hbar$ being the transition frequency), and the symbol $\langle \ldots \rangle$ denotes the averaging over the perturbing ions ensemble.

For a slow (quasistatic) variation of the ion field, the phase shift in (11.54) may be expanded in a power series: $\int_0^\tau \kappa \, dt \approx \kappa(0)\tau + \dot{\kappa}(0)\tau^2/2 + \dots$. The first member of the expansion $(\kappa(0)\tau)$ determines the static shift of the levels in the ion field; substitution of this member into (11.54) gives the δ-function. The second member $(\dot{\kappa}\tau^2/2)$ determines the phase coherence loss by the atom due to ion thermal motion and it leads to the δ-function "expansion" at the value $\delta\omega \sim \sqrt{|\dot{\kappa}|}$. The phase coherence loss time T_φ is defined by the formulae [11.27, 28]:

$$T_\varphi(F) = \left\langle \frac{1}{\sqrt{|\dot{\kappa}|}} \right\rangle_F = \left\langle \frac{1}{\sqrt{|k C_{2k} F^{k-1} \dot{F}|}} \right\rangle_F , \tag{11.55}$$

where $\langle \dots \rangle_F$ means the averaging for fixed field F. The time $T_\varphi(F)$ may be expressed in terms of the lifetime of the given field $T_F \sim F/\sqrt{\langle \dot{F}^2 \rangle_F}$, calculated by *Chandrasekhar* and *I. von Neumann* [11.29]. The ion quasistatic criterion will be [11.10, 11, 28]:

$$\Delta\omega \, T_\varphi(F)|_{\kappa(F) = \Delta\omega} \gg 1 . \tag{11.56}$$

Let us now consider the case of a strong laser field $\mathscr{E}_0$. Apparently, at large values of $\mathscr{E}_0$ the atomic oscillator phase will depend on the $\mathscr{E}_0$ value, as well, so the generalization of (11.54), may be obtained by substituting the laser field strength $\mathscr{E}_0$ into the equation for the frequency of the atom [11.15]:

$$Q(\Delta\omega, \mathscr{E}_0) \propto (d_{\text{if}} \cdot \mathscr{E}_0)^2 \frac{1}{\pi} \text{Re} \left\{ \int_0^\infty d\tau \left\langle \exp\left[i \int_0^\tau dt \sqrt{[\Delta\omega - \kappa(t)]^2 + 4(d_{\text{if}} \cdot \mathscr{E}_0)^2} \right] \right\rangle \right\} . \tag{11.57}$$

Using the sufficiently slow time variation of the ion field, it is possible to account for only the first term in the expansion under the square root symbol in (11.57). Performing the averaging on the rate of the ion field variation $\dot{F}$ and on the field F itself in (11.57) one obtains the equation

$$Q(\Delta\omega, \mathscr{E}_0) = \left(\frac{F_0}{2F_\omega} \right)^{k-1} \frac{1}{\kappa(F_0)} \mathscr{H}\left(\frac{F_\omega}{F_0} \right) \frac{1}{2} \int_{-1}^{+1} d(\cos\theta_{F \cdot \mathscr{E}_0})$$

$$\times (d_{\text{if}} \cdot \mathscr{E}_0)^2 \left\langle \tilde{I}\left(\frac{4|d_{\text{if}} \cdot \mathscr{E}_0|^2}{|\dot{\kappa}(0)|} \right) \right\rangle . \tag{11.58}$$

Here $F_0 = \lambda e N^{2/3} (\lambda = 2.603 \dots)$ is the "normal" Holtsmark field, N is the ion density, $F_\omega = (\Delta\omega/C_{2k})^{1/k}$ and $\mathscr{H}(x)$ is the Holtsmark function [11.29]. The function $\tilde{I}(x)$ describes the deformation of the Holtsmark profile by the electromagnetic field $\mathscr{E}_0$. It is easy to see the parameter defining the deformation to have the value $|d_{if} \cdot \mathcal{E}_0|\ T_{\varphi}(F_{\omega}) \equiv GT_{\varphi}(F_{\omega})$, introduced in the argument of the function $\tilde{I}$ in (11.58). The function $\tilde{I}(y)$ defined by the repulsion of the terms at the crossing point is obtained by estimating the integral (11.57) near the point [11.24, 30]:

$$\tilde{I}(y) = y \left[\int_{-\infty}^{\infty} dx \cos \left(y \int_{0}^{x} \sqrt{1+z^2}\, dz \right) \right]^2 \simeq \begin{cases} \pi , & y \ll 1 \\ \text{const.} \exp(-\pi y/2) , & y \gg 1 . \end{cases}$$

$$(11.59)$$

For the integral over $\cos \theta_{F \cdot \mathcal{E}_0}$ and further averaging it is useful to take the analytical approximation [11.24] which makes it possible to express the spectral deformation with the help of the single universal function

$$Q(\Delta\omega, \mathcal{E}_0) = Q^{(0)}(\beta) I_{\mu_k}(\beta) , \tag{11.60}$$

where $\beta = F_{\omega}/F_0$, $Q^{(0)}(\beta)$ is the line shape in a small field:

$$Q^{(0)}(\beta) = \frac{1}{(2\beta)^{k-1}} \frac{|d_{if}|^2 \mathcal{E}_0^2}{3\kappa(F_0)} \mathcal{H}(\beta) . \tag{11.61}$$

$I_{\mu_k}(\beta)$ is a factor, defining the line shape deformation

$$I_{\mu_k}(\beta) = I\left(\frac{\mu_k}{\beta^{k-1}} \sqrt{\frac{5\,\mathcal{H}(\beta)}{4\pi\beta^{1/2}I(\beta)}} \right) , \tag{11.62}$$

where

$$I(z) = \frac{3\sqrt{\pi}\,\Phi(\sqrt{z})}{4z^{3/2}} - \frac{3\exp(-z)}{2z} = \begin{cases} 1 , & z \ll 1 , \\ \dfrac{3\sqrt{\pi}}{4z^{3/2}} , & z \gg 1 , \end{cases} \tag{11.63}$$

and $\Phi(x)$ is the error function.

The function $I(\beta)$ introduced by *Chandrasekhar* and I. von Neumann [11.29] is connected to the ion field derivation $\langle|\dot{F}|^2\rangle_F$. The parameter μ_k for the cases of linear ($k = 1$) and quadratic ($k = 2$) Stark effect is equal to:

$$\mu_1 = \frac{|d_{if}|^2 \mathcal{E}_0^2}{C_2 N v} ; \quad \mu_2 = \frac{|d_{if}|^2 \mathcal{E}_0^2}{2\lambda C_4 N^{5/3} v} , \tag{11.64}$$

where v is the ion velocity.

Following from (11.60−64), the line shape deformation is spectrally nonuniform. For small laser fields $GT_\varphi(F_\omega) \ll 1$ the factor $I_\mu(\beta) \simeq 1$ and the line shape coincides with (11.54). In the strong field $GT_\varphi(F_\omega) \gg 1$ it follows from (11.60, 63, 64):

$$Q(\Delta\omega, \mathscr{E}_0) = \frac{\pi^{5/4} k^{3/2}}{5^{3/4} 2^{k-1/2} \lambda^{(3-k)/2}} \frac{(C_{2k})^{1/2} v^{3/2} N^{k/3+1/2}}{|d_{if}| \, \mathscr{E}_0} \beta^{(k-1)/2}$$

$$\times [\mathscr{H}(\beta)]^{1/4} [\beta^{1/2} I(\beta)]^{3/4} \, . \tag{11.65}$$

So the absorbed power in strong E.M. fields decreases when the field grows, the absorption coefficient being proportional to the ion temperature $T^{3/4}(v^{3/2})$. The deformation of the static (Holtsmark) shape becomes significant for the parameter values μ_1, μ_2 (11.64), compared with unity. Allowing atomic units $d_{if} \sim 1$; $C_2 \sim 10$, $C_4 \sim 10^3$, $N \sim 10^{17}$ cm^{-3}, $v \sim 10^{-3}$ one finds the value $\mathscr{E}_0 \equiv \mathscr{E}_0^*$ for which $\mu \sim 1$: $\mathscr{E}_0^* \sim 10^{-5}$ a.u. $\sim 10^5$ V/cm for the linear Stark effect and $\mathscr{E}_0^* \sim 10^{-6}$ a.u. $\sim 10^4$ V/cm for the quadratic Stark effect. Note that for the quadratic Stark effect the value $\mathscr{E}_0^*$ is usually smaller than for the linear case (because of the larger density power, introduced in parameter μ_2), but the applicability domain for the static theory in the case is much smaller [11.10].

From (11.62), describing the absorption in a strong field, it follows that the line shape essentially depends on the value μ, responsible for the transitions between individual sublevels. With an increase of μ, the component maximum becomes lower and the width grows; that is, the component "is burned". From the definition of μ it follows that from several Stark components under identical conditions the ones that are rapidly burned are, first, more intense and, second, localized nearer to the line center (C_{2k} is small). Because the most intensive Stark components are, as a rule, nearer to the line center and the weakes components are near the periphery, the effect mentioned leads to the effective smoothing of the spectrum.

From this consideration it follows that for a concrete line shape calculation, it is necessary to calculate the individual Stark component profiles accounting for their intensities and displacements from the line center and then to sum them. It must be taken into account that the components corresponding to the transitions $m = 0 \to m = \pm 1$ are "burned" more rapidly. For these components, the squared nondiagonal matrix element must be multiplied by a factor of two.

For the concrete spectral line shape calculations of the basis of (11.60) it is necessary to know the function $I_\mu(\beta)$. The results of the function $I_{\mu_1}(\beta)$ calculations are shown in Fig. 11.6 for different values of μ_1 and β. The function $I_{\mu_1}(\beta)$ describes the deformation of both the hydrogenic and nonhydrogenic spectra. In the latter case it is seen from (11.55) that μ_1 must be replaced by μ_2/β so that the new parameter μ depends on β itself.

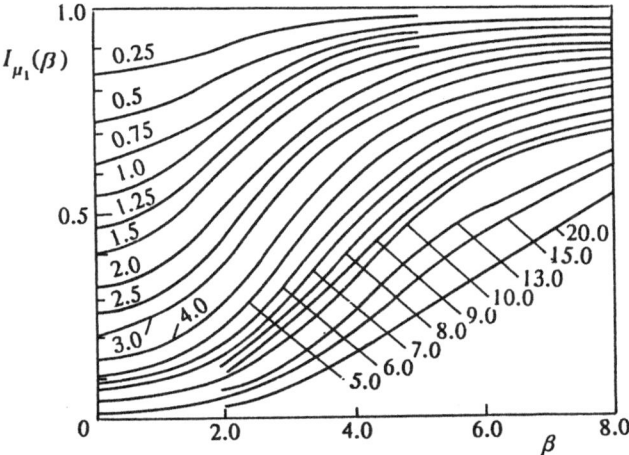

Fig. 11.6. The function $I_{\mu_1}(\beta)$ accounting for spectral line shape deformation in the linear Stark effect case (digits near the lines give the magnitudes of the parameter $\mu_1 = (|d \cdot \mathcal{E}_0|^2 / C_2 N v)$ [11.24]

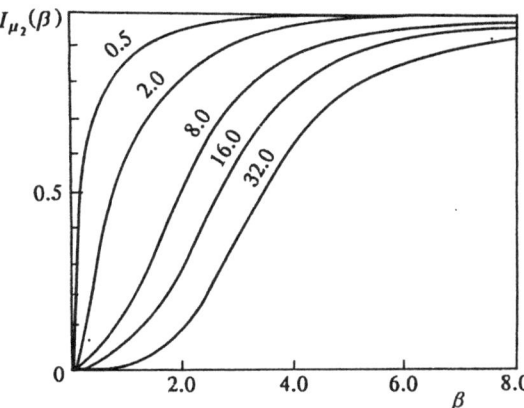

Fig. 11.7. The function $I_{\mu_2}(\beta)$ describing the Holtsmark line shape deformation by laser field for the quadratic Stark effect case [11.25]: $\mu_2 = |d \cdot \mathcal{E}_0|^2 / C_4 v N^{5/3}$

For non-hydrogenic spectral calculations it is more convenient to use the number of curves corresponding to the values $\mu_2 = \text{const.}$ contrary to Fig. 11.6 where for the definite curve the parameter μ_2/β remains constant. The transition procedure with the help of Fig. 11.6 to the number mentioned is obvious. To make the non-hydrogenic spectra deformation types obvious, the values of the function $I_{\mu_2}(\beta)$ for several parameters are shown in Fig. 11.7. It should be remembered that for the quadratic Stark effect $\beta = F_\omega/F_0 = (\Delta\omega/C_4 F_0^2)^{1/2}$, as was pointed out in (11.60).

Let us calculate the hydrogen H_β line shape in a strong laser field with the help of Fig. 11.6. For every of the seven most intense components of this line one finds parameters μ_1 and I_0 (where I_0 is the component's intensity neglect-

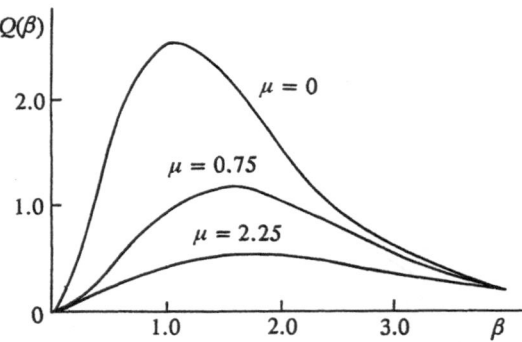

Fig. 11.8. The H_β spectral line shape in the strong electromagnetic field for different values of the parameter μ for the Stark component $102 \rightarrow 001$ [11.25]

ing the deformation) corresponding to the rules mentioned above, the shapes are then summed. The results of the calculation are presented in Fig. 11.8.

It is of interest to consider the reduction of the general formulae (11.65) to the case of pair (binary) collisions. The case response to the limit $\beta \gg 1$ in (11.65) when the frequency shift is due to one ion nearest to the atom. Using the limiting values of the functions $\mathscr{H}(\beta)$ and $I(\beta)$ from [11.29] we obtain for the case of a linear Stark effect ($k = 2$):

$$Q_2 \propto \frac{N v^{3/2} C_2^{3/4}}{V_0 (\Delta \omega^{1/4})} . \tag{11.66}$$

The result may also be obtained by the Landau-Zener method in the binary collisions picture [11.3, 4]. The coincidence of both results demonstrates the equivalence of time and ensemble approaches in the binary domain.

In the center of the line $\beta \ll 1$ ($\Delta \omega \ll C_2 F_0$) it follows from (11.65):

$$Q \sim \frac{\Delta \omega^2 v^2}{(C_2 N v)^{1/2} C_2 V^0} . \tag{11.67}$$

The result (11.67) as the formulae (11.65) as a whole is essentially nonbinary ($Q \sim N^{-1/2}$).

11.4.3 The Change of the Atom Quantization Direction in a Laser Field

In all of the cases considered the transition probability, which critically depends (exponentially) on the field $\mathscr{E}_0$, is changed after averaging over the angle θ between the field F (or internuclear axis R) and field direction $\mathscr{E}_0$ into a smoother field with a power law dependence. The picture is due to the fact that in the collision process the atom is quantized along the ion field (internuclear axes). Here we will consider the example of the atom quantization along the laser field, it resulting in a stronger dependence of the absorption on $\mathscr{E}_0$.

Let us consider the laser radiation absorption by a non-hydrogenic atom at the transitions: $nS \rightleftarrows kP(n > k)$. Suppose that $\kappa_P \ll \kappa_S$, where κ_S and κ_P are the Stark shifts of the S and P (levels) in the ion field. The condition is fulfilled:

$$(\kappa_P)_{\text{eff}} \ll G \ll (\kappa_S)_{\text{eff}} \ . \tag{11.68}$$

The situation when the line shifts are extremely different is typical for the non-hydrogenic system. For helium the term $2P$ displaced under the same electric field is twenty times less than the $3S$ term and approximately two hundred times less than the $4S$ term.

The shift of the spherically symmetric upper S-state which under the condition (11.68) is determined mainly by the line broadening does not depend on the atom quantization axis direction. Let consider in greater detail the influence of the atomic quantization axis change effect on the absorption in the binary case. The absorbed power is proportional to the transition probability from the lower to the higher state. The probability for slow collisions can be calculated by the Landau-Zener method [11.9]. In the binary case $\kappa_S = C_4(\varrho^2 + v^2 t^2)^{-2}$ and the absorbed power Q is obtained by averaging the transition probability w over ϱ and v, calculated with the help of the Landau-Zener formula

$$Q \propto Nv\sigma = 2\pi Nv \int_0^{\varrho_\varphi} \varrho \, d\varrho \, \exp\left(-\frac{2\pi G^2}{|\dot{\kappa}_S|}\right)\left[1 - \exp\left(-\frac{2\pi G^2}{|\dot{\kappa}_S|}\right)\right]$$

$$= Nv\pi \varrho_\omega^2 \Lambda\left(\frac{\pi G^2 C_4^{3/4}}{2v\Delta\omega^{7/4}}\right), \tag{11.69}$$

$$\Lambda(\mathscr{E}) = \int_0^1 dx \exp\left(-\frac{\mathscr{E}}{\sqrt{x}}\right)\left[1 - \exp\left(-\frac{\mathscr{E}}{\sqrt{x}}\right)\right] = \begin{cases} 2\mathscr{E} & , \ \mathscr{E} \ll 1 \\ \dfrac{2}{\mathscr{E}} e^{-\mathscr{E}} , & \mathscr{E} \gg 1. \end{cases} \tag{11.70}$$

The derivative $\dot{\kappa}_S$ is taken at the term crossing point $\kappa = \Delta\omega$, v being the ion velocity. Equation (11.69) (unlike the absorption when hydrogen collides with a proton) exhibits an exponential decay instead of a power law for the cross section. It is easy to understand that the decay is connected to the disappearance of the angular dependence in the exponent.

Equation (11.69) should be averaged over the ion velocities. In line broadening theory it is usual to put the average ion velocity instead of the correct averaging procedure into the corresponding formula. The error in such a method is not too large. But in our case the ion velocity is included in the exponent. So the simple substitution of the average velocity into exponent would be a mistake.

Let us average (11.69) over the Maxwell distribution

$$W(v)\,dv = \frac{1}{(\sqrt{\pi}\,v_0)^3}\exp\left[-\left(\frac{v}{v_0}\right)^2\right]\ ,\quad v_0 = (2\,T/M)^{1/2}\ .$$

It is convenient to rewrite the (11.69) in the form

$$Q \propto 2\pi N\varrho_\omega^2 v^2 \exp\left(-A/v\right)\ ,\tag{11.71}$$

where

$$A = \frac{\pi\,G^2 C_4^{3/4}}{2\varDelta\,\omega^{7/4}}\ .$$

The region of variables where the exponential decay of the absorbed power is observed is of interest. So we let

$$A \gg v_0\ .\tag{11.72}$$

The average power under the condition (11.72) can be easily calculated. The result is

$$\langle Q(v)\rangle = 2\pi N\varrho_\omega^2\frac{4v_0^2}{\sqrt{3}}(A/v_0)^{4/3}\exp\left[-3(A/2v_0)^{2/3}\right]\ .\tag{11.73}$$

Note that the result also strongly depends on v_0.

The absorption considered above is completely connected with the ion thermal motion and it is determined, as already mentioned, by the parameter $GT_\varphi \sim G|\dot{\kappa}|^{-1/2}$.

In fact, there also exists a relaxation in a plasma, connected with spontaneous decay γ_S and electron impact broadening γ_C. Both types of relaxation ensure absorption even in the absence of ion motion, which is determined by the so called nonhomogeneous broadening parameter $\gamma_S\gamma_C/G \ll 1$. For the observation of the effects described here, the absorption due to the relaxation of the levels γ_S, γ_C should not suppress the effects due to ion thermal motion. The condition for this will be

$$\gamma_S\gamma_C T_\varphi^2(\omega) \sim \frac{\gamma_C\gamma_S}{|\dot{\kappa}|} \sim \frac{\gamma_C\gamma_S}{\varDelta\,\omega v/\varrho_{\text{eff}}} \ll 1\ ,\tag{11.74}$$

which apparently is easily fulfilled. For calculating both the ion thermal motion effects and the level relaxation, it is necessary to solve the corresponding kinetic equations [11.30]. One further aspect of the problem is that the dependence of the absorbed power in an intense laser field on $\langle|\dot{\kappa}|^{3/2}\rangle_F$ gives the unique possibility of experimentally investigating the dynamics of ion microfields relative to the ion field F. Remember that up to this point we considered an atom to be quantized along the ion field. The Stark shift of the P-state leads to the nondiagonality of the perturbation matrix in the coordinate system with the axis $0z \parallel \mathscr{E}_0$. The diagonalization is achieved, by the rotation of the coordinate system so that the $0z$ axis will coincide with the vector F. After that the angle $\theta_{F\mathscr{E}_0}$ between the vectors F and $\mathscr{E}_0$ will be introduced into the finite equation for the absorbed power. But under the condition (11.68) the angular dependence falls out of the final results. The process may be described more clearly in the following way: the atomic dipolar moment d interacts with two electric fields: ion field F and laser field $\mathscr{E}_0$. The interaction with the field F determines the broadening and is much larger than the interaction with the field $\mathscr{E}_0$.

Thus, the correct wave functions will be the functions in the coordinate system with the axis $0z \parallel F$ where the interaction of the atom with the ion field is diagonal. This means that there is the angle $\theta_{F\mathscr{E}_0}$ between the matrix element of the vector d_{if}, calculated from these wave functions and the vector $\mathscr{E}_0$. The averaging over the angle $\theta_{F\mathscr{E}_0}$ in (11.58) substantially changes the dependence of Q on $\mathscr{E}_0$. But in the case connected to the spherically symmetric s-state broadening, the interaction κ_S does not depend on the choice of the coordinate system. Thus, under the condition $\kappa_P \ll G \ll \kappa_S$, the atom quantization direction is determined by the laser field $\mathscr{E}_0$. In this case the angle between d_{if} and $\mathscr{E}_0$ is fixed and averaging over the angle $\theta_{F\mathscr{E}_0}$ in (11.58) is absent. This leads to a stronger (exponential) dependence of the absorbed power Q on the laser field $\mathscr{E}_0$.

The condition $\kappa_P \ll G$ is easily fulfilled even for small laser fields due to the small value of the quadratic Stark effect constants for low states. On the other hand, if one selects an upper state that is sufficiently excited, the second inequality in (11.68) will be fulfilled. The necessary laser fields which can be easily estimated by comparing $\kappa_{\mathrm{eff}} \sim C_4 N^{4/3}$ and $G \sim d_{\mathrm{if}} \mathscr{E}_0$ lies in the range of 10^3 V/cm, which is easily obtained in laser physics. The main difficulty is to choose a laser having a frequency in resonance with that of the atomic transition.

11.5 Radiative Collisions

Radiative collisions are the atomic collision processes in an external (laser) radiation field, where the "energy defect" Δ of an atomic transition is compensated by the frequency of the laser field, $\hbar\omega \approx \Delta$. As a result cross sections of the process arise sharply. This process was considered for the first time by *Gudzenko* and *Yakovlenko* [11.2] for the reaction of excitation exchange between two different atoms X and Y which the energetic terms are far from the resonance (Fig. 11.9).

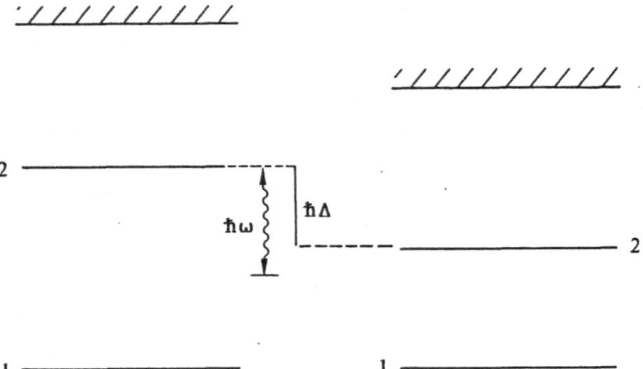

Fig. 11.9. The scheme of radiative collisions of the atoms X and Y

$$X(1)+Y(2)+\hbar\omega \leftrightarrow X(2)+Y(1) , \quad \hbar\omega \simeq \Delta . \tag{11.75}$$

As is known, the probability of excitation transport for slow nonresonant collisions are exponentially small over the adiabaticity parameter $\varrho_{\text{eff}}\Delta/v \gg 1$ (the so called Messy parameter). Switching on the laser field makes the process resonant, and the probability of excitation transport arises at many orders. The analogous effect takes place for the nonresonant charge exchange of an atom at a "different" ion [11.5, 6]. The exponential smallness of the usual cross section $\exp(-\Delta/v)$ in the case may be dramatically extended by switching on the laser field.

The reactions mentioned may be considered as processes with the participation of quasi-energetic atomic levels falling into resonance. Note that radiative collisions, e.g., radiative charge exchanges are possible for spontaneous quantum radiation [11.31] as well, however, the corresponding probabilities are small as a rule over the parameter $e^2/\hbar c$. At the same time these processes may play a noticeable role for multicharged ions with large value of Z [11.32]. The generation system of equations describing the processes of radiative collisions (RC) follow from the Schrödinger equation for the pair of atoms X and Y under consideration with the interaction Hamiltonian:

$$V(t) = \frac{\gamma_{ij}}{R^3(t)} d_i^x d_j^y + (d_i^x+d_i^y) \mathscr{E}_{0i} \cos \omega t , \tag{11.76}$$

where d_i^x, d_j^y are the components of dipole moment operators of atoms X and Y, $\gamma_{i,j} \equiv \delta_{ij}-3 e_i e_j$ is the angular part of the dipole-dipole interaction, $R(t)$ is the distance between colliding atoms, and $\mathscr{E}_{0i}$ are components of laser field strength with the frequency ω.

For state amplitudes a_1 and a_2 describing the states of the system "atom + electromagnetic (E.M.) field":

$$1 \equiv \{X(2), Y(1) , \quad n_\omega\} , \quad 2 \equiv \{X(1), Y(2) , \quad n_\omega + 1\} , \tag{11.77}$$

it is followed by the equation system [11.2]:

$$\left.\begin{aligned} i\dot{a}_1 &= U_1(t)a_1 + V(t)e^{i\Delta\omega t}a_2 , \\ i\dot{a}_2 &= U_2(t)a_2 + V(t)e^{-i\Delta\omega t}a_1 . \end{aligned}\right\} \tag{11.78}$$

Here $U_{1,2} = C_{1,2}R^{-6}(t)$ are Van der Waal's atomic interaction potentials, taking into account the polarizabilities of atomic states in laser field, which are included in the constants $C_{1,2}$; $\Delta\omega = \omega - \Delta$, and $V(t) = B_3\mathscr{E}_0 R^{-3}(t)$ is the transition matrix element describing the excitation transport from the quasi-energetic sublevel of the atom Y to the energetic level 2 of the atom X (or reversed transition). The perturbation $V(t)$ is determined by both the laser field magnitude $\mathscr{E}_0$, defining the amplitude of the quasi-energetic state, and the dipole-dipole interaction, defining the excitation transport probability. The structure of the contants $C_{1,2}$ and B is determined by the dynamic polarization of atoms X and Y as well as by the constant of their dipole-dipole interaction.

The structure of the system (11.78) is typical as in the case of an optical collision (Sect. 11.4) because the atomic collision is due to the dependence of variables in the system on the parameters of the laser field ($\Delta\omega$, $\mathscr{E}_0$). The solution of the system (11.78) under the initial condition $a_1(-\infty) = 0$ determines the probability and the cross section of RC:

$$\sigma^{RC} = \int 2\pi\varrho\,d\varrho\,w(\varrho, \mathscr{E}_0, \Delta\omega) = \int 2\pi\varrho\,d\varrho\,|a_2(\infty)|^2 . \tag{11.79}$$

It is seen from (11.79) that one can say in the case of RC as well as in the line broadening theory about some shape of RC cross section determined by it's dependence on frequency detuning $\Delta\omega = \omega - \Delta$ as well as about the nonlinear effects for increasing laser field strength $\mathscr{E}_0$.

In the linear approximation in $\mathscr{E}_0$ the calculation of the RC cross section shape is performed by the solution of (11.78) with the help of perturbation theory. The corresponding results are analogous to ones obtained upper for line shapes (Sect. 11.2). Among nonlinear effects an interesting case concerns the precise resonance ($\Delta\omega = 0$) when the solution of the system (11.78) takes the form [11.2]:

$$w = \sin^2(2B_3\mathscr{E}_0/v\varrho^2) ; \quad \sigma^{RC} = \pi B_3\mathscr{E}_0/v . \tag{11.80}$$

It is seen that the dependence of the field strength $\mathscr{E}_0$ is linear, in contrast with the squared dependence exhibited in the weak field case.

The case of the spectral line wing $\Delta\omega \gg \Omega_w$ is important, as discussed in Sect. 11.4. Here the main contribution in the RC process is due to the crossing point R_0 of the term:

$$U_1(R_0) - U_2(R_0) = \Delta\omega . \tag{11.81}$$

Corresponding investigation of the RC cross section differentiates in the case by nothing as compared with that mentioned in Sect. 11.4. The main conclusion is that the cross section decrease for the field strength $\mathscr{E}_0 \gg \mathscr{E}_0^*$ due to the repulsion of atomic terms at their crossing point.

The interesting example of RC is the radiative charge exchange of an atom at an ion [11.5,6]. The process is analogous to a crossing point type (11.81) where the potentials $U_{1,2}(R)$ are of the exchange types. The process cross section calculated by the Landau-Zener method has the typical structure (11.21,22), where the point R_0 is determined from (11.81). The limiting values of the cross section average over the angles between the dipole moment d and the field $\mathscr{E}_0$ are of the forms:

$$\langle\sigma\rangle \simeq \pi R_0^2 \begin{cases} \dfrac{4}{3}\delta_0 \propto \mathscr{E}_0^2/v , & \delta_0 \ll 1 , \\[2mm] \dfrac{4}{5\pi^{1/4}}\delta_0^{-1/2} \propto v^{1/2}/\mathscr{E}_0 , & \delta_0 \gg 1 . \end{cases} \tag{11.82}$$

Here v is the reactive velocity of charge exchanging particles, and δ_0 is the typical parameter of Landau-Zener theory:

$$\delta_0 = \frac{\pi}{2}(d\,\mathscr{E}_0)^2 \left(\left| \frac{\partial U_1}{\partial R} - \frac{\partial U_2}{\partial R} \right|_{R_0} v \right)^{-1} . \tag{11.83}$$

The density of light energy flow for typical charge exchange parameters of the atoms at the negative ions are of the order of magnitude of 10^9 W cm^{-2}.

The experimental investigations of RC processes are described in [11.19,20], and RC processes are of the most interest for the initiation of chemical reactions in the gaseous phase [11.22].

11.6 Effect of Electric Microfield
on Resonant Charge Exchange in a Dense Medium

Atomic processes connected with collisions of slow, heavy particles are very sensitive [11.14] to the resonant detuning Δ of electron terms of colliding partners. The resonant detuning may be caused by the interaction of the surrounding dense medium even in the case when the average distances between particles in the medium exceed the effective radius ϱ_0 of the process, e.g., when

$$N\varrho_0^3 \ll 1 , \tag{11.84}$$

where N is the medium density.

An example of such resonant detuning by a medium is the process of nonresonant charge exchange [11.8]. We will provide a simplified theory of the effect. The dependence of the nonresonant charge exchange cross section on resonant detuning Δ takes the form [11.14]

$$d\sigma = 2\pi\varrho\,d\varrho\,w = 2\pi\varrho\,d\varrho\,\cosh^{-2}(\pi\Delta/\gamma v) \ , \tag{11.85}$$

where $\gamma^2/2 \equiv E$ is the energy of electron level, and v is relative velocity.

The foundation of the present model is the model of a medium fluctuating microfield F, causing the resonant detuning in a charge exchange system. It is easy to approach such a model by starting from the interaction potential difference of each of the exchange particles with the medium [11.8]. If R_i is a distance to the ith particle of the medium and r is a distance between the exchange particles $[|r| \approx \varrho_0 \ll N^{-1/3} \approx R_i$ according to (11.84)] then Δ is:

$$\Delta = e\sum_{i=1}^{N} U(R_i) - U(R_i + r) \approx -er\sum_{i=1}^{N} \frac{\partial U}{\partial R_i} \equiv er\cdot\sum_{i=1}^{N} F_i \equiv er\cdot F \ , \tag{11.86}$$

where N is the total number of particles in the medium.

The further calculation may be performed for an arbitrary spherically symmetric field distribution $W(F)\,dF = W(|F|)\,dF$. Substituting (11.86) into (11.85) and performing the averaging over the angle between r and F we immediately obtain:

$$\sigma = \frac{\pi\varrho_0^2}{2}\int_0^\infty dF\,W(F)\frac{\tanh(\pi\varrho_0 eF/2\gamma v)}{\pi\varrho_0 eF/2\gamma v} \ . \tag{11.87}$$

Investigating (11.87) in a general form for a power interaction law: $F_i = Q_k R_i^{-k}$, where Q_k is the parameter defining the type of the interaction (for example, charge, dipole, etc., see below). Let us introduce the dimensionless distribution $\Lambda(\beta)$ defined by the relation:

$$W(F) = 1/F_0\Lambda(F/F_0) \ , \quad \int_0^\infty d\beta\Lambda(\beta) = \int_0^\infty dF\,W(F) = 1 \ , \tag{11.88}$$

where F_0 is the "normal" value of the field F:

$$F_0 = \lambda_k Q_k N^{k/3} \ . \tag{11.89}$$

The numerical factor λ_k depends on the type of interaction. Substituting (11.88, 89) into (11.87) we obtain

$$\sigma = \frac{\pi\varrho_0^2}{2}f\left(\frac{\pi\varrho_0 eF_0}{2\gamma v}\right) \ , \tag{11.90}$$

where the universal function $f(x)$ is of the form

$$f(x) = \int_0^\infty d\beta \Lambda(\beta) \frac{\tanh(\beta x)}{\beta x} . \tag{11.91}$$

The limiting values of the function $f(x)$ are the following:

$$f(x) \approx \begin{cases} 1 , & x \ll 1 , \\ \lambda^{(-1)}/x , & x \gg 1 , \end{cases} \tag{11.92}$$

where $\lambda^{(-1)}$ is the "inverted" moment of the distribution function:

$$\lambda^{(-1)} = \int_0^\infty d\beta \Lambda(\beta)/\beta . \tag{11.93}$$

So, for $F_0 \ll 2\gamma v/\pi e \varrho_0$ the cross section (11.90) makes a transition to the usual charge exchange cross section. For $F_0 \gg 2\gamma v/\pi e \varrho_0$ the cross section takes the form:

$$\sigma = \lambda_k^{(-1)} \frac{\varrho_0 \gamma v}{e F_0} = \frac{\lambda_k^{(-1)}}{\lambda_k} \frac{\varrho_0 \gamma v}{e Q N^{k/3}} . \tag{11.94}$$

The transition from one type of the cross section to another takes place for an increase of the argument of the function f in (11.90). Equating it to unity, one may obtain the density of values N^*, for which the transition takes place:

$$N^* = (2\gamma v / \pi e \varrho_0 \lambda_k Q_k)^{3/k} . \tag{11.95}$$

Let us apply the results of the general consideration to some concrete cases. First of all, let us consider the resonant detuning caused by the polarization α of atoms in a medium, when $k = 5$, $Q_5 = 2\alpha e$ [11.8]. The distribution function $\Lambda_P(\beta)$ is determined with the help of the Markoff method [11.29], and all calculations are analogous to the Coulomb field case. The result takes the form:

$$\Lambda_P(\beta) = \frac{2}{\pi} \beta \int_0^\infty dx\, x \sin x \beta \, e^{-x^{3/5}} \approx \begin{cases} \dfrac{80}{\pi} \beta^2 , & \beta \ll 1 , \\[2ex] \dfrac{16}{5\pi} \Gamma(8/5) \cos \dfrac{\pi}{5} \beta^{-8/5} , & \beta \gg 1 , \end{cases} \tag{11.96}$$

$$\lambda_5 = \left[\frac{5\pi}{6}\Gamma(2/5)\sin\frac{\pi}{5}\right]^{5/3} \approx 7.67 \ ,$$

$$\lambda_5^{(-1)} = \frac{10}{3\pi}\Gamma(5/3) \approx 0.86 \ .$$

To estimate the transition density N^* (11.95) for cesium Cs ($\alpha = 360$), $\gamma \approx \varrho_0 \approx 1$ a.u. and $v \approx 10^{-4}$ a.u., resulting in $N^* \cong 10^{19}$ cm^{-3}.

Now let us consider the case when resonant detuning is due to an ion field ($k = 2, Q_2 = e$). The function $\Lambda(\beta)$ in the case coincides with the well known Holtsmark function $\mathscr{H}(\beta)$ [11.29]. Its parameters are as follows: $\lambda_2 = 2\pi(4/15)^{2/3} \cong 2.603$, and $\lambda_2^{(-1)} = 4\Gamma(2/3)/3\pi \cong 0.575$. Substituting them into (11.94) we find

$$\sigma \approx 0.22 \frac{\varrho_0 \gamma v}{e^2 N^{2/3}} \ . \tag{11.97}$$

The estimation of N^* for the same conditions gives: $N^* \cong 10^{17}$ cm^{-3}.

Let us consider the case when the resonant detuning for the charge exchange of the atom A at the ion A^+ is due to a foreign gas B [11.8]. We will consider the effect for the example of a medium consisting of rigid dipole molecules with a density N_B and a dipole magnitude d_0 ($k = 3, Q_3 = d_0$). The field distribution $\Lambda_d(\beta)$ calculated by Holtsmark for the random dipole orientation takes the form $\Lambda_d(\beta) = \dfrac{4}{\pi}\dfrac{\beta^2}{(1+\beta^2)^2}$, its parameters being: $\lambda_3 \cong 4.54$, $\lambda_3^{(-1)} = 2/\pi$. Their substitution into (11.94) gives

$$\sigma_A \approx 0.14 \frac{\varrho_0 \gamma v}{e^2 d_0 N_B} \ . \tag{11.98}$$

It is of interest to estimate the collision frequency τ_A^{-1} of the exchange particles in the system:

$$\frac{1}{\tau_A} = N_A \langle v\sigma_A\rangle \approx 0.3 \frac{\varrho_0 \gamma}{e^2 d_0} \frac{N_A}{N_B} \frac{T_A}{M_A} \tag{11.99}$$

where T_A and M_A are the temperature and atomic mass, respectively. Letting (in atomic units) $T_A/M_A \approx 10^{-8}$ and $\varrho_0 \gamma/e^2 d_0 \approx 1$ we obtain $\tau_A^{-1} \approx 10^8 N_A/N_B$ s^{-1}.

Note in conclusion that for practical calculations without a high precision, the following simple approximation may be used: let us carry out in (11.91) the function tanh $(\beta x)/\beta x$ in the point of the maximum of the function $\Lambda(\beta)$, that leads to $f(x) \cong$ tanh $(\beta_{max}x)/\beta_{max}x$. For $x \ll 1$, $f(x) \approx 1$, and for $x \gg 1$ the magnitude β_{max}^{-1} is approximately equal to the inverse moment $\lambda^{(-1)}$ (for example, for the function $\mathcal{H}(\beta)$ we have $\lambda^{(-1)} = 0.575$, $\beta_{max}^{-1} = 0.62$, that is, the difference is smaller than 10%). One can replace the magnitude β_{max}^{-1} in the function $f(x)$ by $\lambda^{(-1)}$, and it allows the correct limiting cases for $f(x)$ for $x \gg 1$ and $x \ll 1$ to be obtained.

12. The Influence of Regular and Stochastic Accelerations on Atomic Spectra

12.1 Regular Acceleration. Adiabatic Population Inversion in a Strong Laser Field

A change in the velocity of an atom interacting with the radiation in a plasma or a gas has a considerable influence on the dynamics of the absorption of this radiation, the absorption spectrum, the populations of the excited states, etc. This is due to the fact that one of the main broadening mechanisms is the Doppler effect which causes an effective shift of the frequency of an atomic oscillator by an amount kV ($k = \omega/c$ is the wave vector of radiation whose frequency is k, and V is the velocity of an atom) [12.1 – 3].

When the Doppler broadening is sufficiently large, the main contribution to the absorption is made at the resonance points defined by $V = V_\omega = \Delta\omega/k$ ($\Delta\omega = \omega - \omega_0$ is the frequency shift of the absorbed radiation ω relative to the unperturbed frequency of the atomic transition ω_0). Atoms whose velocity is $V < V_\omega$ are accelerated and can reach the resonance value $V = V_\omega$ and then absorb a photon. Here, the Doppler phase shift kV, which acquires a definite direction under the action of acceleration, characteristically intersects a resonance value $kV_\omega = \Delta\omega$. At the resonance intersection point itself, the mechanism of the transition responsible for the absorption of light is fully analogous to the mechanism of the inelastic Landau-Zener transitions well known from the physics of atomic collisions [12.4]. The parameter p governing the probability of a transition and, consequently, the nonlinear effects in absorption, has a typical Landau-Zener structure: $p = 2\pi V^2/ka$. If $p \ll 1$, then the perturbation theory results are obtained [12.1, 2], whereas for $p \gg 1$ new nonlinear effects of saturation of the absorbed power are observed.

12.1.1 Landau-Zener Nonlinearities in the Spectra of a Two-Level System Subjected to Acceleration

The system of equations for the elements of the density matrix of a spatially homogeneous gas of two-level atoms interacting with a resonant traveling electromagnetic wave and subjected to a constant acceleration a has the following form in the steady state case ([12.2], p. 272):

$$(\gamma + ad/dV)N(V) = \Delta q(V) + 2iG(\varrho - \varrho^*) \tag{12.1}$$

$$(\gamma + ad/dV)\varrho(V) = iGN - i(kV - \Delta\omega)\varrho . \tag{12.2}$$

Here, $N(V) = \varrho_{mm} - \varrho_{nn}(V)$; $\Delta q(V) = (q_m - q_n)W(V)$ is the pumping; $\varrho = \varrho_{mn}(V)$; $G = d_{mn}\mathscr{E}_0/\hbar$ is a matrix element of the dipole interaction of an atom with a wave field $\mathscr{E}_0$; $\Delta\omega = \omega - \omega_{mn}$; and k is the wave vector of the wave (it is assumed that the directions of acceleration and wave propagation are the same).

The absorbed power is given by

$$P(\omega) = q\hbar\omega \int\limits_{-\infty}^{\infty} dV_0 W_{12}(V_0)f(V_0) .$$

The system (12.1,2) describing the amplitudes of the atomic states b_1 and b_2 ($N = |b_1|^2 - |b_2|^2$, $\varrho = b_1^* b_2$ is used below) is [12.5]:

$$\left.\begin{aligned} ia\frac{\partial b_1}{\partial V} &= (kV - \Delta\omega)b_1 + Gb_2 , \\ ia\frac{\partial b_2}{\partial V} &= Gb_1 . \end{aligned}\right\} \tag{12.3}$$

The system (12.3) is identical, apart from the notation, with the system obtained in considering the inelastic Landau-Zener transition in the problem of level crossing [12.3–6]. In fact, the time derivatives become in our case the derivatives with respect to V, and the linearly varying terms become the Doppler frequency shift kV (or kat). The level crossing condition in the Landau-Zener theory corresponds clearly to the resonance condition $\Delta\omega = kV$ in our case.

The system (12.3) does not include the relaxation parameter γ. In investigating the physical singularities of the problem it would be interesting to consider the case of sufficiently small values of γ (or large values of a), when $\gamma = 0$. We shall express the absorbed power in terms of the amplitudes $b_{1,2}$. Employing the symbol $\langle \ldots \rangle_W$ for the averaging over a Maxwellian distribution $W(V_o)$, we write down (12.3) in the form

$$P(\omega) = q\hbar\omega \, \Phi\left(\frac{\Delta\omega}{k}\right) \langle |b_2(-\infty, +\infty)|^2 \rangle_W , \tag{12.4}$$

where

$$\Phi(V) = \int\limits_{-\infty}^{V} f(V')dV' . \tag{12.5}$$

Thus, in the absence of relaxation the absorption in a system of two-level systems is governed by the asymptotic behavior of the amplitude b_2 (or b_1).

In the case of arbitrary but finite values of the field intensity G and high values of $(k/2a)^{1/2}(V-V_\omega)$, we find that

$$|b_2|^2_{(V_0-V_\omega)\to\pm\infty} = \theta(V_\omega-V_0)[1-\exp(-p)] , \quad p = 2\pi G^2/ka . \tag{12.6}$$

According to (12.4), $|b_2|^2$ can be averaged over the Maxwellian distribution $W(V_0)$. We can assume that $|b_2|^2$ has an abrupt jump at $V_0 = V_\omega$. The absorbed power is then

$$P(\omega) = q\hbar\omega\,\Phi\left(\frac{\Delta\omega}{k}\right)[1-\exp(-2\pi G^2/ka)] . \tag{12.7}$$

If $G^2/ak \ll 1$, (12.7) reduces to the perturbation theory result given by (12.4). The nonlinear dependence of the absorbed power on the intensity of light is related to the equalization of the level populations. Moreover, in the case of an atomic system without damping there are no average steady state level populations at all.

12.1.2 Adiabatic Inversion of the Populations of Atomic Levels

The expression for the transition probability W_{12} can also be obtained directly from (12.4) if we bear in mind that $b_1 + b_2 = 1$.

$$\langle W_{12}(V_0, V)\rangle = \int_{-\infty}^{\infty} dV_0 f(V_0) \int_{-\infty}^{\infty} \frac{dV}{a} |b_2(V_0, V)|^2 \approx \Phi\left(\frac{\Delta\omega}{kV_T}\right)(1-e^{-p}) . \tag{12.8}$$

The physical meaning of this result is easily understood using the distribution of levels in the Landau-Zener model. In fact, for high values of $G(p \gg 1)$ the system moves between "adiabatic" terms (separated at the crossing point by G), which correspond to transitions with the probability of 1 between the initial noncrossing ("diabatic") terms corresponding to levels 1 and 2 of the original atom. Consequently, an atom initially entirely in state 1 finds itself in the vacant state 2. State 1 is then emptied completely. On the basis of the above discussion, it is natural to call this effect an adiabatic population inversion.

The main consequence of the above analysis is the conclusion that acceleration gives rise to new nonlinear saturation effects in the absorbed power governed by the Landau-Zener parameter $p = 2\pi G^2/ka$. These effects are manifested in the range $\gamma G/ka \ll 1$, since if $\gamma G/ka \gg 1$, only the conventional inhomogeneous broadening is observed. One of the manifestations of these effects is an adiabatic inversion of populations in a two-level system subjected to resonant illumination and acceleration.

The effects under discussion are of direct interest for the kinetics of ion lasers. In fact, if we assume that $G^2/ka \gtrsim 1$ and $\gamma G/ka \lesssim 1$, we find that the necessary acceleration is $\gamma^2/ka \lesssim 1$. Next, assuming that in the case of optical transitions we have $\gamma \sim 10^{-8}$ a.u., $k = \omega_0/c \sim 10^{-3}$ a.u., we find that $a \sim 10^{-13}$ a.u. $\sim 10^{12}$ cm/s^2. For example, in the case of an argon-ion laser subjected to an electric field ~ 10 V/cm the parameter γ^2/ka estimated in accordance with [12.7−9] is 1/3.

12.2 Model of Brownian Motion and Optical Phenomena. Path Integral Method

Doppler broadening is known to be one of the most important broadening mechanisms in plasmas and gases. The dynamics of the formation of the Doppler line shape is closely related to the nature of the atomic velocity relaxation [12.2, 6]. Different simplified models are used to describe the dynamics, one of the most vivid being the Brownian motion model for a radiating atom. Our purpose is to establish the relation between the atomic state dynamics in a laser field and the relaxation of the Doppler shifts of atomic levels. Stochastically accelerated atoms will be considered in the frame of the Brownian model. For example this model is adequate in the case of a heavy absorbing particle moving in light buffer gas [12.2]. In some cases the Brownian motion model is valid also for the movement of an absorbing ion in plasmas [12.12]. The absorption (emission) of light by molecules moving with Brownian motion is well known in the framework of perturbation theory over an electromagnetic field [12.6, 11]. The problem in this case is reduced to the averaging of the Gaussian exponential function (coordinate of an atom), and the final result includes both the Doppler line shape ($\exp[-(\Delta\omega/k v_T)^2]$) in the absence of the collisions and the narrowed line shape due to Dicke's effect with the width $\gamma_D \sim (k v_T)^2/\nu$ (under the condition $\nu/k v_T \gg 1$). Detailed consideration of Doppler spectra is provided in [12.2, 6].

In the present work we shall devote most of our attention to the model in which only two parameters interest us: the velocity relaxation parameter ν (the elastic collision frequency) and the parameter characterizing the interaction with a laser field $G = d_{12} E_0$ in a two-level system.

12.2.1 The State Amplitude Method and the Path Integration

The system of equations for the state amplitudes of the two-level atom interacting with the resonant radiation has the form

$$i\dot{a}_1 = [kV(t) - \Delta\omega]a_1 + Ga_2 , \quad i\dot{a}_2 = Ga_1 , \quad a_2(V = V_0) = 0 . \qquad (12.9)$$

This initial condition corresponds to pumping into level 1. The final expression for the absorbed power must be averaged over the pumping distribution

$W(V_0)$, for example over the Maxwell distribution. The absorbed power P is expressed with the aid of the transition probability in a unit of time

$$P(\omega) = N\hbar\omega\langle W_{12}\rangle = N\hbar\omega G^2 I(\omega) \ . \tag{12.10}$$

The angular brackets designate the averaging over the random trajectory and over the pumping.

In our case the absorbing particle motion law or the dependence $V = V(t)$ is defined by the Langevin equation

$$\frac{dV}{dt} + vV = n(t) \ . \tag{12.11}$$

Here v is the damping parameter, and $n(t)$ is the delta-correlated accidental force.

The system (12.9) includes the stochastic parameter and, therefore, is a system of stochastic differential equations.

For the averaging over the all-stochastic process in (12.10) it is necessary to start from a general solution for the system (12.9), not connected with any conditions about the nature of the change of $V(t)$ in time. This general solution is easy to obtain in the case of a small field strength (the perturbation theory). For an arbitrary G no general analytical solution exists. However, there is an approximate analytical solution given by *Vainstein*, *Presnyakov* and *Sobelman* (VPS) [12.10], which well describes the known limiting cases and coincides well with the digital solution of the system (12.9) for different potentials. Using the VPS approximation to find the probability results in the following:

$$I(\omega) = \left\langle \frac{1}{\pi} \text{Re} \int_0^\infty dt \exp\left(i \int_0^t \Omega(\tau)d\tau \right) \right\rangle ,$$

$$\Omega(\tau) = \{[kV(\tau) - \Delta\omega]^2 + 4G^2\}^{1/2} \ . \tag{12.12}$$

Equation (12.12) can be considered to be a natural generalization of the varying frequency oscillator model [12.11] in the strong field strength case where the oscillator phase becomes dependent on the value of G (as a result of the dependence of G on the Rabi frequency $\Omega(\tau)$).

The averaging of (12.12) over a random process can be performed as follows. Let us divide the integration interval over τ in (12.12) into N identical parts. Then the average over the trajectories can be defined as

$$\left\langle \exp\left(i \int_0^t \Omega(\tau)d\tau \right) \right\rangle = \lim_{N\to\infty} \int \ldots \int dV_0 \ldots dV_N W(V_0) W\left(V_0, V_1, \frac{t}{N} \right) \ldots$$

$$\ldots W\left(V_k, V_{k+1}, \frac{t}{N} \right) \ldots W\left(V_{N-1}, V_N, \frac{t}{N} \right) \prod_{k=0}^N \exp\left(i(t/N)\Omega[V_k(\tau)]\right) \ . \tag{12.13}$$

Here $W(V_0)$ is the initial velocity distribution, and the transition probability $W(V_k, V_{k+1}, t/N)$ for the time t/N, in the Brownian motion model (12.11), is equal to

$$W(V_0) = \frac{1}{\sqrt{\pi V_T^2}} \exp\left[-(V_0/V_T)^2\right] ,$$

$$W(V_k, V_{k+1}, \Delta t) = \frac{1}{\sqrt{\pi V_T^2(1-e^{-2\nu\Delta t})}} \exp\left(-\frac{(V_{k+1}-V_k e^{-\nu\Delta t})^2}{V_T^2(1-e^{-2\nu\Delta t})}\right) .$$

(12.14)

The limiting transition in (12.13) defines the so called path integral in Gaussian measure [12.14], defined by the transition probability (12.14). The straightforward calculation of integrals in (12.13) is impossible because of the nonlinear dependence on V. The calculation can be performed only for the small field strength case $G \to 0$, where $\Omega(\tau) = kV(\tau) - \Delta\omega$. A straightforward calculation of the Gaussian integrals leads to the result

$$\Phi(\tau) = \langle e^{ik\int_0^\tau V(t)dt} \rangle = e^{-1/2 k^2 \langle r^2(\tau) \rangle} = \exp\left[-\frac{k^2 \bar{V}^2}{2\nu^2}(\nu t - 1 + e^{-\nu t})\right] . \quad (12.15)$$

The value in the exponential is the average squared displacement $\langle r^2(t) \rangle$, obtained using the Brownian motion model [12.13]. The characteristic dependence $\Phi(\tau)$ within the limit of small ($\nu\tau \ll 1$) and large ($\nu\tau \gg 1$) times takes the form

$$\Phi(\tau) \approx \begin{cases} \exp\left(-\frac{k^2 \bar{V}^2 \tau^2}{4}\right) , & \nu\tau \ll 1 , \\[3mm] \exp\left(-\frac{k^2 \bar{V}^2}{2\nu}\tau\right) , & \nu\tau \gg 1 . \end{cases} \quad (12.16)$$

Taking the Fourier transformation of (12.15) one obtains for the line shape of (12.16)

$$I(\omega) \approx \begin{cases} \frac{1}{\sqrt{\pi}kV_T} e^{-\Delta\omega^2/(kV_T)^2} , & \frac{\nu}{kV_T} \ll 1 , \\[3mm] \frac{1}{\pi} \frac{\gamma_D}{\gamma_D^2 + \Delta\omega^2} , & \frac{\nu}{kV_T} \gg 1 . \end{cases} \quad (12.17)$$

Thus, the Doppler line shape for $v/kV \ll 1$ is changed by the Lorentzian line shape for $v/kV \gg 1$ with the linewidth decreased as compared with Doppler line-width by a factor of $kV/v \ll 1$. This effect is called the Dicke narrowing effect, see [12.11].

Now we consider the strong field case. The calculations become much more complicated because of the nonlinear relationship between Ω and the stochastic variable V. Therefore later we shall consider only the case $v/kV \ll 1$, corresponding to a small relaxation. In this case the whole Fokker-Planck measure (12.14) can be replaced by a simpler one, which follows from (12.14) under the condition $vt \ll 1$:

$$W(V_k, V_{k+1}, \Delta t) = \frac{1}{\sqrt{\pi V_T^2}} \exp \left(-\frac{(V_{k+1} - V_k)^2}{2 v V_T^2 \Delta t} \right) . \tag{12.18}$$

The transition probability (12.18) sets the measure, called the *Wiener* measure [12.14, 15], and it obviously determines the diffusion of atoms in velocity space. For $v t_{\text{eff}} \ll 1$ the velocity variations caused by diffusion are greater than the regular velocity variations due to viscous friction $v V$ in the Langevin equation (12.12). Indeed let us estimate the mean squared velocity variation $\langle (V_k - V_{k+1})^2 \rangle$ for a time Δt due to both of the effects mentioned. The variation due to the diffusion $\langle \Delta V^2 \rangle_{\text{diff}}(\Delta t)$ obtained by the straightforward calculation with the help of probability (12.18) gives

$$\Delta V_{\text{diff}}^2 = \langle (V_k - V_{k+1})^2 \rangle \sim \bar{V}^2 v t , \tag{12.19}$$

while the regular variation [obtained by the expansion of the exponents e^{-vt} gives

$$\Delta V_{\text{reg}}^2 = (V_k - V_{k+1})_{\text{reg}}^2 \sim \bar{V}^2 (vt)^2 . \tag{12.20}$$

From a comparison of (12.19) and (12.20) it follows that

$$\Delta V_{\text{diff}}^2 / \Delta V_{\text{reg}}^2 \sim (vt)^{-1} \gg 1 , \tag{12.21}$$

making possible the use of the measure (12.18).

The Katz result [12.14] is of great importance for our outline. It expresses the relationship between the path integral with the Wiener measure (12.18) in terms of a fundamental solution of the parabolic type of differential equation. It was M. Katz, who showed that (for proof, [12.14])

$$\left\langle \exp \left\{ -\int_0^t V[X(\tau)] d\tau \right\} \right\rangle = \int_{-\infty}^{\infty} U(X, t) dX , \tag{12.22}$$

$$X(0) = 0 .$$

The function $U(X, t)$ is defined in the following way

$$\frac{\partial U(X,t)}{\partial t} = \frac{1}{2}\frac{\partial^2 U}{\partial X^2} - V(X)U(X,t) ,$$

$$\lim_{x \to \pm \infty} U(X,t) = 0 , \quad \lim_{t \to 0} U(X,t) = \delta(X) .$$
(12.23)

Taking $\sigma(X) \equiv \int_0^\infty U(X,t)e^{-pt}dt$, the Fourier transform, we obtain

$$\frac{1}{2}\frac{d^2\sigma}{dX^2} - [p+V(X)]\sigma(X) = 0 ,$$

$$\lim_{x \to \pm \infty} \sigma(X) = 0 , \quad \sigma'(-0) - \sigma'(0) = 2 .$$
(12.24)

These results about path integration are sufficient for use in the problem discussed.

Using (12.12) let us write the shape $I(\omega)$ in the form

$$I(\omega) = \int_{-\infty}^{\infty} \frac{dV_0 e^{-(V_0/V_T)^2}}{\sqrt{\pi V_T^2}} \frac{1}{\pi} \mathrm{Re} \left\{ \int_0^\infty dt \left\langle \exp\left(i\int_0^t \Omega[V(\tau)]d\tau \right) \right\rangle_{V_0} \right\} ,$$

$$\Omega(\tau) = \{[\Delta\omega - kV_0 - kV(\tau)]^2 + 4G^2\}^{1/2} .$$
(12.25)

Here the brackets $\langle \ldots \rangle_{V_0}$ mean the path integration for a given value V_0.
Let us introduce the following designations:

$$A = \frac{k\bar{V}}{v} , \quad \frac{\omega - kV_0}{k\bar{V}} = Z , \quad \frac{V}{\bar{V}} = X , \quad X(0) = 0 ,$$

$$\tilde{\omega} = \frac{\omega}{k\bar{V}} , \quad \frac{V_0}{\bar{V}} = Y , \quad \frac{2G}{k\bar{V}} = a .$$
(12.26)

Below the condition of the frequency smallness $v \ll kV$ is used, that is $A \gg 1$. Using (12.26), we write (12.25) in the form

$$I = \frac{1}{v\pi^{3/2}} \mathrm{Re} \left\{ \int_{-\infty}^{\infty} dZ e^{-(\tilde{\omega}-Z)^2} \int_0^\infty dt \left\langle \exp\left(iA\int_0^t \sqrt{(x-Z)^2+a^2}\,d\tau \right) \right\rangle \right\} .$$
(12.27)

According to M. Katz (12.22–24)

$$\int_0^\infty dt\langle\ldots\rangle = \int_{-\infty}^{\infty} dX\sigma(X) ,$$
(12.28)

where $\sigma(x)$ is defined by the equation solution

$$\sigma'' - 2iA\sqrt{(X-Z)^2 + a^2}\,\sigma = 0 \; ,$$

$$[\sigma(0)]_+^- = 0 \; , \quad \sigma(X) \xrightarrow[x \to \pm\infty]{} 0 \; , \tag{12.29}$$

$$[\sigma'(0)]_+^- = 2 \; , \quad [\sigma(0)]_+^- \equiv \sigma(-0) - \sigma(+0) \; .$$

Equation (12.29) can also be obtained from the kinetic equations for the atomic density matrix [12.8, 14], the function from (12.29) in the amplitude approach corresponds to the density matrix elements of the mixed system "atom+E.M. field", which can be used in the case of a strong field. The relationship between the amplitude approach and the kinetic density matrix equation method is about the same as the relationship between the *Feynman* method [12.15] and the Schrödinger method in quantum mechanics. The solution of (12.29) after substitution into (12.27, 28) determines the line shape in a strong E.M. field.

12.3 Investigation of Nonlinear Effects in Absorption Due to Brownian Fluctuations of Atomic Velocity

To solve (12.29), let us make the substitution $X_1 = X - Z$, which leads to

$$\sigma'' - 2iA\sqrt{X_1^2 + a^2}\,\sigma = 0 \; ; \quad \sigma \xrightarrow[X_1 \to \pm\infty]{} 0 \; , \quad [\sigma'(Z)]_+^- = 2 \; ,$$

$$\tag{12.30}$$

$$I = \frac{1}{\nu\pi^{3/2}} \int_{-\infty}^{\infty} dZ\,e^{-(Z-\bar{\omega})^2} \int_{-\infty}^{\infty} dX_1\,\sigma(X_1) \; .$$

Let us also take $p(t) = (t^2 + a^2)^{1/2}$. The function $p(t)$ is defined in the complex plane with a cut from $t = -ia$ to $t = +ia$, and on the real axis [sign $p(t) = \text{sign}(t)$].

This choice of the branch of the analytical function insures the correct transition to the small field ($a = 0$, $p(t) = t$). The large parameter in (12.30) makes it possible to use quasiclassical methods for its solution. Equations (12.30) are solved by reducing them to the standard Airy equation [12.16] by the introduction of the function

$$\sigma(X_1) = [\eta/p(X_1)]^{1/4}\,v(\eta) \; ,$$

$$\eta = \left[\frac{3}{2} \int_0^{X_1} p^{1/2}(t)\,dt \right]^{2/3} \; . \tag{12.31}$$

The equation for the function $v(\eta)$ has the form

$$\frac{d^2 v}{d\eta^2} - 2iA(\eta v) = 0 \; ,$$

$$[v(\eta_0)]_+^- = 0 \; , \quad \left[\frac{dv}{d\eta}\right]_+^- = 2\{\eta_0/p[Z(\eta_0)]\}^{1/4} \; , \tag{12.32}$$

$$v(\eta) \xrightarrow[\eta \to \pm \infty]{} 0$$

$$\eta_0 = \left[\frac{3}{2}\int_0^Z p^{1/2}(t)dt\right]^{2/3} \; . \tag{12.33}$$

The solutions of the equation for $v(\eta)$ are expressed in terms of the Airy functions Ai (x) and Bi (x). After substitution of this solution into (12.30) and a number of transformations the equation for the spectrum is obtained in the form [12.8]:

$$I(G, \omega) = I(0, \omega)J(G) \; , \tag{12.34}$$

where $I(0, \omega)$ is the Doppler line shape in the small field case $(G \to 0)$:

$$I(0, \omega) = \frac{1}{\sqrt{\pi}} e^{-\bar{\omega}^2} \; , \tag{12.35}$$

$J(G)$ is the deformation factor which is not dependent on the frequency for the main part of the lineshape:

$$J(G) = \text{Re}\left\{\left[\int_\infty^\infty d\sigma \, \text{Ai}(\sigma)\varphi(\sigma)\right]^2\right\}$$

$$+ 2\,\text{Im}\left\{\int_\infty^\infty d\sigma \varphi(\sigma) \, \text{Bi}(\sigma) \int_0^\infty d\sigma' \varphi(\sigma') \, \text{Ai}(\sigma')\right\} \; , \tag{12.36}$$

where the function $\varphi(\sigma)$ is connected with the solution of (12.32)

$$\varphi(\sigma) = \left[\frac{C\sigma}{[Z^2(C\sigma)+a^2]^{1/2}}\right]^{3/4} \; , \tag{12.37}$$

is the root of the equation following from (12.33)

$$\frac{2}{3}(C\sigma)^{3/2} = \int_0^Z (t^2+a^2)^{1/4}dt \; , \quad (C \equiv (2A)^{-1/3}e^{-i\pi/6}) \; . \tag{12.38}$$

The limiting values of the function $\varphi(\sigma)$ are

$$\varphi(\sigma) \simeq \begin{cases} \left(\dfrac{C\sigma}{a}\right)^{3/4}, & C\sigma \ll a, \\[2em] 1, & C\sigma \gg a. \end{cases} \tag{12.39}$$

It can be seen that the function $\varphi(\sigma)$ determining the line shape deformation by the laser field is equal to one for small fields ($a \to 0$) and to the small factor $(C\sigma_{\text{eff}}/a)^{3/4}$ for strong fields. As can be seen from the integrals in (12.36), the effective values of the variables are of the order of $\sigma_{\text{eff}} \sim 1$, the deformation factor is defined by the ratio of the laser field G to a critical field G_{cr} determined by

$$G_{\text{cr}} = k\bar{V}|C| \simeq k\bar{V}\left(\frac{v}{k\bar{V}}\right)^{1/3} \equiv \delta\omega_{\text{eff}}. \tag{12.40}$$

An example for the interrelation between variables in (12.40) is shown in Fig. 12.1.

It follows from (12.34, 39) that the laser field line shape deformation factor is of the order of the square of the module of the deformation function $|\varphi(\sigma_{\text{eff}})|^2$ values $\sigma_{\text{eff}} \sim 1$, which contributes the main value to the integral (12.36). For field strengths $G \gg G_{\text{cr}}$, the deformation turns out to be:

$$J(0, G \gg G_{\text{cr}}) \sim (G_{\text{cr}}/G)^{3/2}. \tag{12.41}$$

The results confirm the qualitative estimations of Sect. 12.2.1.

Note that the above consideration is true for nearly all of the important integral frequencies, that is for the entire absorption line shape except the wings.

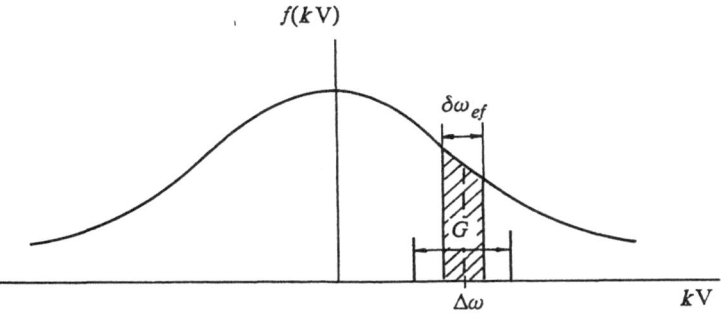

Fig. 12.1. The Doppler line shape formation scheme for atoms with the velocity distribution function $f(k V)$ in a strong laser field $G = |d\,\mathscr{E}_0|$. The domain $\delta\omega_{\text{eff}}$ defines the resonance of the atoms moving with Brownian motion with a laser field

This circumstance allows us to obtain the simple equation (12.3.6) for the distortion factor $J(0, G)$, linking $J(0, G)$ with the integral absorbing power over the frequency spectrum.

In conclusion, we will discuss the results of the theory for the strong relaxation case $v/kV \gg 1$. Here in the limit of weak fields the Dicke narrowing effect [12.11] is realized, as already mentioned. For the strong field case the answer is obvious beforehand: namely, the result of the usual spectral line saturation theory must be obtained, with the elastic collision width coinciding with the collision width $\gamma_D = kV(kV/v)$ figuring in Dicke effect. The detailed consideration on the basis of the kinetic equation confirms the result.

12.4 An Electron in a Planck Radiation Field. "Infrared Catastrophe"

The important example of the stochastic energy exchange of an electron in Coulomb field is the electron collisions with photon gases characterized by a Planck temperature T. The problem plays an important role in astrophysical applications. Following [12.17, 18], we will consider one of the principal aspects of the problem connected with the induced production of soft photons in an external Planck radiation field. The problem of multiple creation of soft quanta (e.g., "infrared catastrophe") has been considered by F. Bloch and A. Nordsieck [12.19] (also [12.18]) far before the discovery of "relict" photons. Therefore they have restricted their analysis to the case of spontaneous quanta creation. The presence of a relict photon background with a temperature of $T = 3\,\mathrm{K}$ makes it important for the problem of the "infrared catastrophe" under induced photon production in an external Planck radiation field. The problem is analyzed below.

12.4.1 Classical Current Approximation

The main approximation for low frequency (soft) photon creation is the classical current method, based on taking into account the creation induced by a classical current produced by a charged particle moving along a given classical trajectory with a velocity $V(t)$.

The Hamiltonian of the interaction between the particle and the magnetic field is of the form

$$\hat{V} = -\frac{e}{c}A \cdot V(t) \;, \quad A = c(2\pi\hbar/\omega\Omega)^{1/2} \sum_{k,\lambda} e_{k,\lambda}(\hat{a}_{k,\lambda} + \hat{a}_{k,\lambda}^+) \;, \qquad (12.42)$$

where A is the second quantized vector potential of the field, expressed in terms of the creation or annihilation operator of the various field modes characterized by the wave vectors k (frequencies $\omega = c|k|$) and polarization vectors $e_{k,\lambda}$. The Schrödinger equation for the wave function Φ of one mode with the wave vector k includes the interaction representation in the form

$$i\hbar \frac{\partial \Phi}{\partial t} = -e \left(\frac{2\pi\hbar}{\omega\Omega}\right)^{1/2} (e_{k,\lambda}, V(t))(\hat{a}_{k,\lambda} + \hat{a}_{k,\lambda}^+)\Phi \ . \tag{12.43}$$

The scattering matrix S that connects the initial and final states of the system is

$$\hat{S}(-\infty, +\infty) = \exp\left(-\frac{i}{\hbar}\int_{-\infty}^{+\infty} \hat{V}dt\right) \ . \tag{12.44}$$

In the classical current representation, the action of the matrix $\hat{S}$ on the operators $\hat{a}_{k,\lambda}$, $\hat{a}_{k,\lambda}^+$ is given by the relations [12.20]

$$\hat{S}^{-1}\hat{a}_{k,\lambda}\hat{S} = \hat{a}_{k,\lambda} + ie\left(\frac{2\pi}{\hbar\omega\Omega}\right)^{1/2} \int_{-\infty}^{+\infty} (e_{k,\lambda}\cdot V(t))e^{-i\omega t}dt \ , \tag{12.45}$$

$$\hat{S}^{-1}\hat{a}_{k,\lambda}^+\hat{S} = \hat{a}_{k,\lambda}^+ - ie\left(\frac{2\pi}{\hbar\omega\Omega}\right)^{1/2} \int_{-\infty}^{+\infty} (e_{k,\lambda}\cdot V(t))e^{i\omega t}dt \ . \tag{12.46}$$

An exact solution of (12.43) is known for one mode, [12.8]. The expression for the probability W_{ls} of the transition from an initial state containing s photons (i.e., from a state corresponding to the sth level of the oscillator) to the final state containing l photons is

$$W_{ls} = \begin{cases} x_{k,\lambda}^{s-l} \dfrac{l!}{s!} [L_l^{s-l}(x_{k,\lambda})]^2 \exp(-x_{k,\lambda}) \ , & s \geq l \ , \\[2ex] x_{k,\lambda}^{l-s} \dfrac{s!}{l!} [L_s^{l-s}(x_{k,\lambda})]^2 \exp(-x_{k,\lambda}) \ , & s < l \ , \end{cases} \tag{12.47}$$

$$x_{k,\lambda} = \frac{2\pi e^2}{\hbar\omega\Omega} \left| \int_{-\infty}^{+\infty} e_{k,\lambda}\cdot V(t)e^{-i\omega t}dt \right|^2 \ , \tag{12.48}$$

where L_p^k is a Legendre polynomial of degree k.

In the case of a monochromatic field, the number s of the photons can be expressed in terms of the electromagnetic field strength:

$$\mathscr{E}_0^2\Omega/8\pi = s\hbar\omega \quad (s,\Omega \to \infty) \ , \tag{12.49}$$

after which the parameter $x_{k,l}$, which we shall denote simply by x_ω takes the form

$$x_\omega = \frac{\mathscr{E}_0^2 e^2}{4s(\hbar\omega)^2} \left| \int_{-\infty}^{\infty} e_{k,\lambda} \cdot V(t) e^{-i\omega t} dt \right|^2 . \tag{12.50}$$

For multimode spontaneous emission, the occupation numbers $n_{k,\lambda} \equiv l$ of each of the modes are small, and the transition probability $W_{ls}(s=0, l=n_{n,\lambda})$ for each of the oscillators of the field is greatly simplified, taking the form of a Poisson distribution

$$W_{n_{k,\lambda}} = \frac{1}{n_{k,\lambda}!} (x_{k,\lambda})^{n_{k,\lambda}} \exp(-x_{k,\lambda}) . \tag{12.51}$$

The probability that N field oscillators with frequencies $\omega_1, \omega_2, \ldots, \omega_N$ will simultaneously emit a definite number $n = \sum_{i=1}^{N} n_{k_i \lambda_i}$ of photons is

$$W_n = \sum_{n = n_{k_1,\lambda_1} + \ldots + n_{k_N}, \lambda_N} \prod_{i=1}^{N} W_{n_{k_i \lambda_i}} . \tag{12.52}$$

From the known addition theorem for the Poisson probabilities we have

$$W_n = \frac{1}{n!} \left(\sum_{k,\lambda} x_{k,\lambda} \right)^n \exp\left(-\sum_{k,\lambda} x_{k,\lambda} \right) . \tag{12.53}$$

Changing in the usual manner from summation over the oscillators to integration over the frequencies and angles, we obtain a known expression for the probability of multiphoton spontaneous emission in a finite frequency interval (ω_1, ω_2) [12.19, 20]

$$W_n = \frac{1}{n!} \left(\int_{\omega_1}^{\omega_2} y_\omega d\omega \right)^n \exp\left(-\int_{\omega_1}^{\omega_2} y_\omega d\omega \right) , \tag{12.54}$$

$$y_\omega = \frac{1}{2\pi^2 c^3} \int_{-\pi}^{\pi} x_k \omega^2 \sin^2 \Theta d\Theta . \tag{12.55}$$

In the case of a Coulomb field the parameter x_k that determines the probabilities of multiphoton processes can be calculated in an explicit analytic form.

12.4.2 Multiphoton Induced Processes in a Planck Radiation Field

We consider multiphoton processes in a non-single-mode monochromatic external radiation field. As an example of physical interest, we use a Planck thermal radiation field.

Known results for spontaneous multiphoton processes were already cited in Sect. 12.4.1. The process can occur because the smallness of the QED parameter $e^2/\hbar c$ is offset by the logarithmically large average number of emitted photons. This can be seen directly from the structure of the probability W_n of the emission of low frequency phonons. In fact, at $\omega \tau_{collis} \ll 1$ we have from (12.54):

$$W_n(\omega_1, \omega_2) = \frac{(\omega_2/\omega_1)^{-\beta}}{n!} \left[\beta \ln \left(\frac{\omega_2}{\omega_1} \right) \right]^n ,$$

$$\beta \sim e^2/\hbar c, \quad \bar{n} = \sum_{n=1}^{\infty} n W_n(\omega_1, \omega_2) = \beta \ln \left(\frac{\omega_2}{\omega_1} \right) .$$

(12.56)

Since the emission and absorption probabilities for an individual oscillator of the continuous spectrum are proportional to $(x_\omega/\Omega)^n$, the emission (absorption) will be one single photon for this oscillator, since $W_{n+1}/W_n \to 0$ as $\Omega \to \infty$.

However, the emission from the aggregate of all of the oscillators in a given frequency interval need not necessarily be the emission of one single photon. This was already seen with (12.54) as an example: depending on the number of field oscillators considered (on the width of the frequency interval), the value $n_{\max}$ at which W_n is a maximum can take on different values.

Equilibrium Planck radiation is characterized by a Gibbs probability distribution W_s^G of finding s photons (prior to scattering) in an oscillator of frequency ω at an emission temperature T_P:

$$W_s^G = [1 - \exp(-\hbar\omega/T_P)] \exp(-s\hbar\omega/T_P) .$$

The probability W_n^e of emission by an oscillator is obtained by weighting the probabilities (12.47) of exciting this oscillator over this distribution:

$$\tilde{W}_n^e = \sum_{s=0}^{\infty} W_{s,n}^e W_s^G = [1 - \exp(-\hbar\omega/T_P)] \sum_{s=0}^{\infty} \frac{s!}{(s+n)!}$$

$$\times e^{-\hbar\omega/T_P} \left(\frac{x_k}{\Omega} \right)^n \left[L_s^n \left(\frac{x_k}{\Omega} \right) \right]^2 \exp \left(-\frac{x_k}{\Omega} \right)$$

$$= [x_k \exp(-\hbar\omega/T_P)]^{-n} I_n \left(\frac{2 x_k \exp(-\hbar\omega/T_P)}{\Omega(1 - \exp(\hbar\omega/T_P))} \right)$$

$$\times \exp \left(-\frac{1 + \exp(-\hbar\omega/T_P)}{1 - \exp(-\hbar\omega/T_P)} \frac{x_k}{\Omega} \right) ,$$

where $W^e_{s,n}$ is the probability that the field oscillator will emit n photons if their initial number is s. For a distinct separation of the parts that are small as $\Omega \to \infty$, we take x_k hereafter to mean (12.48) multiplied by Ω. The same transformation applies also to y, defined earlier by (12.55).

As $\Omega \to \infty$, we obtain

$$\tilde{W}^e_n = \frac{1}{n!}\left(\frac{x_\omega}{\Omega(1-\exp(-\hbar\omega/T_P))}\right)^n \exp\left(-\frac{1-\exp(-\hbar\omega/T_P)}{1+\exp(-\hbar\omega/T_P)}\right),$$

(12.57)

i.e., it is a Poisson distribution. In analogy with the derivation of (12.54), we obtain the photon-emission probability $W^e_n(\omega_1, \omega_2)$ for the frequency interval (ω_1, ω_2):

$$W^e_n(\omega_1, \omega_2) = \frac{1}{n!}\left(\int_{\omega_1}^{\omega_2}\frac{y_\omega d\omega}{1-\exp(-\hbar\omega/T_P)}\right)^n$$

$$\times \exp\left(-\int_{\omega_1}^{\omega_2}\frac{1+\exp(-\hbar\omega/T_P)}{1-\exp(-\hbar\omega/T_P)}y_\omega d\omega\right).$$

(12.58)

Similarly, for absorption,

$$\tilde{W}^a_n = \sum_{s=n}^{\infty}W^a_{l,n}W^G_s = (1-e^{-\hbar\omega/T_P})\sum_{s=n}^{\infty}e^{-s\hbar\omega/T_P}\frac{(s-n)!}{s!}$$

$$\times\left(\frac{x_k}{\Omega}\right)^n\left[L^n_{s-n}\left(\frac{x_k}{\Omega}\right)\right]^2 e^{-x_k/\Omega}$$

$$= [1-\exp(-\hbar\omega/T_P)]e^{-n\hbar\omega/T_P}\sum_{s=0}^{\infty}e^{-s\hbar\omega/T_P}\frac{s!}{(s+n)!}$$

$$\times\left(\frac{x_k}{\Omega}\right)^n\left[L^n_s\left(\frac{x_k}{\Omega}\right)\right]^2 e^{-x_k/\Omega} = e^{-n\hbar\omega/T_P}\tilde{W}^e_n,$$

(12.59)

where $W^a_{l,n}$ is the probability that the oscillator will absorb n photons out of a finite number $l = s-n$. Taking (12.57) into account, we obtain from (12.59)

$$\tilde{W}^a_n = \frac{1}{n!}\left(\frac{x_k\exp(-\hbar\omega/T_P)}{\Omega(1-\exp(-\hbar\omega/T_P))}\right)^n \exp\left(-\frac{1+\exp(-\hbar\omega/T_P)}{1-\exp(-\hbar\omega/T_P)}\right).$$

(12.60)

The absorption probability integrated over the interval (ω_1, ω_2) is

$$W_n^a(\omega_1, \omega_2) = \frac{1}{n!} \left(\int_{\omega_1}^{\omega_2} \frac{y_\omega \exp(-\hbar\omega/T_P)}{1 - \exp(-\hbar\omega/T_P)} d\omega \right)^n$$

$$\times \exp\left(-\int_{\omega_1}^{\omega_2} \frac{1 + \exp(-\hbar\omega/T_P) d\omega}{1 - \exp(-\hbar\omega/T_P)} \right). \tag{12.61}$$

Now, when the initial occupation numbers of the oscillators are not zero, both emission and absorption is possible. The observed n-photon process (we refer for the sake of argument to emission) is an aggregate consisting of the emission of $n+k$ and the absorption of k photons, where k assumes all possible values from zero to infinity. Since emission and absorption in a frequency interval are statistically independent, the probabilities w_n^e and w_n^a actually observed are determined by the set of two equations

$$w_n^e = \sum_{k=0}^{\infty} W_{n+k}^e(\omega_1, \omega_2) W_k^a(\omega_1, \omega_2) \left[w_0 + \sum_{p=1}^{\infty} (w_p^e + w_p^a) \right]^{-1}, \tag{12.62}$$

$$w_n^a = \sum_{k=0}^{\infty} W_{n+k}^a(\omega_1, \omega_2) W_k^e(\omega_1, \omega_2) \left[w_0 + \sum_{p=1}^{\infty} (w_p^e + w_p^a) \right]^{-1}, \tag{12.63}$$

with the appropriate normalization, where $w_0 = w_0^e = w_0^a$.

The number of terms that contribute to (12.62, 63) depends at a given (ω_1, ω_2) on the ratio $\hbar\omega/T_P$. As $T_P \to 0$, a contribution is made only by single-photon processes corresponding to the first terms of these sums. As $T_P \to \infty$ the contribution of the multistep-induced emission or absorption processes that determine mainly the process increases sharply. This cause the resultant probability to deviate from a purely Poisson probability. Indeed, solution of this system yields

$$w_n^e = Y^n I_n(X) \exp[-1/2 X(Y + Y^{-1})], \tag{12.64}$$

$$w_n^a = Y^{-n} I_n(X) \exp[-1/2 X(Y + Y^{-1})], \tag{12.65}$$

where I_n is a modified Bessel function of order n, and the arguments X and Y are given by

$$X = \left(\int_{\omega_1}^{\omega_2} \frac{y_\omega \exp(-\hbar\omega/T_P) d\omega}{1 - \exp(-\hbar\omega/T_P)} \int_{\omega_1}^{\omega_2} \frac{y_\omega d\omega}{1 - \exp(-\hbar\omega/T_P)} \right)^{1/2},$$

$$\tag{12.66}$$

$$Y = \left[\int_{\omega_1}^{\omega_2} \frac{y_\omega \exp(-\hbar\omega/T_P) d\omega}{1 - \exp(-\hbar\omega/T_P)} \left(\int_{\omega_1}^{\omega_2} \frac{y_\omega d\omega}{1 - \exp(-\hbar\omega/T_P)} \right)^{-1} \right]^{1/2}.$$

These expressions are a new result, viz., a generalization of the Poisson equation for spontaneous bremsstrahlung radiation to include the case of induced emission and absorption in an external equilibrium thermal radiation field. The Poisson equations are obtained from (12.64) in the limit as $T_P \to 0$ (there are no thermal photons), while (12.65) yields zero in this case, as it should.

Using the known relation

$$\sum_{n=1}^{\infty} y^n I_n(x) = \frac{1}{2} \left\{ \exp\left[\frac{x}{2}(y+y^{-1})\right] - I_0(x) \right\} , \tag{12.67}$$

we can also calculate such quantities as the mean number $\bar{n}$ and the mean squared number $\bar{n}^2$ of the emitted photons, and the variance that characterizes the given distribution

$$\bar{n} = \sum_{n=1}^{\infty} n w_n^e = \sum_{n=1}^{\infty} n Y^n I_n(X) \exp\left[-\frac{X}{2}(Y+Y^{-1})\right]$$

$$= Y e^{-X(Y+Y^{-1})} \frac{d}{dY}\left[\sum_{n=1}^{\infty} Y^n I_n(X)\right] .$$

Using (12.67) we obtain

$$\bar{n} = \frac{X}{2}(Y-Y^{-1}) . \tag{12.68}$$

Similarly,

$$\overline{n^2} = \sum_{n=1}^{\infty} n^2 w_n^e = \left(\frac{X}{2}\right)^2 (Y-Y^{-1}) + \frac{X}{2}(Y+Y^{-1}) . \tag{12.69}$$

From this we get an expression for the variance $D(n) = \overline{(n-\bar{n})^2} = \overline{n^2} - \bar{n}^2$:

$$D(n) = \frac{X}{2}(Y-Y^{-1}) . \tag{12.70}$$

We rewrite $\bar{n}$, $\overline{n^2}$, and $D(n)$ in standard notation:

$$\bar{n} = D(n) = \frac{1}{2} \int_{\omega_1}^{\omega_2} \frac{1+\exp(-\hbar\omega/T_P)d\omega}{1-\exp(-\hbar\omega/T_P)} , \tag{12.71}$$

$$\overline{n^2} = \frac{1}{2} \int_{\omega_1}^{\omega_2} \frac{1+\exp(-\hbar\omega/T_P)}{1-\exp(-\hbar\omega/T_P)} y_\omega d\omega \left(1 + \frac{1}{2} \int_{\omega_1}^{\omega_2} \frac{1+\exp(-\hbar\omega/T_P)}{1-\exp(-\hbar\omega/T_P)} y_\omega d\omega\right) . \tag{12.72}$$

As $T_P \to 0$, the variance, and with it also n (12.71), become like the usual results for spontaneous emission [12.19]. As $T_P \to \infty$, the distribution width increases strongly:

$$D(n) \approx T_P \int\limits_{\omega_1}^{\omega_2} \frac{y_\omega d\omega}{\hbar \omega} \; , \tag{12.73}$$

obviously corresponding to an increase of the probability of producing many photons. However, the normalization of the distribution remains the same as before.

12.4.3 Calculation of the Absorbed Energy

We now calculate the change in one particle-scattering act of the total energy of a field having a continuous spectrum. It is expedient to use the S-matrix formalism for this calculation, as in Sect. 12.4.1. Using the operator $\hat{S}$, we express the field energy E_f after scattering in the form

$$E_f = \left\langle i \left| \hat{S}^{-1} \left(\sum_{k,\lambda} \hbar \omega \hat{a}_{k,\lambda}^+ \hat{a}_{k,\lambda} \right) \hat{S} \right| i \right\rangle \; . \tag{12.74}$$

With the aid of relations (12.45, 46), we obtain

$$E_f = \left\langle i \left| \sum_{k,\lambda} \hbar \omega \left[\hat{a}_{k,\lambda}^+ \hat{a}_{k,\lambda} + ie \left(\frac{2\pi}{\hbar \omega \Omega} \right)^{1/2} \int\limits_{-\infty}^{+\infty} (eV(t)) e^{-i\omega t} dt \right. \right. \right.$$

$$- ie \left(\frac{2\pi}{\hbar \omega \Omega} \right)^{1/2} \int\limits_{-\infty}^{+\infty} eV(t) e^{i\omega t} dt$$

$$\left. \left. \left. + \frac{2\pi e^2}{\omega \hbar \Omega} \left| \int\limits_{-\infty}^{+\infty} eV(t) e^{-i\omega t} dt \right|^2 \right] \right| i \right\rangle \; . \tag{12.75}$$

Elementary operations with $\hat{a}_{k,\lambda}^+$ and $\hat{a}_{k,\lambda}$ yield the calculation result:

$$E_f = \sum_{k,\lambda} \hbar \omega n_{k,\lambda} + \sum_{k,\lambda} \frac{2\pi e^2}{\Omega} \left| \int\limits_{-\infty}^{+\infty} (eV(t)) e^{i\omega t} dt \right|^2 \; . \tag{12.76}$$

The first term in (12.76) is, obviously, the initial energy E_i of the field, and the second is the sought difference in energy before and after the scattering, $\Delta E = E_f - E_i$, and equals the already known spontaneous emission energy. In particular, this result also holds for a Planck radiation field, as can be verified directly by calculating the sum

$$\sum_{n=-\infty}^{n=+\infty} n\hbar\omega\,(w_n^e - w_n^a)$$

with the aid of (12.67).

To determine the energy acquired by the electron, we must take the kinetics of the processes into account. In the case of the scattering of many classical electrons having a distribution function $f(\varepsilon)$, the expression for the energy increment ΔE is

$$\Delta E = N\hbar\omega\,Vd\varepsilon d\sigma_{\mathrm{res}} \sum_{n=1}^{\infty} n[w_n^e f(\varepsilon - n\hbar\omega) - w_n^a f(\varepsilon + n\hbar\omega)]\Delta t\,, \quad (12.77)$$

where $\varepsilon = \dfrac{mv^2}{2}$, $\varepsilon \gg \hbar\omega$.

Here w_n^e and w_n^a are taken for that frequency interval in which ΔE is calculated. It must not be very large, viz., such that the integrands in the expressions for w_n^e and w_n^a do not change too much in it, so that photon emission from even the opposite ends of the interval (ω_1, ω_2) has high probabilities, in accordance with the meaning of the differential description of ΔE. To this end we must have

$$\omega_2 - \omega_1 \ll g(\omega)/g'(\omega)|_{\omega = \omega_1}\,, \tag{12.78}$$

where g is one of the functions

$$y_\omega/[1 - \exp(-\hbar\omega/T_P)]\,, \quad y_\omega \exp(-\hbar\omega/T_P)/[1 - \exp(-\hbar\omega/T_P)]$$

for which the ratio $g(\omega)/g'(\omega)|_{\omega = \omega_1}$ is a minimum.

At high temperatures the conditions (12.78) cease to depend on T_P, and the expression in the argument of the modified Bessel function becomes proportional to T_P. If the interval (ω_1, ω_2) that can be considered is fixed (is not too small), then w_n^e and w_n^a are no longer Poisson probabilities. In other words, in accordance with the preceeding discussion, they do not describe processes that are multiphoton in (ω_1, ω_2).

Expanding $f(\varepsilon - n\hbar\omega)$ up to the terms of the first order of smallness, substituting the probabilities, and regrouping the corresponding terms, we have

$$dE = \left[N\hbar\omega f(\varepsilon) \exp\left\{ -\frac{X}{2}(Y + Y^{-1}) \right\} \sum_{n=1}^{\infty} [n\,Y^n I_n(X) - n\,Y^{-n} I_n(X)] \right.$$

$$\left. - Nf(\varepsilon)\exp\left[-\frac{X}{2}(Y + Y^{-1}) \right] (\hbar\omega)^2 \frac{\partial f}{\partial \varepsilon} \sum_{n=1}^{\infty} n^2 I_n(X)(Y^n + Y^{-n}) \right]$$

$$\times Vd\varepsilon d\sigma \Delta t\,. \tag{12.79}$$

Calculating the sums in (12.79) by elementary methods and returning to the original notation, we obtain

$$\Delta E = N\hbar\omega\, V d\varepsilon\, d\sigma_{\text{res}} \left\{ f(\varepsilon) \int_{\omega_1}^{\omega_2} y_\omega\, d\omega + \frac{\partial f}{\partial\varepsilon}\, \hbar\omega \right.$$

$$\left. \times \left[\int_{\omega_1}^{\omega_2} \frac{1+e^{-\hbar\omega/T_P}}{1-e^{-\hbar\omega/T_P}}\, y_\omega\, d\omega + \frac{1}{2}\left(\int_{\omega_1}^{\omega_2} y_\omega\, d\omega \right)^2 \right] \right\} \Delta t\ . \tag{12.80}$$

Changing to a small interval $d\omega$, we have

$$\frac{dE}{d\omega} = NV d\varepsilon\, d\sigma_{\text{res}} \left[f(\varepsilon)\hbar\omega\, y_\omega + \hbar^2\frac{\partial f}{\partial\varepsilon}\frac{1+e^{-\hbar\omega/T_P}}{1-e^{-\hbar\omega/T_P}}\, y_\omega\, \omega^2 \right] dt\ . \tag{12.81}$$

At large $T_P \gg \hbar\omega$ and also $\omega\tau_{\text{collis}} \ll 1$,

$$\frac{dE}{d\omega\, dt} = \left(\frac{Ze^2}{mV^2}\right)^2 V d\varepsilon\, d\sigma \left(f(\varepsilon) + 2T_P\frac{\partial f}{\partial\varepsilon} \right) \frac{2e^2 V^2}{3\pi c^3}\ . \tag{12.82}$$

This result agrees with the classical increase in energy of an electron in a Planck field [12.17].

References

Chapter 1

1.1 H. A. Bethe, E. E. Salpeter: *Quantum Mechanics of One- and Two-Electron Atoms* (Springer, Berlin, Heidelberg 1958)

1.2 L. D. Landau, E. M. Lifshitz: *Quantum Mechanics: Non-Relativistic Theory* (Pergamon, Oxford 1977)

1.3 I. I. Sobelman: *Introduction to the Theory of Atomic Spectra* (Pergamon, Oxford 1973)

1.4 I. I. Sobelman, L. A. Vainshtein, Yu. A. Yukov: *Excitation of Atoms and Broadening of Spectral Lines*, Springer Ser. Chem. Phys., Vol. 7 (Springer, Berlin, Heidelberg 1981)

1.5 H. R. Griem: *Spectral Line Broadening by Plasmas* (Academic, New York 1974)

1.6 R. F. Stebbings, F. B. Dunning (eds.): *Rydberg States of Atoms and Molecules* (Cambridge Univ. Press, Cambridge 1983)

1.7 S. G. Rautian, G. I. Smirnov, A. N. Shalagin: *Nonlinear Resonances in Spectra of Atoms and Molecules* (Nauka, Moscow 1979) (in Russian)

1.8 M. A. Leontovich, B. B. Kadomtsev (eds.): *Reviews of Plasma Physics*, Vol. 12 (Consultants Bureau, New York, London 1987)

1.9 R. D. Cowan: *The Theory of Atomic Structure and Spectra* (California Press, Los Angeles 1981)

1.10 A. A. Nikitin, Z. B. Rudzikas: *Foundation of Spectra Theory of Atoms and Ions* (Nauka, Moscow 1983) (in Russian)

1.11 V. S. Lisitsa: Usp. Fiz. Nauk *153*, 379 (1987) [English transl.: Sov. Phys.-Usp. *30*, 927 (1987)]

1.12 V. I. Kogan, A. B. Kukushkin, V. S. Lisitsa: Phys. Rep. *213*, 1 (1992)

1.13 M. Born: *The Mechanics of the Atom* (Bell, London 1927)

1.14 V. S. Lisitsa: Usp. Fiz. Nauk *122*, 449 (1977) [English transl.: Sov. Phys.-Usp. *20*, 603 (1977)]

1.15 V. I. Kogan, V. S. Lisitsa, G. V. Sholin: *Spectral-Line Broadening in a Plasma.* In: *Reviews of Plasma Physics*, ed. by B. B. Kadomtsev, Vol. 13 (Consultant Bureau, New York, London 1988) p. 213

1.16 Yu. A. Klimontowitch: *Kinetic Theory of Electromagnetic Processes* (Nauka, Moscow 1980) (in Russian)

1.17 I. C. Percival, M. J. Seaton: Phil. Trans. R. Soc. London A *251*, 113 (1958)

1.18 K. Burnett: Phys. Rep. *118*, 339 (1985)

1.19 A. V. Anufrienko, A. L. Godunov, A. V. Demura, Yu. K. Zemtzov, V. S. Lisitsa, A. N. Starostin, M. D. Taran, V. A. Schipakov: Zh. Eksp. Teor. Fiz. *98*, 1304 (1990) [English transl.: Sov. Phys.-JETP *71* (4), 728 (1990)]

Chapter 2

2.1 J. Jackson: *Classical Electrodynamics* (Wiley, New York 1975)

2.2 L. D. Landau, E. M. Lifshitz: *The Classical Theory of Fields* (Pergamon, Oxford 1975)

2.3   V. I. Gervids, V. I. Kogan: Pisma Zh. Eksp. Teor. Fiz. *22*, 308 (1975) [English transl.: JETP Lett. *22*, 142 (1975)]
2.4   V. I. Kogan, A. B. Kukushkin: Zh. Eksp. Teor. Fiz. *87*, 1164 (1984) [English transl.: Sov. Phys. JETP *60*, 665 (1984)]
2.5   V. I. Gervids, V. I. Kogan: *Electron Bremsstrahlung in a Static Potential.* Review (CNII Atominform-IAE, Moscow 1988). English transl.: Polarization Bremsstrahlung of Particles and Atoms, ed. by V. N. Tsytovich, I. M. Oiringel (Plenum, New York 1991), p.
2.6   V. I. Kogan, A. B. Kukushkin, V. S. Lisitsa: Phys. Rep. *213*, 1 (1992)
2.7   H. A. Kramers: Phil. Mag. *46*, 836 (1923)
2.8   J. W. B. Hughes: Proc. Phys. Soc. *91*, 810 (1967)
2.9   L. D. Landau, E. M. Lifshitz: *Quantum Mechanics: Non-Relativistic Theory* (Pergamon, Oxford 1977)
2.10   U. I. Safronova, V. S. Senashenko: *Theory of Multicharged Ions Spectra* (Energoizdat, Moscow 1983) (in Russian)
2.11   R. K. Janev, L. P. Presnyakov, V. P. Shevelko: *Physics of Highly Charged Ions*, Springer Ser. Electroph., Vol. 13 (Springer, Berlin, Heidelberg 1985)
2.12   H. A. Bethe, E. E. Salpeter: *Quantum Mechanics of One- and Two-Electron Atoms* (Springer, Berlin, Heidelberg 1958)
2.13   G. W. Erickson: Phys. Rev. Lett. *27*, 780 (1971)
2.14   A. G. Zhidkov, A. N. Tkachev, S. I. Yakovlenko: Zh. Eks. Teor. Fiz. *91*, 445 (1986) [English transl.: Sov. Phys.-JETP *64*, 261 (1986)]
2.15   G. Luders: Ann. Phys. Leipzig *8*, 301 (1951)
2.16   E. Kh. Ahmedov, A. L. Godunov, Yu. K. Zemtsov, V. A. Makhrov, A. N. Starostin, M. D. Taran: Zh. Eksp. Teor. Fiz. *89*, 470 (1985) [English transl.: Sov. Phys.-JETP *62*, 266 (1985)]
2.17   S. Klarsfeld: Phys. Lett. *30* A, 382 (1969)
2.18   S. Suckewer, Phys. Scr. *23*, 772 (1981)
2.19   V. A. Brysgunov, S. Yu. Luk'anov, M. T. Pakhomov: Zh. Eks. Teor. Fiz. *82*, 1904 (1982) [English transl.: Sov. Phys.-JETP *55*, 1095 (1982)]
2.20   W. Heitler: *Quantum Theory of Radiation* (Clarendon, Oxford 1954)
2.21   U. Fano: Phys. Rev. *124*, 1866 (1961)
2.22   A. S. Kompaneets: Zh. Eks. Teor. Fiz. *54*, 974 (1968) [English transl.: Sov. Phys.-JETP *27*, 519 (1968)]
2.23   R. F. Stebbings, F. B. Dunning (eds.): *Rydberg States of Atoms and Molecules* (Cambridge Univ. Press, Cambridge 1983)

Chapter 3

3.1   S. P. Goreslavsky, N. B. Delone, V. P. Krainov: Zh. Eks. Teor. Fiz. *82*, 1789 (1982); *87*, 1164 (1984) [English transl.: Sov. Phys.-JETP *55*, 1032 (1982); *60*, 665 (1984)]
3.2   R. A. Gantsev, N. F. Kazakova, V. P. Krainov: *Radiation Transition Rates in Hydrogen-Like Plasmas.* In: *Plasma Chemistry*, ed. by B. M. Smirnov, Vol. 12 (Energoatomizdat, Moscow 1985) p. 96 (in Russian)
3.3   L. D. Landau, E. M. Lifshitz: *Quantum Mechanics: Nonrelativistic Theory* (Pergamon, Oxford 1977)
3.4   P. F. Naccache: J. Phys. B *5*, 1308 (1972)
3.5   V. I. Kogan, A. B. Kukushkin: Zh. Eks. Teor. Fiz. *87*, 1164 (1984) [English transl.: Sov. Phys.-JETP *60*, 665 (1984)]
3.6   V. I. Kogan, A. B. Kukushkin, V. S. Lisitsa: Phys. Rep. *213*, 1 (1992)
3.7   L. D. Landau, E. M. Lifshitz: *Mechanics* (Pergamon, Oxford 1976)
3.8   D. R. Bates, A. Damgaard: Phil. Trans. *242*, 101 (1949)
3.9   I. I. Sobelman: *Introduction to the Theory of Atomic Spectra* (Pergamon, Oxford 1979)
3.10   V. A. Davydkin, B. A. Zon: Opt. Spectrosk. *51*, 25 (1981) [English transl.: Opt. Spectrosc. USSR *51*, 13 (1981)]

3.11 H. A. Bethe, E. E. Salpeter: *Quantum Mechanics of One- and Two-Electron Atoms* (Springer, Berlin, Heidelberg 1958)
3.12 L. D. Landau, E. M. Lifshitz: *Classical Field Theory* (Pergamon, Oxford 1975)
3.13 J. Jackson: *Classical Electrodynamics* (Wiley, New York 1975)
3.14 L. Kim, R. H. Pratt: Phys. Rev. A *36*, 45 (1987)
3.15 A. B. Kukushkin, V. S. Lisitsa: Zh. Eks. Teor. Fiz. *88*, 1570 (1985) [English transl.: Sov. Phys.-JETP *61*, 937 (1985)]
3.16 M. J. Seaton: Mon. Not. R. Astron. Soc. *119*, 90 (1959)
3.17 R. M. Pendelly: Mon. Not. R. Astron. Soc. *127*, 145 (1964)
3.18 H. P. Summers: Mon. Not. R. Astron. Soc. *178*, 101 (1977)
3.19 V. A. Abramov, F. F. Baryshnikov, A. I. Kazanskii, I. V. Komarov, V. S. Lisitsa, M. I. Chibisov: Charge Exchange of Atoms on Multiply Charged Ions. In: *Reviews of Plasma Physics*, ed. by M. A. Leontovich, B. B. Kadomtsev, Vol. 12 (Consultants Bureau, New York, London 1987) p. 123
3.20 I. I. Sobelman, L. A. Vainshtein, Yu. A. Yukov: *Excitation of Atoms and Broadening of Spectral Lines*, Springer Ser. Chem. Phys., Vol. 7 (Springer, Berlin, Heidelberg 1981)
3.21 B. M. Smirnov: *Excited Atoms* (Energoizdat, Moscow 1982) (in Russian)
3.22 I. L. Beigman, E. D. Mikhaltchi: J. Quant. Spectrosc. Radiat. Transfer *91*, 365 (1969)
3.23 I. L. Beigman, I. M. Gaisinsky: Preprint Lebedev Inst. of Physics, No. 181 (Moscow 1979)
3.24 S. T. Belyaev, G. I. Budker: *Multiquantum Recombination in Ionized Gases*. In: *Plasma Physics and the Problem of Controlled Thermonuclear Reactions*, ed. by M. A. Leontovich, Vol. 3 (Pergamon, Oxford 1958) p. 250

Chapter 4

4.1 V. I. Kogan, A. B. Kukushkin, V. S. Lisitsa: Phys. Rep. *213*, 1 (1992)
4.2 E. Fermi: Z. Physik. *29*, 315 (1924)
4.3 V. N. Tsytovich, I. M. Oiringel (eds.): *Polarization Bremsstrahlung of Particles and Atoms* (Nauka, Moscow 1987) [English transl.: Plenum, New York 1991]
4.4 L. D. Landau, E. M. Lifshitz: *Classical Field Theory* (Pergamon, Oxford 1975)
4.5 J. Jackson: *Classical Electrodynamics* (Wiley, New York 1975)
4.6 V. B. Berestetsky, E. M. Lifshitz, L. P. Pitaevsky: *Quantum Electrodynamics* (Nauka, Moscow 1989)
4.7 V. A. Bazylev, M. I. Chibisov: Usp. Fiz. Nauk. *133*, 617 (1981) [English transl.: Sov. Phys.-Usp. *24*, 276 (1981)]
4.8 J. N. Gan, R. J. Henry: Phys. Rev. A *16*, 968 (1977)
4.9 M. N. Gailitis: In *Atomic Collisions* (Latvian State University, Riga 1963) p. 93
4.10 W. Heitler: *Quantum Theory of Radiation* (Clarendon, Oxford 1954)
4.11 I. L. Beigman, L. A. Vainshtein, B. N. Chichkov: Zh. Eks. Teor. Fiz. *80*, 964 (1981) [English transl.: Sov. Phys.-JETP *53*, 490 (1981)]
4.12 I. I. Sobelman, L. A. Vainshtein, Yu. A. Yukov: *Excitation of Atoms and Broadening of Spectral Lines*, Springer Ser. Chem. Phys., Vol. 7 (Springer, Berlin, Heidelberg 1981)
4.13 V. M. Buimistrov, L. I. Trakhtenberg: Zh. Eksp. Teor. Fiz. *69*, 108 (1975) [English transl.: Sov. Phys.-JETP *42*, 54 (1975)]
4.14 B. A. Zon: Zh. Eksp. Teor. Fiz. *73*, 128 (1977) [English transl.: Sov. Phys.-JETP *46*, 65 (1977)]
4.15 V. A. Astapenko, V. M. Buimistrov, Yu. A. Krotov, L. K. Mikhailov, L. I. Trakhtenberg: Zh. Eksp. Teor. Fiz. *88*, 1560 (1985) [English transl.: Sov. Phys.-JETP *61*, 930 (1985)]
4.16 M. Ya. Amusya: *Bremsstrahlung Radiation* (Energoatomizdat, Moscow 1990) (in Russian)
4.17 N. F. Mott, H. S. W. Massey: *The Theory of Atomic Collisions* (Clarendon, Oxford 1965)

4.18   I.C. Persival, M.J. Seaton: Phil. Trans. R. Soc. *251* (3), 113 (1958)
4.19   A. Jablonski: Phys. Rev. *68*, 78 (1945)
4.20   A.B. Kukushkin, V.S. Lisitsa: *Polarization Radiation Mechanism in Atomic Colli-sions.* In: *Polarization Bremsstrahlung of Particles and Atoms,* ed. by V.N. Tsyto-vich, I.M. Oiringel (Nauka, Moscow 1987) Chap. 11 [English transl.: Plenum, New York 1991)

Chapter 5

5.1   H.A. Bethe, E.E. Salpeter: *Quantum Mechanics of One- and Two-Electron Atoms* (Springer, Berlin, Heidelberg 1958)
5.2   L.D. Landau, E.M. Lifshitz: *Quantum Mechanics: Non-Relativistic Theory* (Perga-mon, Oxford 1977)
5.3   I.I. Sobelman: *Introduction to the Theory of Atomic Spectra* (Pergamon, Oxford 1973)
5.4   B.M. Smirnov: Usp. Fiz. Nauk *131*, 577 (1980) [English transl.: Sov. Phys.-Usp. *23*, 450 (1980)]
5.5   V.S. Letokhov, V.I. Mishin, A.A. Puretskii: *Plasma Chemistry 4*, 224 (1977) (in Rus-sian)
5.6   R.J. Elliot, R. London: J. Phys. Chem. Solids *15*, 196 (1960)
5.7   H. Hasegawa, R.E. Howard: J. Phys. Chem. Solids *21*, 179 (1961)
5.8   B.B. Kadomtsev: Zh. Eksp. Teor. Fiz. *58*, 1765 (1970) [English transl.: Sov. Phys.-JETP *31*, 945 (1970)]
5.9   B.B. Kadomtsev, V.S. Kudryavtsev: Zh. Eksp. Teor. Fiz. *62*, 144 (1972) [English transl.: Sov. Phys.-JETP *35*, 76 (1972)]
5.10   R.H. Garstang: Rep. Prog. Phys. *40*, 105 (1977)
5.11   H.R. Griem: *Spectral Line Broadening by Plasmas* (Academic, New York 1974)
5.12   V.S. Lisitsa: Usp. Fiz. Nauk *122*, 449 (1977) [English transl.: Sov. Phys.-Usp. *20*, 603 (1977)]
5.13   K.J. Gordon, C.P. Gordon, F.J. Lockman: Astrophys. J. *192*, 337 (1974)
5.14   S.A. Gulyaev: Astron. Zh. *53*, 1010 (1976) [English transl.: Sov. Astron. 20, 573 (1976)]
5.15   M.J. Seaton: Rep. Prog. Phys. *46*, 167 (1983)
5.16   N.B. Delone, V.P. Krainov: *Atoms in a Strong Optical Field* (Energoatomizdat, Moscow 1984) (in Russian)
5.17   R.F. Stebbings, E.B. Dunning (eds.): *Rydberg States of Atoms and Molecules* (Cam-bridge Univ. Press, Cambridge 1983)
5.18   G.F. Drukarev: Zh. Eksp. Teor. Fiz. *75*, 473 (1978); *82*, 1388 (1982) [English transl.: Sov. Phys.-JETP *48*, 237 (1978); *55*, 806 (1982)]
5.19   L.A. Bureeva: Astron. Zh. *45*, 1215 (1968) [English transl.: Sov. Astron. *12*, 962 (1969)]
5.20   S.P. Goreslavskii, N.B. Delone, V.P. Krainov: Zh. Eksp. Teor. Fiz. *82*, 1789 (1982) [English transl.: Sov. Phys.-JETP *55*, 1032 (1982)]
5.21   L.D. Landau, E.M. Lifshitz: *The Classical Theory of Fields* (Pergamon, Oxford 1962)
5.22   V.I. Kogan, A.B. Kukushkin: Zh. Eksp. Teor. Fiz. *87*, 1164 (1984) [English transl.: Sov. Phys.-JETP *60*, 665 (1984)]
5.23   A.B. Kukushkin, V.S. Lisitsa: Zh. Eksp. Teor. Fiz. *88*, 1570 (1985) [English transl.: Sov. Phys.-JETP *61*, 937 (1985)]
5.24   D.R. Herrick: Phys. Rev. A *26*, 323 (1982)
5.25   J.R. Hiskes, C.B. Tarter, D.A. Moody: Phys. Rev. *133*, A 424 (1964)
5.26   S.A. Gulyaev: Astron. Zh. *55*, 1002 (1978) [English transl.: Sov. Astron. *22*, 572 (1978)]
5.27   B.M. Smirnov, M.I. Chibisov: Zh. Eksp. Teor. Fiz. *49*, 841 (1965) [English transl.: Sov. Phys.-JETP *22*, 585 (1966)]

5.28   R. Ya. Damburg, V. V. Kolosov: J. Phys. B *11*, 1921 (1978)
5.29   R. Ya. Damburg, V. V. Kolosov: *Asymptotic Approach to the Stark Problem in the Case of the Hydrogen Atom* (Zinatue, Riga 1977) (in Russian)
5.30   A. Aliajh, J. T. Broad, J. Hinze: J. Phys. B *19*, 2617 (1986)
5.31   B. B. Kadomtsev, V. I. Kogan, B. M. Smirnov, V. D. Shafranov: Usp. Fiz. Nauk *124*, 547 (1978) [English transl.: Sov. Phys.-Usp. *21*, 272 (1978)]
5.32   D. Banks, J. G. Leopold: J. Phys. B *11*, 37 (1978)
5.33   D. F. Zaretskii, V. P. Krainov: Zh. Eksp. Teor. Fiz. *67*, 1301 (1974) [English transl.: Sov. Phys.-JETP *40*, 647 (1975)]
5.34   M. B. Kadomtsev, B. M. Smirnov: Zh. Eksp. Teor. Fiz. *80*, 1715 (1981) [English transl.: Sov. Phys.-JETP *53*, 885 (1981)]
5.35   C. Lanczos: Z. Phys. *62*, 518 (1930)
5.36   N. Froman, P. O. Froman: *JWKB Approximation* (North-Holland, Amsterdam 1965)
5.37   D. S. Bailey, J. R. Hiskes, A. C. Riviere: Nucl. Fusion *5*, 41 (1965)
5.38   N. Hoe, B. D. Etat, G. Couland: Phys. Lett. A *85*, 327 (1981)
5.39   M. L. Zimmerman, M. G. Littman, M. M. Kash, D. Kleppner: Phys. Rev. A *20*, 2251 (1979)

Chapter 6

6.1   H. A. Bethe, E. E. Salpeter: *Quantum Mechanics of One- and Two-Electron Atoms* (Springer, Berlin, Heidelberg 1958)
6.2   L. D. Landau, E. M. Lifshitz: *Quantum Mechanics: Non-Relativistic Theory* (Pergamon, Oxford 1977)
6.3   R. J. Elliot, R. Loudon: J. Phys. Chem. Solids *15*, 196 (1960)
6.4   H. Hasegawa, R. E. Howard: J. Phys. Chem. Solids *21*, 179 (1961)
6.5   A. Galindo, P. Pascual: Nuovo Cimento B *34*, 155 (1976)
6.6   H. C. Praddaude: Phys. Rev. A *6*, 1321 (1972)
6.7   J. Simola, J. Virtamo: J. Phys. B *11*, 3309 (1978)
6.8   D. Cabib, E. Fabri, G. Fiorio: Nuovo Cimento B *10*, 185 (1972)
6.9   A. G. Zhilich, B. S. Monozon: Fiz. Tverd. Tela *8*, 3559 (1966) [English transl.: Sov. Phys.-Solid State *8*, 2846 (1967)]
6.10   A. Aliajh, J. Hinze, J. T. Broad: J. Phys. B *23*, 45 (1990)
6.11   M. L. Zimmerman, M. M. Kash, D. Kleppner: Phys. Rev. Lett. *45*, 1092 (1980)
6.12   C. W. Clark, K. T. Taylor: J. Phys. B *13*, L737 (1980)
6.13   E. A. Solov'ev: Zh. Eksp. Teor. Fiz. *82*, 1762 (1982) [English transl.: Sov. Phys.-JETP *55*, 1017 (1982)]
6.14   D. R. Herrick: Phys. Rev. *26*, 323 (1982)
6.15   M. Robnik: J. Phys. A *14*, 3195 (1981)
6.16   A. I. Baz', Ya. B. Zel'dovich, A. M. Perelomov: *Scattering, Reactions and Decay in Nonrelativistic Quantum Mechanics* (Israel Program for Scientific Translations, Jerusalem; Wiley, New York 1969)
6.17   P. A. Braun: Zh. Eksp. Teor. Fiz. *84*, 850 (1983) [English transl.: Sov. Phys.-JETP *57*, 492 (1983)]
6.18   A. P. Kazantsev, V. L. Pokrovsky, J. Bergou: Phys. Rev. A *28*, 3659 (1983)
6.19   H. Forster, W. Strupat, W. Rosner, G. Wunner, H. Ruder, H. Herold: J. Phys. B *17*, 1301 (1984)
6.20   J. B. Delos, S. K. Knudson, D. W. Noid: Phys. Rev. A *30*, 1208 (1987)
6.21   A. H. Lichtenberg, M. A. Liberman: *Regular and Stochastic Motion*, Springer Ser. Appl. Math. Sci., Vol. 38 (Springer, Berlin, Heidelberg 1983)
6.22   G. Wunner, H. Ruder, H. Herold: Phys. Lett. A *85*, 430 (1981)
6.23   V. B. Pavlov-Verevkin, B. I. Zhilinskii: Phys. Lett. A *75*, 279 (1980)
6.24   H. Friedrich: Phys. Rev. A *26*, 1827 (1982)
6.25   H. Herold, H. Ruder, G. Wunner: J. Phys. B *14*, 751 (1981)
6.26   G. Wunner, H. Ruder, H. Herold: Phys. Lett. A *85*, 430 (1981)

6.27 W. Rosner, G. Wunner, H. Herold, H. Ruder: J. Phys. B *17*, 29 (1984)
6.28 N. Hoe, H. W. Drawin, L. Herman: J. Quant. Spectrosc. Radiat. Transfer 7, 429 (1967)
6.29 A. V. Demura, V. S. Lisitsa: Zh. Eksp. Teor. Fiz. *62*, 2161 (1972) [English transl.: Sov. Phys.-JETP *35*, 1130 (1972)]
6.30 L. P. Gor'kov, I. E. Dzyaloshinskii: Zh. Eksp. Teor. Fiz. *53*, 717 (1967) [English transl.: Sov. Phys.-JETP *26*, 449 (1968)]
6.31 C. P. Slichter: *Principles of Magnetic Resonance with Examples from Solid State Physics* (Harper and Row, New York 1963)
6.32 T. Ishimura: J. Phys. Soc. Jpn. *23*, 422 (1967)
6.33 V. S. Lisitsa: Opt. Spektrosk. *31*, 862 (1971) [English transl.: Opt. Spectrosc. USSR *31*, 468 (1971)]
6.34 N. L. Manakov, L. P. Rapoport: Zh. Eksp. Teor. Fiz. *69*, 842 (1975) [English transl.: Sov. Phys.-JETP *42*, 430 (1975)]
6.35 V. S. Lisitsa, G. V. Sholin: Zh. Eksp. Teor. Fiz. *61*, 912 (1971) [English transl.: Sov. Phys.-JETP *34*, 484 (1972)]
6.36 Yu. N. Demkov, V. N. Ostrovskii, E. A. Solov'ev: Zh. Eksp. Teor. Fiz. *66*, 125 (1974) [English transl.: Sov. Phys.-JETP *39*, 57 (1974)]
6.37 M. Born: *The Mechanics of the Atom* (Bell, London 1927)
6.38 G. L. Kotkin, V. G. Serbo: *Collection of Problems on Classical Mechanics* (Nauka, Moscow 1977) (in Russian)
6.39 Yu. N. Demkov, B. S. Monozon, V. N. Ostrovskii: Zh. Eksp. Teor. Fiz. *57*, 1431 (1969) [English transl.: Sov. Phys.-JETP *30*, 775 (1970)]
6.40 E. A. Solov'ev: Zh. Eksp. Teor. Fiz. *85*, 109 (1983) [English transl.: Sov. Phys.-JETP *58*, 63 (1983)]
6.41 A. V. Turbiner: Zh. Eksp. Teor. Fiz. *84*, 1329 (1983) [English transl.: Sov. Phys.-JETP *57*, 770 (1983)]
6.42 L. A. Burkova, I. E. Dzyaloshinskii, G. F. Drukarev, B. S. Monoson: Zh. Eksp. Theor. Fiz. *71*, 526 (1976) [English transl.: Sov. Phys.-JETP *44*, 276 (1976)]

Chapter 7

7.1   I. I. Sobelman, L. A. Vainshtein, Yu. A. Yukov: *Excitation of Atoms and Broadening of Spectral Lines*, Springer Ser. Chem. Phys., Vol. 7 (Springer, Berlin, Heidelberg 1981)
7.2   S. G. Rautian, G. I. Smirnov, A. N. Shalagin: *Nonlinear Resonances in Spectra of Atoms and Molecules* (Nauka, Novosibirsk 1979) (in Russian)
7.3   Ia. B. Zeldovitch: Usp. Fiz. Nauk *110*, 139 (1973) [English transl.: Sov. Phys.-Usp. *16*(3), 427 (1973)]
7.4   N. A. Blochinzew: Phys. Z. Sov. Union *4*, 501 (1933)
7.5   E. V. Lifshitz: Zh. Eksp. Teor. Fiz. *53*, 943 (1967) [English transl.: Sov. Phys.-JETP *26*, 570 (1968)]
7.6   E. A. Oks: *Plasma Spectroscopy: The Influence of Microwave or Laser Fields*, Springer Ser. Atom. Plasm., Vol. 9 (Springer, Berlin, Heidelberg 1991)
7.7   T. Ishimura: J. Phys. Soc. Jpn. *23*, 422 (1967)
7.8   V. S. Lisitsa: Opt. Spectrosk. *XXXI*, 862 (1971) [English transl.: Opt. Spectrosc. USSR *31*, 468 (1971)]
7.9   D. F. Zaretzkii, V. P. Krainov: Zh. Eksp. Teor. Fiz. *66*, 537 (1974) [English transl.: Sov. Phys.-JETP *39*, 257 (1974)]
7.10 M. Baranger, B. Mozer: Phys. Rev. *123*, 25 (1961)
7.11 I. Sh. Averbukh, N. F. Perelman: Zh. Eksp. Teor. Fiz. *88*, 1131 (1985) [English transl.: Sov. Phys.-JETP *61*(4), 665 (1985)]
7.12 M. Abramowitz, I. A. Stegun (eds.): *Handbook of Mathematical Functions* (National Bureau of Standards, Applied Math. Series *55*, 1964)
7.13 J. E. Bayfield, L. D. Gardner, Y. Z. Gulkok, S. D. Sharma: Phys. Rev. A *24*, 138 (1981)

7.14   I. Ia. Bersons: Zh. Eksp. Teor. Fiz. *85*, 70 (1983); *86*, 860 (1984) [English transl.: Sov. Phys.-JETP *58*, 40 (1983); *59*, 502 (1984)]

Chapter 8

8.1   E. A. Oks, G. V. Sholin: Zh. Eksp. Teor. Fiz. *68*, 975 (1975) [English transl.: Sov. Phys.-JETP *41* (3), 482 (1975)]
8.2   A. I. Zhuzhunashvili, E. A. Oks: Zh. Eksp. Teor. Fiz. *73*, 2142 (1977) [English transl.: Sov. Phys.-JETP *46* (6), 1122 (1977)]
8.3   E. A. Oks, V. A. Rantsev-Kartinov: Zh. Eksp. Teor. Fiz. *79*, 99 (1980) [English transl.: Sov. Phys.-JETP *52* (1), 50 (1980)]
8.4   C. C. Gallagher, M. A. Levine: Phys. Rev. Lett. *30*, 897 (1973); J. Quant. Spectrosc. Radiat. Transfer *15*, 275 (1975)
8.5   W. R. Rutgers, H. de Kluiver: Z. Naturforsch. *29a*, 42 (1974)
8.6   H. A. Blochinzew: Phys. Z. Sov. Union *4*, 501 (1933)
8.7   M. Baranger, B. Mozer: Phys. Rev. *123*, 25 (1961)
8.8   V. P. Gavrilenko, E. A. Oks: Zh. Eksp. Teor. Fiz. *80*, 2150 (1981) [English transl.: Sov. Phys.-JETP *53*, 1122 (1981)]
8.9   V. P. Gavrilenko, E. A. Oks: Fiz. Plazmy *13*, 39 (1987) [English transl.: Sov. J. Plasma-Phys. *13*, 22 (1987)]
8.10   E. Wigner: *Group Theory and its Application to the Quantum Mechanics of Atomic Spectra* (Academic, New York 1959)
8.11   I. S. Gradshtein, I. M. Ryzhik: *Table of Integrals, Series and Products* (Academic Press, New York 1966)
8.12   M. D. Anosov: Opt. Spektrosk. *47*, 209 (1979) [English transl.: Opt. Spectrosc. USSR *47*, 121 (1979)]
8.13   D. G. Yakovlev, V. Yu. Iasevich: Zh. Eksp. Teor. Fiz. *92*, 74 (1987) [English transl.: Sov. Phys.-JETP *65* (1), 40 (1987)]
8.14   R. V. Jensen: Comments on At. Mol. Phys. *25*, 119 (1990)
8.15   B. I. Meerson, E. A. Oks, P. V. Sasorov: J. Phys. B *15*, 3599 (1982); Sov. Phys.-JETP Lett. *29*, 72 (1979)
8.16   B. V. Chirikov: Phys. Rep. *52*, 263 (1979)

Chapter 9

9.1   M. A. Leontovich, L. I. Mandelstamm: Z. Phys. *47*, 131 (1928)
9.2   To the 50th Anniversary of M. A. Leontovich and L. I. Mandelstamm Work, "On the Schrödinger Equation Theory": Usp. Fiz. Nauk. *124*, 547 (1978) [Engl. transl.: Sov. Phys.-Usp. *21*, 272 (1978)]
9.3   G. A. Gamov: Z. Phys. *51*, 204 (1928); *52*, 510 (1928)
9.4   G. Breit, E. Wigner: Phys. Rev. *49*, 519, 642 (1936)
9.5   U. Fano: Nuovo Cimento *12*, 156 (1935)
9.6   U. Fano: Phys. Rev. *124*, 1866 (1961)
9.7   A. S. Kompaneets: Zh. Eksp. Teor. Fiz. *54*, 974 (1968) [English transl.: Sov. Phys.-JETP *27*, 519 (1968)]
9.8   W. Heitler: *Quantum Theory of Radiation* (Clarendon, Oxford 1954)
9.9   L. I. Gudzenko, S. I. Yakovlenko: Zh. Eksp. Teor. Fiz. *62*, 1686 (1972) [English transl.: Sov. Phys.-JETP *35*, 877 (1972)]
9.10   V. S. Lisitsa, S. I. Yakovlenko: Zh. Eksp. Teor. Fiz. *66*, 1981 (1974) [English transl.: Sov. Phys.-JETP *39*, 975 (1974)]
9.11   N. F. Mott, H. S. Massey: *Theory of Atomic Collisions* (Oxford Univ. Press, Oxford 1965)

9.12 G. V. Sholin, A. V. Demura, V. S. Lisitsa: Zh. Eksp. Teor. Fiz. *64*, 2097 (1973) [English transl.: Sov. Phys.-JETP *37*, 1057 (1973)]

9.13 M. L. Strekalov, A. I. Burshtein: Zh. Eksp. Teor. Fiz. *61*, 101 (1971) [English transl.: Sov. Phys.-JETP *34*, 53 (1972)]

9.14 E. M. Purcell: Astrophys. J. *116*, 457 (1952)

9.15 V. I. Kogan, V. S. Lisitsa, A. D. Selidovkin: Zh. Eksp. Teor. Fiz. *65*, 152 (1973) [English transl.: Sov. Phys.-JETP *38*, 78 (1974)]

9.16 I. L. Beigman, V. A. Boiko, S. A. Pikus, A. Yu. Faenov: Zh. Eksp. Teor. Fiz. *71*, 975 (1976) [English transl.: Sov. Phys.-JETP *44*, 511 (1976)]

9.17 A. B. Kukushkin, V. S. Lisitsa: *Polarization Radiation Mechanism in Atomic Collisions*. In: *Polarization Bremsstrahlung Radiation of Particles and Atoms*, ed. by V. Tsytovitch, I. Oiringel (Nauka, Moscow 1987, pp. 273–303 (in Russian); Plenum, New York 1992)

9.18 E. B. Kleiman, I. M. Oiringel: *Photon-Plasmon Transitions as a Mechanism of Polarizational Destruction of Metastable Atomic Levels in Plasmas*. In: *Polarization Bremsstrahlung Radiation of Particles and Atoms*, ed. by V. Tsytovitch, I. Oiringel (Nauka, Moscow 1987, pp. 304–318 (in Russian); Plenum, New York 1992)

9.19 K. Barnard, J. Cooper, E. W. Smith: J. Quant. Spectrosc. Radiat. Transfer *14*, 1025 (1974)

9.20 A. V. Vinogradov, E. A. Yukov: Fiz. Plasmy *1*, 860 (1975) [English transl.: Sov. J.-Plasma Phys. *1*, 472 (1975)]

9.21 E. A. Oks: *Plasma Spectroscopy: The Influence of Microwave or Laser Fields*, Springer Ser. Atom. Plasm., Vol. 9 (Springer, Berlin, Heidelberg 1991)

9.22 L. M. Biberman, V. S. Vorob'ev, I. T. Yakubov: *Kinetics of Nonequilibrium Low-Temperature Plasmas* (Nauka, Moscow 1982) (in Russian)

9.23 V. Ts. Gurovich, V. S. Engelsht: Zh. Eksp. Teor. Fiz. *72*, 444 (1977) [English transl.: Sov. Phys.-JETP *45*, 232 (1977)]

9.24 M. I. Chibisov: Pizma Zh. Eksp. Teor. Fiz. *24*, 56 (1976) [English transl.: Sov. Phys.-JETP Lett. *24* (2), 46 (1976)]

9.25 V. A. Abramov, F. F. Baryshnikov, A. I. Kazanskii, I. V. Komarov, V. S. Lisitsa, M. I. Chibisov: *Charge Exchange of Atoms on Multiply Charged Ions*. In: *Reviews of Plasma Physics*, ed. by M. A. Leontovich, B. B. Kadomtsev, Vol. 12 (Consultants Bureau, New York, London 1987) p. 123

9.26 B. M. Smirnov: Asymptotic Methods in Atomic Collision Theory (Atomizdat, Moscow 1976) (in Russian)

9.27 L. A. Woltz, V. L. Jacobs, C. F. Hooper, R. C. Manchini: Phys. Rev. A *44*, 1281 (1991)

9.28 A. K. Grigoriadi, O. I. Fisun: Fiz. Plasmy *8*, 776 (1982) [English transl.: Sov. J.-Plasma Phys. *8*, 440 (1982)]

9.29 K. Sakimoto: J. Phys. B *19*, 3011 (1986)

Chapter 10

10.1 V. S. Lisitsa, G. V. Sholin: Zh. Eksp. Teor. Fiz. *61*, 912 (1971) [English transl.: Sov. Phys.-JETP *34*, 484 (1972)]

10.2 Yu. N. Demkov, V. N. Ostrovskii, E. A. Solov'ev: Zh. Eksp. Teor. Fiz. *66*, 125 (1974) [English transl.: Sov. Phys.-JETP *39*, 57 (1974)]

10.3 E. Wigner: *Group Theory and its Applications to the Quantum Mechanics of Atomic Spectra* (Academic, New York 1959)

10.4 V. A. Abramov, F. F. Baryshnikov, V. S. Lisitsa: Zh. Eksp. Teor. Fiz. *74*, 897 (1978) [English transl.: Sov. Phys.-JETP *47*, 469 (1978)]

10.5 V. A. Abramov, F. F. Baryshnikov, A. I. Kazanskii, I. V. Komarov, V. S. Lisitsa, M. I. Chibisov: *Charge Exchange of Atoms on Multiply Charged Ions*. In: *Reviews of Plasma Physics*, ed. by M. A. Leontovich, B. B. Kadomtsev, Vol. 12 (Consultants Bureau, New York, London 1987) p. 123

10.6   N. F. Mott, H. S. Massey: *Theory of Atomic Collisions* (Oxford Univ. Press, Oxford 1965)

10.7   Yu. N. Demkov, V. I. Osherov: Zh. Eksp. Teor. Fiz. *53*, 1589 (1967) [English transl.: Sov. Phys.-JETP *26*, 916 (1968)]

10.8   V. A. Abramov, F. F. Baryshnikov, V. S. Lisitsa: Pizma Zh. Eksp. Teor. Fiz. *27*, 494 (1978) [English transl.: Sov. Phys.-JETP Lett. *47*, 469 (1978)]

10.9   L. D. Landau, E. M. Lifshitz: *Quantum Mechanics: Nonrelativistic Theory* (Pergamon, Oxford 1977)]

10.10   F. F. Baryshnikov, V. S. Lisitsa: Zh. Eksp. Teor. Fiz. *72*, 1797 (1977) [English transl.: Sov. Phys.-JETP *45*, 943 (1977)

10.11   H. Griem: *Plasma Spectroscopy* (McGraw-Hill, New York 1966)

10.12   V. S. Lisitsa: Usp. Fiz. Nauk *122*, 449 (1977) [English transl.: Sov. Phys.-Usp. *20*, 603 (1977)]

10.13   M. J. Seaton: Proc. Phys. Soc. London *77*, 174 (1961)

10.14   I. I. Sobelman: *Introduction to the Theory of Atomic Spectra* (Pergamon, Oxford 1973)

10.15   I. S. Gradshtein, I. M. Ryzhik: *Table of Integrals, Series and Products* (Academic, New York 1966)

10.16   M. Abramovitz, I. A. Stegun (eds.): *Handbook of Mathematical Functions* (National Bureau of Standards, Applied Math. Series 55, 1964)

10.17   H. Bateman, A. Erdelyi: *Higher Transcendental Functions* (McGraw-Hill, New York 1953)

10.18   W. W. Gargaro, D. S. Onley: Phys. Rev. C4, 1032 (1971)

10.19   F. F. Baryshnikov, V. S. Lisitsa: Zh. Eksp. Teor. Fiz. *80*, 926 (1981) [English transl.: Sov. Phys.-JETP *54*, 267 (1981)]

10.20   V. I. Gervids, V. I. Kogan, V. S. Lisitsa: *Multicharged Ions and Plasma Radiation*. In: *Plasma Chemistry*, ed. by B. M. Smirnov (Energoatomizdat, Moscow 1982) pp. 3 – 85 (in Russian)

10.21   F. F. Baryshnikov, V. S. Lisitsa: Fiz. Plasmy *4*, 1177 (1978) [English transl.: Sov. Phys.-Plasma *4*, 660 (1978)]

10.22   A. I. Gurevich, V. M. Dubovik, L. M. Satarov: In: *Questions of Atomic Collisions Theory*, ed. by Yu. A. Vdovin (Atomizdat, Moscow 1970) p. 87 (in Russian)

Chapter 11

11.1   A. I. Gurevich: Zh. Eksp. Teor. Fiz. *64*, 1199 (1973) [English transl.: Sov. Phys.-JETP *37*, 610 (1973)]

11.2   L. I. Gudzenko, S. I. Yakovlenko: Zh. Eksp. Teor. Fiz. *62*, 1686 (1972) [English transl.: Sov. Phys.-JETP *35*, 877 (1972)]

11.3   V. S. Lisitsa, S. I. Yakovlenko: Zh. Eksp. Teor. Fiz. *66*, 1550 (1974) [English transl.: Sov. Phys.-JETP *39*, 759 (1974)]

11.4   V. S. Lisitsa, S. I. Yakovlenko: Zh. Eksp. Teor. Fiz. *68*, 479 (1975) [English transl.: Sov. Phys.-JETP *41*, 233 (1975)]

11.5   L. I. Gudzenko, S. I. Yakovlenko: Zh. Tekh. Fiz. *45*, 274 (1975) [English transl.: Sov. Phys.-Tech. Phys. *20*, 150 (1975)]

11.6   R. Z. Vitlina, A. V. Chaplik, M. V. Entin: Zh. Eksp. Teor. Fiz. *67*, 1667 (1974) [English transl.: Sov. Phys.-JETP *40*, 829 (1974)]

11.7   A. M. Bouch-Bruevich, S. G. Przhubelskii, A. A. Fedorov, V. V. Khromov: Zh. Eksp. Teor. Fiz. *71*, 1733 (1976) [English transl.: Sov. Phys.-JETP *43*, 230 (1976)]

11.8   R. Z. Vitlina, A. M. Dykhne: Zh. Eksp. Teor. Fiz. *64*, 510 (1973) [English transl.: Sov. Phys.-JETP *37*, 260 (1973)]

11.9   L. D. Landau, E. M. Lifshitz: *Quantum Mechanics: Nonrelativistic Theory* (Pergamon, Oxford 1977)

11.10   H. R. Griem: *Spectral Line Broadening by Plasmas* (Academic, New York 1974)

11.11 I. I. Sobelman: *Introduction to the Theory of Atomic Spectra* (Pergamon, Oxford 1973)
11.12 L. Spitzer: Phys. Rev. *58*, 348 (1940)
11.13 V. I. Kogan, V. S. Lisitsa, J. Quant. Spectrosc. Radiat. Transfer *12*, 881 (1972)
11.14 N. F. Mott, H. S. Massey: *Theory of Atomic Collisions* (Oxford Univ. Press, Oxford 1965)
11.15 S. G. Rautian, G. I. Smirnov, A. N. Shalagin: *Nonlinear Resonances in Spectra of Atoms and Molecules* (Nauka, Novosibirsk 1979) (in Russian)
11.16 I. I. Sobelman, L. A. Vainshtein, E. A. Yukov: *Excitation of Atoms and Broadening of Spectral Lines*, Springer Ser. Chem. Phys., Vol. 7 (Springer, Berlin, Heidelberg 1981)
11.17 S. P. Andreev, V. S. Lisitsa: Zh. Eksp. Teor. Fiz. *72*, 73 (1977) [English transl.: Sov. Phys.-JETP *45*, 38 (1977)]
11.18 A. G. Zhidkov: Zh. Eksp. Teor. Fiz. *88*, 372 (1985) [English transl.: Sov. Phys.-JETP *61*, 218 (1985)]
11.19 T. A. Vartan'an, Yu. N. Maximov, S. G. Przhibelskii, V. V. Khromov: Pizma Zh. Eksp. Teor. Fiz. *29*, 281 (1979) [English transl.: Sov. Phys.-JETP Lett. *29*, 252 (1979)]
11.20 D. B. Lidow, R. W. Falcone, J. F. Young, E. Harris: Phys. Rev. Lett. *36*, 462 (1976)
11.21 S. I. Yakovlenko: Usp. Fiz. Nauk *136*, 593 (1982) [English transl.: Sov. Phys.-Usp. *25*, 216 (1982)]
11.22 S. I. Yakovlenko: *Radiative-Collisional Phenomena* (Energoatomizdat, Moscow 1984) (in Russian)
11.23 V. S. Lisitsa: Zh. Eksp. Teor. Fiz. *65*, 879 (1973) [English transl.: Sov. Phys.-JETP *38*, 435 (1974)]
11.24 V. S. Lisitsa: Zh. Eksp. Teor. Fiz. *69*, 195 (1975) [English transl.: Sov. Phys.-JETP *42*, 109 (1975)]
11.25 F. F. Baryshnikov, V. S. Lisitsa: Fiz. Plasmy *3*, 101 (1977) [English transl.: Sov. Phys.-Plasma *3*, 398 (1977)]
11.26 V. S. Lisitsa: Acta Phys. Pol. A *55*, 87 (1979)
11.27 H. K. Wimmerl: J. Quant. Spectrosc. Radat. Transfer *1*, 1 (1961)
11.28 V. S. Lisitsa: Usp. Fiz. Nauk *122*, 449 (1977) [English transl.: Sov. Phys.-Usp. *20*, 603 (1977)]
11.29 S. Chandrasekhar: Rev. Mod. Phys. *1*, 15 (1943)
11.30 V. S. Lisitsa: *Effect of Strong Electromagnetic Fields on Atomic Spectral Line Shapes in Plasmas and Gases.* In: *Spectral Line Shapes*, Vol. 5 (Ossolineum, Wroclaw, Poland 1989) p. 305
11.31 G. F. Drukarev, A. A. Michailov: Opt. Spectrosk. *37*, 384 (1974) [English transl.: Opt. Spectrosc. USSR *37*, 220 (1974)]
11.32 E. L. Duman, L. I. Menshikov: Dok Akad. Nauk SSR *244*, 1342 (1979) [English transl.: Sov. Phys. Dokl. *244*, 116 (1979)]

Chapter 12

12.1 S. G. Rautian, G. I. Smirnov: Zh. Eksp. Teor. Fiz. *74*, 1295 (1978) [English transl.: Sov. Phys.-JETP *47*, 678 (1978)]
12.2 S. G. Rautian, G. I. Smirnov, A. M. Shalagin: *Nonlinear Resonances in Spectra of Atoms and Molecules* (Nauka, Novosibirsk 1979) (in Russian)
12.3 F. F. Baryshnikov, V. S. Lisitsa, S. A. Sukhin: Pizma Zh. Eksp. Teor. Fiz. *33*, 199 (1981) [English transl.: Sov. Phys.-JETP Lett. *33*, 188 (1981)]
12.4 N. F. Mott, H. S. Massey: *Theory of Atomic Collisions* (Oxford Univ. Press, Oxford 1965)
12.5 F. F. Baryshnikov, V. S. Lisitsa, S. A. Sukhin: Zh. Eksp. Teor. Fiz. *81*, 497 (1981) [English transl.: Sov. Phys.-JETP *54*, 267 (1981)]
12.6 L. A. Vainshtein, I. I. Sobelman, E. A. Yukov: *Atom Exutation and Spectral Line Broadening* (Springer, Berlin, Heidelberg 1981)

12.7 A. P. Kolchenko, G. I. Smirnov: Zh. Eksp. Teor. Fiz. *71*, 925 (1976) [English transl.: Sov. Phys.-JETP *44*, 486 (1976)]

12.8 F. F. Baryshnikov, V. S. Lisitsa: Zh. Eksp. Teor. Fiz. *82*, 1058 (1982) [English transl.: Sov. Phys.-JETP *56*, 618 (1982)]

12.9 V. S. Lisitsa: *Effect of Strong Electromagnetic Fields on Atomic Spectral Line Shapes in Plasmas and Gases.* In: *Spectral Line Shapes*, Vol. 5 (Ossolineum, Wroclaw, Poland 1989) p. 305

12.10 L. A. Vainshtein, L. P. Presnyakov, I. I. Sobelman: Zh. Eksp. Teor. Fiz. *43*, 518 (1962) [English transl.: Sov. Phys.-JETP *18*, 1383 (1962)]

12.11 I. I. Sobelman: *Introduction to the Theory of Atomic Spectra* (Pergamon, Oxford 1973)

12.12 S. Chandrasekhar: Rev. Mod. Phys. *1*, 15 (1943)

12.13 M. I. Podgoretskii, A. V. Stepanov: Zh. Eksp. Teor. Fiz. *40*, 561 (1961) [English transl.: Sov. Phys.-JETP *13*, 393 (1961)]

12.14 M. Katz: *Probability and Related Topics in Physical Sciences* (McGraw-Hill, New York 1965)

12.15 R. P. Feynman, A. R. Hibbs: *Quantum Mechanics and Path Integrals* (McGraw-Hill, New York 1964)

12.16 M. Abramovitz, I. A. Stegun (eds.): *Handbook of Mathematical Functions* (National Bureau of Standards, Applied Math. Series *55*, 1964)

12.17 A. S. Kompaneets: Zh. Eksp. Teor. Fiz. *54*, 974 (1968) [English transl.: Sov. Phys.-JETP *27*, 519 (1968)]

12.18 V. S. Lisitsa, Yu. A. Savel'ev: Zh. Eksp. Teor. Fiz. *92*, 484 (1987) [English transl.: Sov. Phys.-JETP *65*, 273 (1987)]

12.19 A. I. Akhiezer, V. B. Berestetskii: *Quantum Electrodynamics* (Wiley Interscience, New York 1965)

12.20 J. D. Bjorken, S. D. Drell: *Relativistic Quantum Mechanics* (McGraw-Hill, New York 1965)

Subject Index

Absorption
- by an atom in plasmas 2
- due to Brownian fluctuations 269
- line shape 271
- nonlinear effects in 237
- of equivalent photons 55, 57, 60
- of light
-- by atomic particles 230
-- by the medium 230
Acceleration 261–263
- near turning points 9
- of electron in attractive potential 28
Action
- -angle variables 40, 174
- variables 83
Adiabatic approximation 135, 199
Adiabatic collision theory 245
Adiabatic (energy) levels 135, 141
Adiabatic lifetime 200
Adiabatic perturbation 129
Adiabatic population inverion 263
- in a strong laser field 262
Adiabatic term 263
Adiabatic theory in strong magnetic field 98, 108
Adiabatic wave functions 134, 138
Airy
- functions 148, 269, 270
- equation 269
Allowed transitions 16, 20
Atom
- in crossed F–B fields 120
- in electric and strong magnetic fields 125
- in a laser field 235
- in a magnetic field 93, 98, 127
- in a nonresonant oscillating electric field 128
- in a resonant oscillating electric field 152
- quantization direction 253
-- change effect 251
-- in a laser field 250, 251
Atomic collisions in electromagnetic field 240

Atomic dipole moment 236, 253
Atomic energy 1, 70
Atomic level population 48
Atomic spin polarization 228
Atomic state
- mixing 2, 20, 201
- polarization 193
Atomic transitions in strong magnetic field 108
Axis of quantization 207
Autoionization
- decay 1, 127, 176, 201, 203
- levels 184
- lines 22, 184
- of the atom 182
- rate 62, 203, 204
- state 22
- transitions 203
- width 23, 203

Balmer series 161
Baranger-Mozer 137
- satellites 160
Barrier penetrability 107
Bethe
- -Born formulae 227
- rule 11
-- defect (BRD) 32, 34, 46, 48, 49
Bifurcation 117
Binary collisions 245, 250
- of an atom with broadening particles 244
Bloch F., Nordsieck A. 272
Blokhintsev 129, 130, 160
- spectrum 129, 130, 132, 134, 152, 158, 165, 169
- satellites 131, 142, 150
- theory 150
- wave functions 169
Bohr
- magneton 70, 93
- radius 70, 109, 206
- -Sommerfeld
-- quantization conditions 58, 98, 104, 107

Springer-Verlag
and the Environment

$\mathbf{W}$e at Springer-Verlag firmly believe that an international science publisher has a special obligation to the environment, and our corporate policies consistently reflect this conviction.

$\mathbf{W}$e also expect our business partners – paper mills, printers, packaging manufacturers, etc. – to commit themselves to using environmentally friendly materials and production processes.

$\mathbf{T}$he paper in this book is made from low- or no-chlorine pulp and is acid free, in conformance with international standards for paper permanency.